RECENT ADVANCES IN RELIABILITY AND MAINTENANCE MODELING

Recent Advances in Reliability and Maintenance Modeling contains the papers presented at the 11th Asia-Pacific International Symposium on Advanced Reliability and Maintenance Modeling (APARM 2024, Nagoya, Japan, 26–30 August 2024). The contributions discuss and explore solutions to the various reliability challenges facing society. Reliability and maintenance is the technology required in various fields such as (but not limited to):

- Power systems
- Communication networks
- Transportation
- Cloud computing
- Electronic systems
- Buildings and infrastructure
- Medical and healthcare
- Aviation and railway systems.

Recent Advances in Reliability and Maintenance Modeling is of interest to academics and professionals interested or involved in the above mentioned areas.

PROCEEDINGS OF THE 11TH ASIA-PACIFIC INTERNATIONAL SYMPOSIUM ON ADVANCED RELIABILITY AND MAINTENANCE MODELING, 26–30 AUGUST 2024, NAGOYA, JAPAN

Recent Advances in Reliability and Maintenance Modeling

Edited by

Hiroyuki Okamura
Hiroshima University, Japan

Shinji Inoue
Kansai University, Japan

Xiao Xiao
Tokyo Metropolitan University, Japan

CRC Press
Taylor & Francis Group
Boca Raton London New York Leiden

CRC Press is an imprint of the
Taylor & Francis Group, an **informa** business
A BALKEMA BOOK

First published 2024
by CRC Press/Balkema
4 Park Square, Milton Park, Abingdon, Oxon, OX14 4RN

and by CRC Press/Balkema
2385 NW Executive Center Drive, Suite 320, Boca Raton FL 33431

CRC Press/Balkema is an imprint of the Taylor & Francis Group, an informa business

British Library Cataloguing-in-Publication Data
A catalogue record for this book is available from the British Library

Library of Congress Cataloging-in-Publication Data
A catalog record has been requested for this book

ISBN: 978-1-032-79098-5 (hbk)
ISBN: 978-1-032-79267-5 (pbk)
ISBN: 978-1-003-49130-9 (ebk)

DOI: 10.1201/9781003491309

Typeset in Times New Roman
by MPS Limited, Chennai, India

Recent Advances in Reliability and Maintenance Modeling – Okamura, Inoue & Xiao (Eds)
© 2025 the Editor(s), ISBN: 978-1-032-79098-5

Table of Contents

 ISBN: 978-1-032-79098-5

Preface

The 11th Asia-Pacific International Symposium on Advanced Reliability and Maintenance Modeling (APARM 2024), which continues a series of successful workshops – AIWARM 2004 (Hiroshima, Japan), AIWARM 2006 (Busan, Korea), AIWARM 2008 (Taichung, Taiwan), APARM 2010 (Wellington, New Zealand), APARM 2012 (Nanjing, China), APARM 2014 (Sapporo, Japan), APARM 2016 (Seoul, Korea), APARM 2018 (Qingdao, China), APARM 2020 (Vancouver, Canada), APARM 2022 (Hsinchu, Taiwan) – will be held on August 26–30, 2024, in Nagoya, Japan. APARM 2024 is dedicated to disseminating cutting-edge research and practical applications in reliability engineering, maintenance, and related fields within the Asia-Pacific region. The symposium aims to bring together both young and senior researchers, scientists, and practitioners to discuss significant and challenging issues in reliability and maintainability, fostering a strong research network across the Asia-Pacific region. Covering all key topics in reliability, maintainability, and safety engineering, the symposium will provide an in-depth representation of both theoretical and practical advancements in these areas.

All submitted full papers were reviewed by two Program Committee members. After the review process, 34 full papers were accepted for inclusion in this proceeding. We would like to acknowledge the efforts of all the Technical Program Committee members for reviewing the submitted papers. The members of the Technical Program Committee are Prof. T. Akiba, *Chiba Institute of Technology, Japan,* Prof. J. Cai, *Nanjing University of Aeronautics & Astronautics, China,* Prof. Q. Hu, *Chinese Academy of Science, China,* Prof. M. Imaizumi, *Aichi Gakusen University, Japan,* Prof. S. Inoue, *Kansai University, Japan,* Prof. L. Jin, *University of Electro-Communications, Japan,* Prof. M. Kimura, *Aichi University, Japan,* Prof. C. Lv, Xian *University of Architecture and Technology, China,* Prof. Y. Minamino, *Tottori University, Japan,* Prof. S. Mizutani, *Aichi Institute of Technology, Japan,* Prof. T. Nakamura, *Tokai University, Japan,* Prof. H. Okamura, *Hiroshima University, Japan,* Prof. S. Ota, *Kanagawa University, Japan,* Prof. J. Shen, *Nanjing University of Science and Technology, China,* Prof. N. Takahashi, *National Defense Academy of Japan, Japan,* Prof. N. Tamura, *Hosei University, Japan,* Prof. Y. Tamura, *Yamaguchi University, Japan,* Prof. J. Wang, *China Jiliang University, China,* Prof. X. Xiao, *Tokyo Metropolitan University, Japan,* Prof. Z. Ye, *Zhengzhou University, China,* Prof. X. Zhao, *Nanjing University of Aeronautics & Astronautics, China,* Prof. J. Zheng, *Hiroshima University, Japan*. Organizing APARM 2024 and editing this proceeding has been a privilege for us. We greatly enjoyed working with the authors and reviewers of the full papers.

We express our gratitude to Aichi University, Nagoya, for providing generous support, including the symposium venue. We would also like to thank our conference sponsors: Aichi University, the SECOM Science and Technology Foundation, and the Reliability Engineering Association of Japan for their support. Additionally, we appreciate CRC Press for their assistance with this publication.

In conclusion, we would like to extend our appreciation to the following individuals for their contributions: General Co-chair: Prof. Zhisheng Ye, *Zhengzhou University, China*, Program Co-Chairs: Prof. Mitsutaka Kimura, Aichi University, Japan, Prof. Xufeng Zhao, *Nanjing University of Aeronautics & Astronautics, China*, Publicity Co-Chairs: Prof. Taishin Nakamura, *Tokai University, Japan*, Prof. Yuka Minamino, *Tottori University, Japan*, Web Co-Chairs: Prof. Yoshinobu Tamura, *Yamaguchi University, Japan*, Prof. Shuhei Ota, *Kanagawa University, Japan*, Local Organization Co-Chairs: Prof. Satoshi Mizutani, *Aichi Institute of Technology, Japan*, Prof. Mitsuhiro Imaizumi, *Aichi Gakusen University, Japan*,

Conference Secretariat: Prof. Junjun Zheng, *Hiroshima University, Japan.* We also extend our heartfelt thanks to their families for their support.

Editors and Publication Co-Chairs

Hiroyuki Okamura	*Hiroshima University, Japan*
Shinji Inoue	*Kansai University, Japan*
Xiao Xiao	*Tokyo Metropolitan University, Japan*

Editor Biographies

Hiroyuki Okamura received the B.S.E., M.S., and D.Eng. degrees in engineering from Hiroshima University, Higashihiroshima, Japan, in 1995, 1997, and 2001, respectively. In 1998, he joined Hiroshima University as an Assistant Professor, where he has been an Associate Professor with the Department of Information Engineering, Graduate School of Engineering, since 2003. He is now a Processor with the Department of Information Engineering, Graduate School of Engineering, since 2018. His research interests include performance evaluation, dependable computing, and applied statistics. He is a regular member of IEEE, ACM, Institute of Electrical, Information and Communication Engineers, the Information Processing Society of Japan, the Reliability Engineering Association of Japan, and a fellow member of the Operations Research Society of Japan.

Shinji Inoue was born in Fukuoka Prefecture, Japan, in 1978. He received the B.S.E., M.S., and Ph.D. degrees from Tottori University, Japan, in 2001, 2003, and 2006, respectively. He has been working as Full Professor of the Faculty of Informatics, Kansai University, Osaka, Japan, since April 2021 after he had worked as Assistant Professor of the Graduated School of Engineering, Tottori University, Tottori, Japan (April 2006–March 2017) and as Associate Professor of the Faculty of Informatics, Kansai University (April 2017–March 2021). His research interests include software reliability engineering, quality control, and functional safety. He is a regular member of the IEEE, the Operations Research Society of Japan, the Japanese Society for Quality Control, the Information Processing Society of Japan, Reliability Engineering Association of Japan, and the IEICE. Dr. Inoue received The IEEE Reliability Society Japan Chapter 2006 Outstanding Young Scientist Award, Outstanding Paper Award at IEEE-IEEM 2008, Amity Global Academic Excellence Award (Amity University, Noida, Uttar Pradesh, India) in 2017, and Best Paper Award at APARM 2018.

Xiao Xiao received the B.Eng., M.Eng., and Ph.D.Eng. from Hiroshima University, Hiroshima, Japan, in 2008, 2010, and 2013, respectively. She worked as an Assistant Professor in the Department of Information Engineering, Graduate School of Engineering, Hiroshima University from April 2013 to September 2013. In October 2013, she joined the Department of Management Systems Engineering, Graduate School of Systems Design, Tokyo Metropolitan University, as an assistant professor, and was promoted to associate professor in April 2022. Her research areas include software reliability engineering, system performance evaluation, time series analysis, and network design. She is a regular member of ACM, IEEE, IEICE, IPSJ, JIMA, ORSJ, and REAJ.

Recent Advances in Reliability and Maintenance Modeling – Okamura, Inoue & Xiao (Eds)
© 2025 the Author(s), ISBN: 978-1-032-79098-5

On a queueing model with two kinds of working vacations

A. Aissani & S. Lassel
Department of Artificial Intelligence and Data Science, Faculty of Computer Science, RIIMA Laboratory, University of Science and Technology Houari Boumediene (USTHB), Bab Ezzouar, Algeria

ABSTRACT: Vacation queueing models are useful for those systems in which the server wants to utilize his idle time for different purposes: maintenance or computer security actions, saving energy etc. In classical models, the server is unavailable to customers during a vacation period and the server stop completely the original work and cannot come back to the regular busy period until the vacation period ends. Here we consider a Non-Markovian model with two types of vacations (called differentiated, for example long and short) and in which the server is available to customers but at different rates than normal mode (working vacations). We obtain the joint probability distribution of the server state and the queue size. Several mean performance measures of interest are derived. Numerical illustrations are given in order to show the effect of some parameters on these performances measures.

1 INTRODUCTION

Vacation queueing models are useful for those systems in which the server (or the mobile) wants to utilize his idle time for different purposes: maintenance or computer security actions, saving energy in call centers or WSN etc. Examples and applications can be found in the articles and surveys by Doshi 1986, Guo *et al.* 2015, Hall *et al.* 2015, Isci *et al.* 2013, Kempa 2015, Levy *et al.* 1975, Lu *et al.* 2013, and their references. Many papers consider models with several kinds of vacations (called differentiated), for example a first kind is long and the second one's is short, Ait Braham *et al.* 2023, Phung-Duc 2015, Zhang *et al.* 2017. In classical models, the server is unavailable to customers during a vacation period and the server stop completely the original work and cannot come back to the regular busy period until the vacation period ends. Working vacation is one of vacation policy under which the server provides service at lower speed (thus at lower consumption energy) during the vacation period rather than stopping service completely. Here we consider a model with two kinds of vacations (differentiated) and in which the server is available to customers but at a different rate than normal mode (working vacations). Such a model of working vacation has been visibely introduced by Servi *et al.* 1975, and discussed further by Phung-Duc 2015 and Zhang *et al.* 2017. Surveys about working vacations can be found in Chandrasekaran *et al.* 2016, Dyviat *et al.* 2024. The next section is devoted to the model description. In section 3 we derive the system of equations describing the evolution of our system. The resolution of this system is provided in section 4 using Fourier transform: Generating function in the discrete component and Laplace transform in the continuous ones. In section 5 we provide some performance measures of interest agremented by illustrations demonstrating how a practitionner can use these results to show the effect of system parameters on performance measures. Section 6 is devoted to a short discussion about the statistical identification of the model in practice.

DOI: 10.1201/9781003491309-1

2 MODEL FORMULATION

Consider an $M/G/1$ queueing model in which server takes vacations each time it becomes idle. Customers arrive at the system according to a Poisson process with rate λ and request for service takes a random duration S with general probability distribution $H(x) = P(S < x)$, Laplace-Stieltjes transform $h(s)$ and first two moments h_1 and h_2. The server starts a first kind (long) of vacation when the server becomes idle. The length of this vacation follows an exponential distribution with mean $1/\gamma_1$. On returning from this vacation, if the server is still idle, it takes a second kind (short) vacation whose duration follows an exponential distribution with mean $1/\gamma_2$. Kind 2 vacations are repeated as long the system is empty upon the completion of a vacation. On returning from either kind 1 or 2 vacation, if there are some customers in the system, the server immediately starts serving customers until the system is empty again. All the considered random variables are assumed to be mutually independent and all moments are assumed to be finite. This vacation model was introduced in the context of Power-Saving mode in the work by Phung-Duc 2015 using Markovian assumption. Ait Braham *et al.* 2023 revisited this question in a Non Markovian context when the service time in normal mode is arbitrary distributed. In the above works, the server is unavailable to serve customers until the vacation periods end. The service can start only when a new busy period begins. In this paper, the server is allowed to serve customers during vacation period of any kinds but at a lower constant rates μ_1 and μ_2.

3 SYSTEM STATE OF EQUATIONS

Let

$$S(t) = \begin{cases} 0, \text{ if server is busy by a customer at time t,} \\ 1, \text{ if server is on vacation type I,} \\ 2, \text{ if server is on vacation type II.} \end{cases} \tag{1}$$

and $S = lim_{t\to\infty} S(t)$ be the limiting random variable (if it exists) when $t \to \infty$ and $N = lim_{t\to\infty} N(t)$ the limiting variables of $N(t)$, the number of customers in the system at time t. The bi-dimensional discrete stochastic process $\{S(t), N(t)\}$ is Non-Markovian since the probability distribution of service in normal state is arbitrary distributed. The model is Non-Markovian because there is an uncertainty about the service in normal mode which is non exponential. However, we can "markovize" the model by introducing a new continuous random variable $\xi(t)$ describing "uncertainty" about the residual service time. So, the new tri-dimensional stochastic process $\{S(t), N(t]; \xi(t)\}$ is now Markovian. It is possible to arrive to such a "Markovization" by solving the uncertainty with introduction of another continuous random variable describing the elapsed service time rather than the residual one. See for example the work by (Ait Braham *et al.* 2023)which considers a similar vacation problem without working vacations. Let

$$P_i(m) = P\{S = i, N = m, \}, i = 1, 2, m \in \mathbb{N}, x \in \mathbb{R}^+ \tag{2}$$

$$P_0(m, x) = P\{S = 0, N = m, \xi(t) < x\}, m \in \mathbb{N}, x \in \mathbb{R}^+ \tag{3}$$

be the steady-state probabilities. For $i = 1$ we have the following equations

$$(\lambda + \gamma_1 + \mu_1)P_1(1) = \lambda P_1(0) + \mu_1 P_1(2), \tag{4}$$

$$(\lambda + \gamma_1 + \mu_1)P_1(m) = \lambda P_1(m-1) + \mu_1 P_1(m+1), m \geq 2. \tag{5}$$

Similarly, for $i = 2$ we get

$$(\lambda + \gamma_2 + \mu_2)P_2(1) = \lambda P_2(0) + \mu_2 P_2(2), \tag{6}$$

$$(\lambda + \gamma_2 + \mu_1)P_2(m) = \lambda P_2(m-1) + \mu_2 P_2(m+1), m \geq 2. \tag{7}$$

and for $i = 0$,

$$\begin{aligned}\lambda P_0(1,x) = &\frac{\partial P_0(1,x)}{\partial x} - \frac{\partial P_0(1,0)}{\partial x} + \frac{\partial P_0(2,0)}{\partial x} H(x)\\ &+ \gamma_1 P_1(1) H(x) + \gamma_2 P_2(1) H(x),\end{aligned} \tag{8}$$

$$\begin{aligned}\lambda P_0(m,x) = &\frac{\partial P_0(m,x)}{\partial x} - \frac{\partial P_0(m,0)}{\partial x} + \frac{\partial P_0(m+1,0)}{\partial x} H(x),\\ &+ \gamma_1 P_1(m) H(x) + \gamma_2 P_2(m) H(x) + \lambda P_0(m-1,x), m \geq 2.\end{aligned} \tag{9}$$

These equations are obtained in usual way by considering the evolution of the process during an infinitesimal interval of time (a slot). In the next section we obtain the solution of this system of equations using generating function in the discrete component and Laplace transform in continuous one. This allows us to obtain several performance measures of interest.

4 RESOLUTION OF THE SYSTEM

Taking the generating function in the discrete component m for the probability distribution $P_1(m) = P\{S = 1, N = m\}$, we obtain

$$Q_1(z) = \frac{[(\lambda + \gamma_1 + \mu_1)P_1(0) - \mu_1 P_1(1)]z - \mu_1 P_1(0)}{(\lambda + \gamma_1 + \mu_1)z - \lambda z^2 - \mu_1}. \tag{10}$$

The polynome in the denominator $T(z) = (\lambda + \gamma_1 + \mu_1)z - \lambda z^2 - \mu_1$ has two distinct roots $0 < \widehat{z}_1 < 1$ and $z_1 > 1$

$$\widehat{z}_1 = \frac{\lambda + \mu_1 + \gamma_1 - \sqrt{\Delta}}{2\lambda}; z_1 = \frac{\lambda + \mu_1 + \gamma_1 + \sqrt{\Delta}}{2\lambda}$$

where the discriminant $\Delta = (\lambda + \mu_1 + \gamma_1)^2 - 4\lambda\mu_1 > 0$. Putting $z = \widehat{z}_1$ in (10) lead to an indetermination $Q_1(\widehat{z}_1) = \frac{0}{0}$. Since $Q_1(z)$ is an analytical function at least in the disk $\|z < 1\|$, then the zeros of the denominator are also zeros of numerator. So, we have the following equation:

$$[(\lambda + \mu_1 + \gamma_1)P_1(0) - \mu_1 P_1(1)]\widehat{z}_1 - \mu_1 P_1(0) = 0 \tag{11}$$

Thus, we have a relation between the two unknown constants $P_1(1)$ and $P_1(0)$

$$P_1(1) = \frac{(\lambda + \mu_1 + \gamma_1)\widehat{z}_1 - \mu_1}{\mu_1 \widehat{z}_1} P_1(0). \tag{12}$$

After simplification we get

$$Q_1(z) = \frac{\mu_1 P_1(0)}{\lambda \widehat{z}_1 (z_1 - z)} \tag{13}$$

It is well known that the two roots $\widehat{z}_1$ and z_1 of the polynome $T(z)$ satisfy the relations

$$\widehat{z}_1 + z_1 = \frac{\lambda + \mu_1 + \gamma_1}{\lambda}; \widehat{z}_1.z_1 = \frac{\mu_1}{\lambda}.$$

So, $z_1 = \frac{\mu_1}{\lambda \widehat{z}_1}$ and

$$Q_1(z) = \frac{P_1(0)}{1 - \frac{z}{z_1}}; Q_2(z) = \frac{P_2(0)}{1 - \frac{z}{z_2}} \tag{14}$$

From equations (8)–(9), we obtain

$$\begin{aligned}(\lambda - \lambda z)Q_0(z,x) = &\frac{\partial Q_0(z,x)}{\partial x} - \frac{\partial Q_0(z,0)}{\partial x} + \gamma_1(Q_1(z) - P_1(0))H(x) + \\ &+ \gamma_2(Q_2(z) - P_2(0))H(x) + \frac{1}{z}\left(\frac{\partial Q_0(z,0)}{\partial x} - z\frac{\partial P_0(1,0)}{\partial x}\right)H(x).\end{aligned} \tag{15}$$

Let $f_0(z,s)$ be the Laplace transform of the generating function $Q_0(z,x)$ in the continuous component x. Then multiplying equation (15) by e^{-sx} and after integration from $x = 0$ to ∞ we get

$$\begin{aligned}s(s - \lambda + \lambda z)f_0(z,s) = &\frac{\partial Q_0(z,0)}{\partial x} - \gamma_1(Q_1(z) - P_1(0))h(s) \\ &- \gamma_2(Q_2(z) - P_2(0))h(s) - \frac{1}{z}\left(\frac{\partial Q_0(z,0)}{\partial x} - z\frac{\partial P_0(1,0)}{\partial x}\right)h(s)\end{aligned} \tag{16}$$

Rearranging the right hand side lead to

$$\begin{aligned}s(s - \lambda + \lambda z)f_0(z,s) = &(1 - \frac{1}{z}h(s))\frac{\partial Q_0(z,0)}{\partial x} \\ &- \gamma_1(Q_1(z) - P_1(0))h(s) - \gamma_2(Q_2(z) - P_2(0))h(s) \\ &+ \frac{\partial P_0(1,0)}{\partial x}h(s).\end{aligned} \tag{17}$$

The function $f_0(z,s)$ is analytic in the domain $\Re e(s) \geq 0$ and since the left hand side is null at the point $s = \lambda - \lambda z$, then the right hand side must also be zero. So, we have

$$\begin{aligned}\left(1 - \frac{1}{z}h(\lambda - \lambda z)\right)\frac{\partial Q_0(z,0)}{\partial x} = &\gamma_1(Q_1(z) - P_1(0))h(\lambda - \lambda z) + \gamma_2(Q_2(z) \\ &- P_2(0))h(\lambda - \lambda z) - \frac{\partial P_0(1,0)}{\partial x}h(\lambda - \lambda z).\end{aligned} \tag{18}$$

Rearranging this equation, yields

$$\begin{aligned}\frac{\partial Q_0(z,0)}{\partial x} = &\frac{h(\lambda - \lambda z)}{1 - \frac{1}{z}h(\lambda - \lambda z)} \\ &\times \left[\gamma_1(Q_1(z) - P_1(0)) + \gamma_2(Q_2(z) - P_2(0)) - \frac{\partial P_0(1,0)}{\partial x}\right].\end{aligned} \tag{19}$$

Next, replacing (19) in (17) we obtain:

$$\begin{aligned}s(s - \lambda + \lambda z)f_0(z,s) = &\left[(1 - \frac{1}{z}h(s))\frac{h(\lambda - \lambda z)}{1 - \frac{1}{z}h(\lambda - \lambda z)} - h(s)\right] \\ &\times \left(\gamma_1(Q_1(z) - P_1(0)) + \gamma_2(Q_2(z) - P_2(0)) - \frac{\partial P_0(1,0)}{\partial x}\right).\end{aligned} \tag{20}$$

Similarly to (Ait Braham *et al.* 2023), using again the analytical properties of Fourier transform, we get after some algebra

$$Q_0(z,\infty) = \left[\gamma_1(Q_1(z) - P_1(0)) + \gamma_2(Q_2(z) - P_2(0)) - \frac{\partial P_0(1,0)}{\partial x}\right] \times \frac{z[1 - h(\lambda - \lambda z)]}{(\lambda - \lambda z)[z - h(\lambda - \lambda z)]}. \tag{21}$$

At the bound of the system of equations (or the Markovian graph), we have

$$(\lambda + \gamma_1)P_1(0) = \frac{\partial P_0(1,0)}{\partial x} + \mu_1 P_1(1) \tag{22}$$

So,

$$\frac{\partial P_0(1,0)}{\partial x} = (\lambda + \gamma_1)P_1(0) - \mu_1 P_1(1) \tag{23}$$

Next, for $i = 2$,

$$\lambda P_2(0) = \gamma_1 P_1(0) + \mu_2 P_2(1);\ P_2(0) = \frac{\gamma_1}{\lambda} P_1(0) + \frac{\mu_2}{\lambda} P_2(1) \tag{24}$$

At this stage, we have some undetermined constants: $P_1(0)$, $P_2(0)$, $P_1(1)$, $P_2(1)$ and $\frac{\partial P_0(1,0)}{\partial x}$. First, we can try to found $P_1(1)$, $P_2(1)$ by using a property of generating function

$$P_1(1) = Q_1'(0) = \frac{P_1(0)}{z_1};\ P_2(1) = Q_2'(0) = \frac{P_2(0)}{z_2} \tag{25}$$

From equations (23) and (25) we obtain

$$\frac{\partial P_0(1,0)}{\partial x} = (\lambda + \gamma_1)P_1(0) - \mu_1 \frac{P_1(0)}{z_1} \tag{26}$$

and from equation (25),

$$P_2(0) = \frac{\gamma_1}{\lambda} P_1(0) + \frac{\mu_2}{\lambda}\frac{P_2(0)}{z_2};\ P_2(0) = \frac{\gamma_1 z_2}{\lambda z_2 - \mu_2} P_1(0) \tag{27}$$

So, we have only one unknown constant $P_1(0)$. It can be determined by using the normalization condition

$$Q_1(1) + Q_2(1) + Q_0(1,\infty) = 1 \tag{28}$$

But it is not difficult to see that

$$Q_0(1,\infty) = 1 - Q_1(1) - Q_2(1) = \rho = \lambda h_1 \tag{29}$$

So, $P_1(0) = \frac{1-\rho}{\Phi}$, where

$$\Phi = \frac{z_1}{z_1 - 1} + \frac{\gamma_1 (z_2)^2}{(z_2 - 1)(\lambda z_2 - \mu_2)} \tag{30}$$

Since $1 - \rho$ must be strictly positive, we have a necessary condition $\rho < 1$ for stability. We can prove that this condition is also sufficient as in the work by Ait Braham *et al.* 2023.

5 PERFORMANCE EVALUATION

From the results above, we can derive several performance measures. Some illustrations and examples are showing now to illustrate the effect of given parameters on mean performance measures. The probability that the server is in vacation type 1 is given by

$$Q_1(1) = \frac{z_1}{z_1 - 1} P_1(0) \tag{31}$$

and in vacation type 2

$$Q_2(1) = \frac{\gamma_2 z_2}{(\lambda z_2 - \mu_2)(z_2 - 1)} P_1(0) \tag{32}$$

Figure 1 shows the evolution of these probabilities versus vacation rate type 1, γ_1 when $\gamma_2 = 0.4$: Q_1 decreases while Q_2 increases with increasing of γ_1. Figure 4 shows similar evolution versus working service rate type 1 μ_1.

The mean number of customers when server is in vacation kind 1

$$N_1 = E(N/I) = Q_1'(1) = \frac{z_1}{(1 - z_1)^2} P_1(0) \tag{33}$$

and in vacation kind 2,

$$N_2 = E(N/2) = Q_2'(1) = \frac{z_2}{(1 - z_2)^2} P_2(0) \tag{34}$$

Figures 2 and 3 show the evolution of N_1 and N_2 versus vacation γ_1 and working rate type 1 μ_1 when $\gamma_2 = 0.4$.

The mean number of customers in the system

$$E(Q) = Q_0'(1) + Q_1'(1) + Q_2'(1) \tag{35}$$

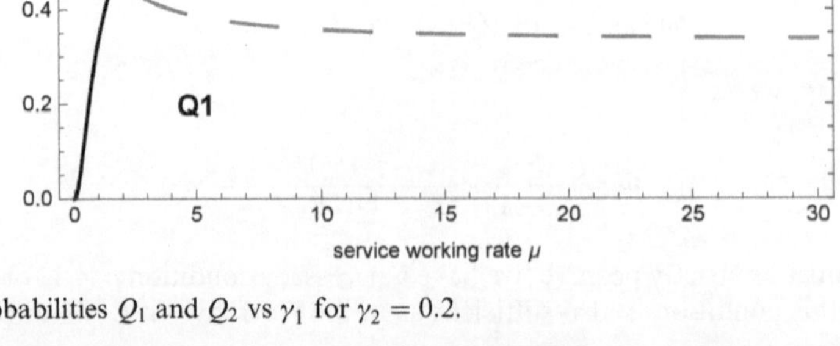

Figure 1. Probabilities Q_1 and Q_2 vs γ_1 for $\gamma_2 = 0.2$.

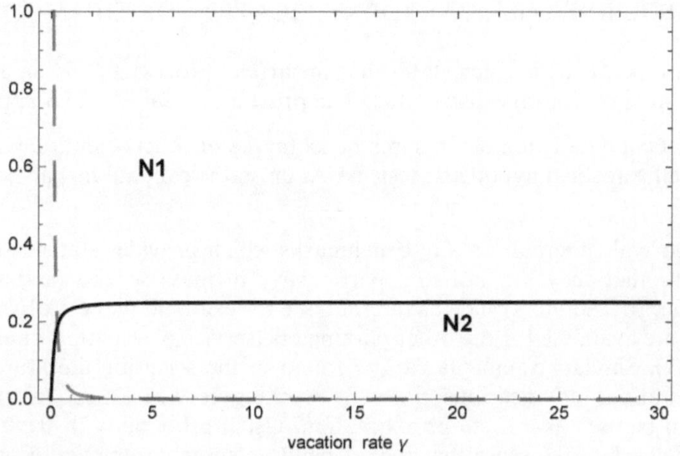

Figure 2. Mean number of customers $N1$ and $N2 = E(N2)$ vs γ_1 for $\gamma_2 = 0.2$.

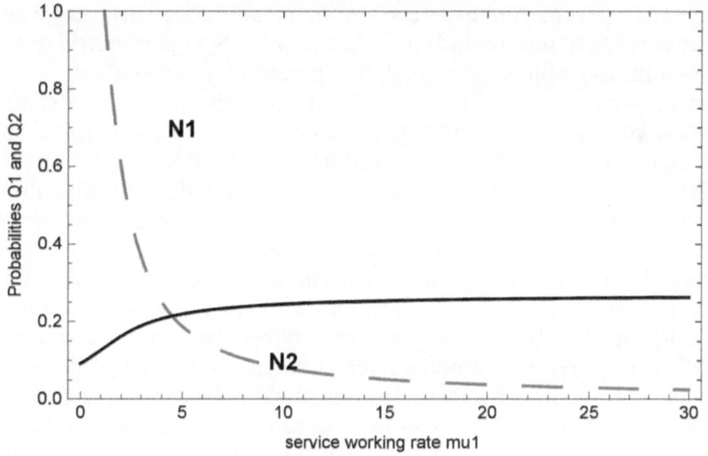

Figure 3. Mean number of customers $N1$ and $N2$ vs μ_1 for $\mu_2 = 0.2$.

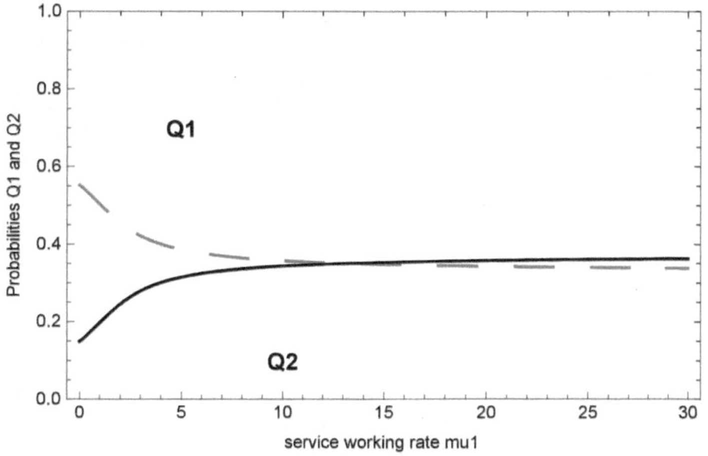

Figure 4. Probabilities $Q1$ and $Q2$ vs μ_1 for $\mu_2 = 0.2$.

6 A NOTE ABOUT STATISTICAL IDENTIFICATION OF THE MODEL

Our model is of type $M/G/1$ which states that the arrival process is Poisson and the service time in normal state is arbitrary distributed. The practionner can use two approachs:

- Refer to statistical data available in some benchmarks or other scientific papers.
- Provide himself statistical hypothesis tests based on statistical data he had collected on the system.

Concerning arrivals, there are many benchmarks which provides statistical argumenting experimentations that they are Poisson, particularly in mass service (actually known as Queueing, Cloud, IoT Services) such as internet, see for example the work by (Foster 2003). Statistical data are available for the other parameters (service or vacation times) in the work by (Moore 2001). Similar arguments can be found in the scientific literature about exponentiality assumptions of vacations times or working service times. In any case, such assumptions can be used as a first approximation. Usual probability distributions are qualified as "parametric" in the sense that they depend on some parameters. For example, the exponential distribution depends on only one parameter (mean or failure rate), the normal distribution depends on two parameters (mean and variance). To identify such parametric distributions there are parametric tests (test on the mean or the variance) when such parameters have such a physical interpretation. There are also Non parametric tests (Khi-Square or Kolmogorov-Smirnov) which test directly the probability distribution.

However, (Benaoudia *et al.* 2023) show example in wich on the same statistical sample, the chi-square test conducts to accept two different distributions: Gamma and Raighley, which are both characterized by Increasing Failure Rate (IFR).

In practice, Engineer is not interested by the concrete parametric probability distribution itself, but simply by some "ageing property" of the component or the system. The simplest example is given by Reliability or Survival analysis (Lai *et al.* 2006; Shaked *et al.* 1994), in which practitioners are interested only about a stage of ageing or rejuvenation; for example "Increasing(Decreasing) Failure Rate" i.e. *IFR*(*DFR*).So, we can consider the class of all probability distributions with such property *IFR*(*DFR*). Such a class is non parametric rather than parametric since it contains a collection of PDFs which have in common a given ageing property. The properties of such classes have also some interest for other stochastic models such as Queueing, Insurance, Medicine or Biological models (Shaked *et al.* 1994). Non parametric tests for non parametric probability distributions $IFR, NBU, IFRA, \ldots$ and their duals can be found in (Lai *et al.* 2006). Now, concerning the statistical identification of the "uncertainty" about the residual service time in normal mode. A first approach is to use probability theory and thus to use traditional methods of statistical inference as discussed above for parametric and/or non parametric distributions. The notion of "uncertainty" have been discussed from different points of view. In particular from the point of view of entropy which comes from the founding paper of (Shannon 1948) about Information theory. Ebrahimi 1997 introduced a new non parametric probability distribution called Decreasing Uncertainty Residual Life Distribution (DURL). He proposed a non fparametric test for this non parametric probability distribution. Recentely, (Benaoudia *et al.* 2023) proposed a new non parametric test for such a non parametric distribution.

7 CONCLUSION

We have provided an insight in the topic of differentiated working vacation vacations and general service in normal mode and exponential working service in vacation. We have obtained some results for the case of exponential working vacations. In an ongoing work we hope to obtain more results by solving the system for general service time distributions in working vacations and exploiting the solution in order to obtain the mean performance measures and use them for the energy trade-off problem.

ACKNOWLEDGEMENT

We are thankful to the anonymous referees for their helpful comments. We appreciate also the fruitful discussions and hospitality with professor Tuan Phung-Duc during my visit at Tsukuba University (Japan) in july 2023. This work was supported in parts by Ministery of High Education and Scientific Research (DGRSDT) through grant PRFRU $N°C00L07UN$160420220004.

REFERENCES

Ait Braham K., Taleb S. and Aissani, A. (2023). Performance analysis of a non-markovian queue with differentiated vacations. *SN Computer Science*, **4**, 512. https://doi.orgL10.1007/s42979-023-02057-9

Benaoudia H. and Aissani A. (2023). Non parametric test for decreasing uncertainty on residual life distribution. *Mathematics*, 11, pp.1–17, https://doi.org/10.3390/math11204227.

Bachmat E. *Mathematical Adventures in Performance Analysis Birkhauser*.

Chandrasekaran *et al.* (2016). A survey on working vacation queueing models. *International J. of Pure and Applied Mathematics*, vol.106, N°6, 33–41.

Dyviat *et al.* A literature survey on queueing model with working vacation. *RT A*, N° 1 (17), vol.19, 2024.

Foster, L., Cao, J., Cleveland, W., Lin, D. and Sun, D. (2003). Internet traffic tends toward Poisson and independent as the load increases. In: Denison, D., Hansen, M., Holmes, C., Mallick, B., Yu, B. (eds.) *Nonlinear Estimation and Classification*. LNCS, vol. 171, pp. 83–110. Springer, Heidelberg

Gandhi, A., Harchol-Balter, M. and Kozuh, M.A. (2012). Are sleep states effective in data centers. *In: Green Computing Conf. IGCC2012*, Eds IEEE Computer 1, pp. 1–10.

Guo, X., Zhou, S. and Niu, Z. (2015). Optimal wake-up mechanism for single base station with sleep mode. *In: 25th Int. Teletraffic Congress (ITC2015)*, 1–8.

Isci C., McIntosh S., J., Das R., Hanson J., Piper S. *et al.* (2013). Efficient virtualization power management with low-latency server power-states. *ACM SIGARCH Comput. Architec. News*,**41**(3), 96–107.

Kempa, W.M. (2013). On buffer overflow duration in WSN with a vacation-type power saving mechanism. In: *Int. 22nd ITC Specialist Seminar on Energy Efficient and Green Networking (SSEEGN)*, Riccarton, pp. 55–55.

Lai, C.D. and Xie, M. (2006). *Stochastic Ageing and Dependance in Reliability*; Springer: Berlin/Heidelberg, Germany.

Lu, X., Aalto, S. and Lassila, P. (2013). Performance-energy trade-off in data centers:imapct of switching delay. In: *Proc. 22nd ITC Specialist Seminar on Energy Efficient and Green Networking (SSEEGN)*, Riccarton, pp. 55–55.

Moore, D., Voelker, G. and Savage, S. (2001). Inferring internet denial-of-service activity. In: *Proceedings of the 2001 USENIX Security Symposium*, pp. 9–22. IEEE Press, New York

Phung-Duc, T. (2015). Systems with power-saving modes. In; International Conference on Analytical and Stochastic Modeling Techniques and Applications ASMTA2015 Eds Springer pp. 158–172.

Sampath, M.I.G.S. and Liu, J. (2020). Impact of customer's impatience on an *M/M/1* queueing systems subject to differentiated vacations with a waiting server. *Qual. Technol. Quant. Manage.*,17**2**, 125–148.

Shaked, M. and Shantikumar, J.G. (1994). *Stochastic Orders and Their Applications*; Academic Press: San Diego, CA, USA.

Upadhyaya, S. (2016). Queueing systems with vacation: An overview. *Int. J. Math. Operat. Res.* **92**, 167–213.

Zhang H. and Zhou G. (2017). *M/M/1* Queue with *m* kinds of differentiated working vacations. *J. Appl. Math. Comput.* 54, 213–227.

Recent Advances in Reliability and Maintenance Modeling – Okamura, Inoue & Xiao (Eds)
© 2025 the Author(s), ISBN: 978-1-032-79098-5

Optimal preventive maintenance policy for a proportional resilience system with Weibull lifetime

Tien-Yu Chan, Ruey-Huei Yeh & Chao-Peng Liu
Department of Industrial Management, National Taiwan University of Science and Technology, Taiwan

ABSTRACT: Systems are often subject to the interaction of internal degradation and external random shocks, making them prone to premature failures. This study aims to explore the impacts of random shocks and resilience (self-healing capabilities) on the maintenance policy for a deteriorating system to minimize the expected total cost rate. This study considers the scenario where resilience diminishes with both the system operational time and the number of random shocks and investigates the impact of preventive maintenance actions on various levels of resilience. Mathematical models for the expected total cost rate under proportional resilience systems are constructed, and the optimal preventive maintenance policy for each system is determined when the lifetime distribution of the system is Weibull. Finally, some conclusions are drawn through numerical analysis for decision-makers in assessing system replacement and preventive maintenance policies.

1 INTRODUCTION

Due to extreme weather conditions, governments and corporations are placing increasing emphasis on sustainable operations, actively developing and promoting materials and products that are recyclable or have extended lifespans. In recent years, products and systems that are recyclable or possess self-healing capabilities have been continually developed to extend the lifespan of products and systems to fulfill the objectives of a circular economy. Merryn *et al.* (2021) proposed that in the Circular Economy (CE), extending the lifespan of products is a key strategy for maximizing the value of products and components within the system, where self-healing technologies are significant methods for enhancing product reliability and lifespan. The concept of self-healing was first introduced by Holling (1937) and applied in the field of biology, defined as "the behavior of ecosystems can be described by two attributes: resilience and stability," where resilience is defined as "the persistence and tolerance of an organism or system to absorb external variations within the system's internal relationships." This concept has been extensively applied in various fields in the following years, including reliability (Liu *et al.* 2017), architecture (Yildirim *et al.* 2015), and materials science (Tan *et al.* 2022).

In general, the failure causes of engineering systems are influenced by internal degradation and external random shocks. Keedy and Feng (2012) summarized that there are mainly two failure modes in systems: the delayed failure mode caused by cyclic stress and the instantaneous failure mode caused by a single impact overload. They integrated these with a probabilistic model to develop a comprehensive reliability model for medical stents. Song *et al.* (2016) categorized system failures into soft failure processes and hard failure processes, considering the magnitude of transmitted impacts and the dependency relationship between impact size and damage.

When a system temporarily fails due to external shocks, maintenance is required to restore it to the operational condition. Efficient maintenance strategies can extend the equipment's

 DOI: 10.1201/9781003491309-2

lifespan and significantly reduce maintenance costs and downtime. Regarding repairable products, Wang (2022) differentiated maintenance strategies into Corrective Maintenance (CM) and Preventive Maintenance (PM), developing distinct maintenance strategies for both single-machine and multi-machine systems and their interconnections. Zheng and Viliam (2020) introduced a condition-based maintenance strategy for systems affected by dependent soft and hard failures, optimizing the maintenance approach to address the specific vulnerabilities introduced by these failures.

However, when the system's tolerance reaches a critical value, the system degradation process will be restarted through replacement. Lim *et al.* (2016) proposed an age replacement policy based on the age of the system failure with a stochastic chance of minor repairs of 1-p or perfect repairs of p and derived the expected cost per unit of time and the optimal age of the replacement. Badía *et al.* (2018) categorized system failures into two types, major-type failures and minor-type failures. When the system is in major failure, it will be replaced, and when the system is in minor failure, a repair strategy will be adopted. Assuming that the failure process follows the Pólya process, and based on the age and a fixed number of failures, the long-term average cost rate between minor repair and perfect repair is derived and the optimal replacement strategy is found.

In the past, reliability studies rarely incorporated the self-healing capabilities of engineering systems into reliability models following external impacts. Therefore, this research focuses on repairable and resilient (self-healing) systems under self-degradation and random shocks. It assumes that the system's degradation follows a Weibull distribution, and its self-healing ability decreases with both the operational time and the number of impacts. To minimize the expected total cost rate, this study seeks to determine the optimal replacement time. Furthermore, it explores the cost-effectiveness and impacts on the optimal replacement strategy under different models of self-healing capabilities.

The structure of this paper is as follows: Section 2 constructs cost models for systems whose lifetimes follow a Weibull distribution and whose self-recovery capabilities decrease with the number of impacts. In Section 3, the study seeks the optimal maintenance policy to minimize the expected total cost rate. Subsequently, Section 4 conducts a numerical analysis. Finally, some conclusions are provided in Section 5.

2 MATHEMATICAL MODEL

In this Section, the system under consideration is described and mathematical notations are defined in Section 2.1 and Section 2.2, respectively. Section 2.3 describes the mathematical models for system degradation states. Finally, Section 2.4 involves the construction of cost models for various types of systems.

2.1 *System description*

Given that the initial state of the repairable system is brand new with a strictly increasing failure rate function denoted by $h(t)$. Both intrinsic degradation and external shocks contribute to failures, where the shock process is modeled by a Poisson Process with a rate of λ. Upon experiencing an external shock, the system instantaneously self-recovers (resilience), and during the recovery period from the shock to full restoration, no additional shocks occur. The magnitude of each shock and the proportion of recovery are denoted by θ and π, respectively, where each shock magnitude follows an independent uniform distribution and the recovery amount is less than the shock magnitude. The value of the self-recovery proportion is π, which ranges from 0 to 1.

When the system fails, minimal repairs are conducted to restore the system's capacity to its pre-failure state such that the failure process follows a non-homogeneous Poisson process. The duration of repairs and PM is relatively short compared to the planned period and thus is considered negligible. PM actions are employed to slow down the rate of failure. A total of m PM actions are performed periodically within the replacement period. Each maintenance operation can reduce the failure rate $h(t)$ by the amount δ which is positive and less than $h(t)$. Since the PM is performed periodically with interval Δt, we have $\Delta t = T/(m+1)$ where T is the replacement time. This leads to the constraint $0 \leq \delta \leq h(\Delta t)$ when $h(t)$ is an increasing function of t.

When the system is replaced, replacement action is performed instantaneously, hence the time required for replacement is negligible. The objective of this paper is to determine the optimal PM and replacement policy such that the expected cost rate is minimized for various degradation systems under external shocks.

2.2 *Notations*

The mathematical notations used in this paper are defined below.

T	System replacement time.
$h(t)$	Failure rate function of the system at time t.
C_m	Cost per minimal repair.
$C_{pm}(\delta)$	PM cost at δ level of maintenance.
C_r	System replacement cost per replacement.
$N(t)$	The number of shocks suffered by the system during the replacement interval $[0,t]$.
m	PM frequency in the interval $[0, T]$.
θ_i	Magnitude of i-th random shock, $i = 1,2, \ldots ,N(t)$. $\theta_i \sim U(0,w)$ with $E[\theta_i] = w/2$.
s_i	Arrival time of the i-th random shock. $i = 1,2, \ldots ,N(t)$.
π	Proportion of resilience after a random shock.
δ	Magnitude of each PM action.
t_k	Time epoch of the k-th PM, $k = 1,2, \ldots ,m$.
λ	The arrival rate of random shocks.
α	Scale parameter of Weibull distribution
β	Shape parameter of Weibull distribution
$N(t)$	Number of random shocks within time interval $[0,t]$.
$M(t)$	Number of failures within time interval $[0,t]$.
$EC(T)$	The expected total cost of the system within the replacement interval $[0,T]$.
$ECR(T)$	Expected total cost rate of the system within the replacement interval $[0,T]$.
$\overline{EC}$	Expected total cost rate without PM.

2.3 *Cost rate model for degradation systems under random shocks*

This section describes the degradation process for systems with various levels of self-recovery capabilities and the expected costs based on the degradation process. Consider that the system lifetime follows the Weibull distribution and hence the failure rate function of the system can be represented as $h(t) = \alpha\beta(\alpha t)^{\beta-1}$, where β is the shape parameter influencing the slope of the system's degradation curve.

When the system fails, minimal repairs are employed to restore it back to operational status, incurring a minimal repair cost C_m. Upon reaching the replacement time T, the system is immediately replaced, incurring a replacement cost C_r. After a replacement, the system is renewed. Suppose that periodical (fixed time interval) PM actions are performed to reduce the number of failures and each PM action with level δ incurs a cost $C_{pm}(\delta) = a + b\delta$. As the system operates over time, the number of external random shocks $N(t)$ gradually increases, and the total number of system failures $M(t)$ also increases, accordingly. Therefore, within the replacement time T, the expected total cost rate of the system can be expressed by

$$ECR(T) = \frac{C_m E[M(T)] + mC_{pm}(\delta) + C_r}{T} \tag{1}$$

2.3.1 *Model for degradation system without resilience*

Within the replacement period $[0,T]$, the system undergoes $N(t)$ random impacts and each impact increases the system's failure rate by θ_i which is uniformly distributed with $E[\theta_i] = w/2$, where $i = 1, 2, \cdots, N(t)$. Additionally, there are k PM actions performed periodically within $[0,T]$ and each PM action reduces the system's failure rate by δ. The degradation process for systems without resilience is illustrated in Figure 1.

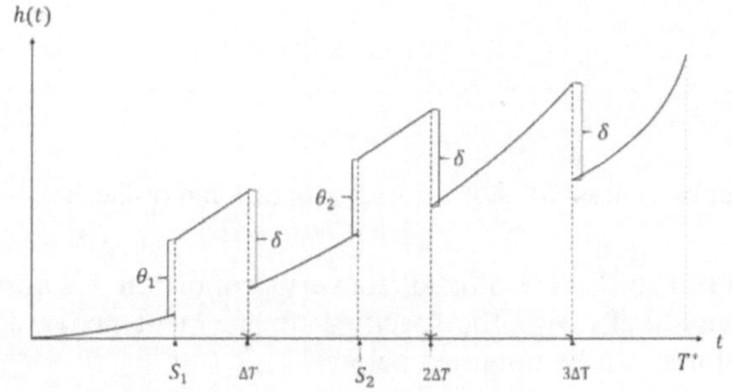

Figure 1. The degradation process for systems without resilience.

Under PM, the failure rate function of the system can be represented as

$$h(t) = \alpha\beta(\alpha t)^{\beta-1} + \sum_{i=1}^{N(t)} \theta_i - k\delta \text{ for } k\Delta t \le t < (k+1)\Delta t \tag{2}$$

Without resilience, the system suffers $N(T)$ random shocks within the replacement interval $[0,T]$. Since minimal repairs are performed upon failures, the failure process follows a non-homogeneous Poisson process and the expected number of system failures can be obtained as $E[M(T)] = \int_0^T h(t)dt$ and the arrival times of random shocks $(s_1, s_2, \cdots, s_{N(T)})$ follow the ordered uniform distribution on [0, T]. Furthermore, there are k PM actions with level δ performed on $t_k = k\Delta t = \frac{kT}{m+1}$. Therefore, the expected number of system failures can be derived as follows:

$$\begin{aligned} E[M(T)] = E[E[M(T)|N(T)]] &= E\left[\int_0^{T_1} \alpha\beta(\alpha t)^{\beta-1}dt + \frac{wT}{4}\cdot N(T)\right] - E[\delta\sum_{k=1}^{m}(T - t_k)] \\ &= (\alpha T)^\beta + \frac{w\lambda T^2}{4} - \frac{m\delta T}{2} \end{aligned} \tag{3}$$

where the expected number of shocks is $E(N(T)) = \lambda T$ since the arrival of shocks follows the Poisson process with intensity λ. The expected total cost rate of the system without resilience can be obtained by substituting Eq. (3) into equation Eq. (1).

2.3.2 *Model for degradation system with proportional resilience*

Within the replacement period $[0,T]$, there are a total of $N(t)$ random shocks arrived, and each shock magnitude is θ_i, and the system's self-resilience varies with the size of the shock in a fixed proportion π. Again, there are k PM actions with level δ performed, the degradation process for systems with resilience can be illustrated in Figure 2. From Figure 2, the failure state of the system can be described by

$$h(t) = \alpha\beta(\alpha t)^{\beta-1} + \sum_{i=1}^{N(t)} \theta_i(1-\pi) - k\delta \text{ for } k\Delta t \le t < (k+1)\Delta t \tag{4}$$

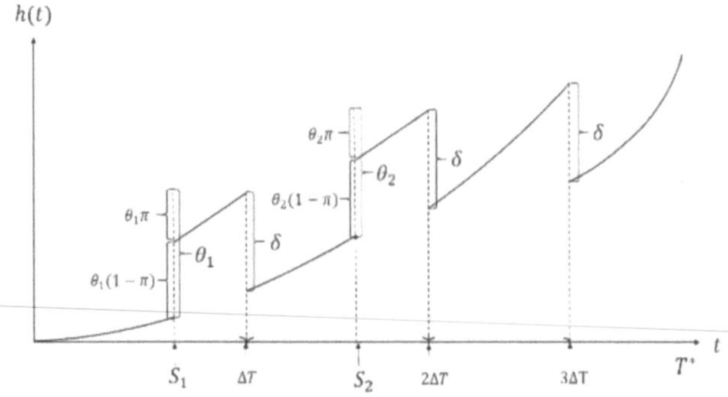

Figure 2. The degradation process for systems with proportional resilience.

According to the results of Eq.(4), a fixed recovery proportion π is applied, and the system endures $N(T)$ random shocks over the specified replacement period [0, T]. The expected number of system failures can be obtained below.

$$E[M(T)] = E[E[M(T)|N(T)]] = E\left[\int_0^{T_1} \alpha\beta(\alpha t)^{\beta-1}dt + \frac{wT(1-\pi)}{4} \cdot N(T)\right] - E\left[\delta\sum_{k=1}^{m}(T-t_k)\right]$$
$$= (\alpha T)^{\beta} + (1-\pi)\frac{\lambda w T^2}{4} - \frac{m\delta T}{2} \tag{5}$$

By substituting Eq.(5) into Eq.(1), the expected total cost rate for the system with a fixed proportion of resilience can be obtained.

3 OPTIMIZATION MAINTENANCE POLICY

In the previous Section, the expected total cost rate models are constructed for both non-resilient and resilient systems whose lifetime follows the Weibull distribution. In this section, the optimal replacement times and PM policy will be derived such that the expected total cost rate is minimized for both systems.

3.1 *Optimal replacement time for systems without resilience*

Based on the expected total cost rate model for the non-resilient system derived in Section 2.3.1, the first and second derivatives of Eq.(1) can be obtained as follows:

$$\frac{\partial ECR(T)}{\partial T} = C_m \cdot \left[a^{\beta}(\beta-1)T^{\beta-2} + \frac{\lambda w}{4}\right] - m(a+b\delta)T^{-2} - C_r T^{-2} \tag{6}$$

and

$$\frac{\partial^2 ECR(T)}{\partial^2 T^2} = C_m \cdot \left[a^{\beta}(\beta-1)(\beta-2)T^{\beta-3}\right] + 2m(a+b\delta)T^{-3} + 2C_r T^{-3} \tag{7}$$

From Eq.(7), it is obvious that $\partial^2 ECR(T)/\partial T^2 > 0$ when $\beta \geq 2$. That means the curve is an upward-opening convex function, indicating a minimum within the region existed. Therefore, by setting Eq.(6) to zero, T^* can be obtained. For a specific value β, the following properties can be derived.

Property 1: When $\beta = 2$ and $C_{pm}(\delta) = a + b\delta$, given any m and δ, the optimal replacement time T^* can be obtained as

$$T^* = 2\sqrt{\frac{C_r + m(a + b\delta)}{C_m(4\alpha^2 + \lambda w)}} \tag{8}$$

By substituting the optimal replacement time from Eq.(8) into Eq.(1), the optimal expected total cost rate can be obtained. For other values of β, we may derive the optimal replacement time by setting Eq.(6) equal to 0 and solving the equation by any search method.

3.2 *Optimal replacement time for systems with proportional resilience*

Based on the expected total cost rate model for the proportional resilience system derived from in Section 2.3.2, the first and second derivatives of Eq.(1) can be obtained as follows:

$$\frac{\partial ECR(T)}{\partial T} = C_m\left[(\beta-1)\alpha^\beta T^{\beta-2+}\frac{w\lambda(1-\pi)}{4}\right] - m(a + b\delta)T^{-2} - C_r T^{-2} \tag{9}$$

and

$$\frac{\partial^2 ECR(T)}{\partial T^2} = C_m[(\beta-1)(\beta-2)\alpha^\beta T^{\beta-3}] + 2m(a + b\delta)T^{-3} + 2C_r T^{-3} \tag{10}$$

Again, from Eq.(10), it is obvious that $\partial^2 ECR(T)/\partial T^2 > 0$ when $\beta \geq 2$. By setting Eq.(9) equal to zero, T^* can be obtained. For a specific value β, the following properties can be found.

Property 2: When $\beta = 2$ and $C_{pm}(\delta) = a + b\delta$, given any m and δ, the optimal replacement time T^* can be obtained as

$$T^* = 2\sqrt{\frac{C_r + m(a + b\delta)}{C_m(4\alpha^2 + \lambda w(1-\pi))}} \tag{11}$$

By substituting the optimal replacement time from Eq.(11) into Eq.(1), the optimal expected total cost rate can be obtained.

3.3 *Optimal level and number of preventive maintenance*

Since the purpose of PM action is to reduce the expected total cost rate within the replacement period, if it does not have any advantage, then PM action is not required, that is, $m = 0$. Hence, there is an upper bound of the expected total cost when $m = 0$, denoted as $\overline{EC} = C_m E[M(T)] + C_r$. For any m>0 and δ>0, the expected total cost can be expressed as follows.

$$EC(T, m, \delta) = \overline{EC} + mC_{pm}(\delta) - C_m\frac{m\delta T}{2} = \overline{EC} + m(a + b\delta) - C_m\frac{m\delta T}{2} \tag{12}$$

Observing Eq. (12), it is clear that the preventive cost parameters a and b play important roles in finding the optimal PM policy. To derive the optimal maintenance number m and maintenance level δ, we first consider the constraint of δ, which is $0 \leq \delta \leq h(\Delta t)$. This means that there is an upper bound of the maintenance level δ for any m>0, which is $\overline{\delta_m} = h\left(\frac{T}{m+1}\right)$. Given any values of T and m, we have the following property.

Property 3: When there is a linear relationship between the cost of PM and the level of maintenance, denoted as $C_{pm}(\delta) = a + b\delta$, $a \geq 0$, $b \geq 0$, given any values of T and m, if $C_m T/2 - b \leq 0$, then $\delta^* = 0$; otherwise, $\delta^* = \overline{\delta_m}$.

From Eq.(12), we have that if $C_m T/2 - b \leq 0$, then it is an increasing function of δ. Hence, the optimal maintenance level is $\delta^* = 0$, which means PM is not required. On the other hand, Eq.(12) is a decreasing function of δ. Therefore, the optimal maintenance level is the upper limit of δ, which is $\delta^* = \overline{\delta_m}$.

4 NUMERICAL EXAMPLES

In this section, we explore the impact of PM action on the systems with and without resilience. The parameter setting is shown in Table 1. That means the preventive cost is $C_{pm}(\delta) = 10 + 10\delta$, the minimal repair cost is $C_m = 20$, and the replacement cost is $C_r = 1000$.

Table 1. Parameters setting.

α	β	w	λ	π	C_m	C_r	a	b
1	2	5	0.5	0.5	20	1000	10	10

Based on the parameter setting given in Table 1, for any $T > 1$, $C_m T/2 - b = 10(T - 1) > 0$. From property 3, it is worthy to perform PM when $T > 1$. Figure 3 shows the expected cost rates for systems without resilience as the replacement time T increases. Table 2 shows the optimal maintenance policies and the corresponding expected cost rates for systems with and without PM. From Figure 3 and Table 2, for systems without resilience, the replacement time increases from 5.55 to 8.91 and $ECR(T^*)$ is significantly reduced from 360.56 to 268.26 by 12 PM actions with maintenance level 1.38.

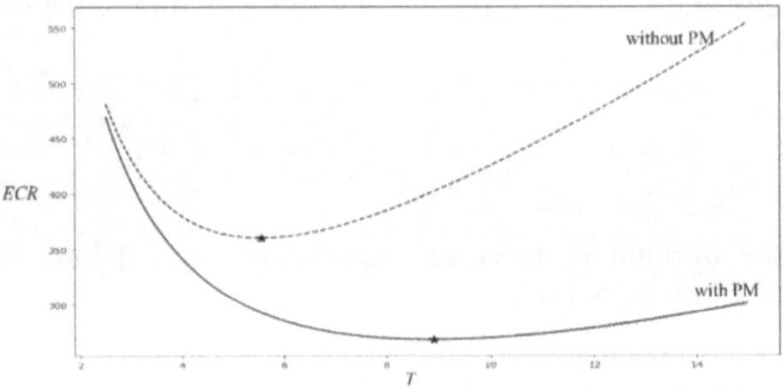

Figure 3. The impact of PM action on the systems without resilience.

Table 2. Optimal maintenance policy without resilience.

	T^*	$ECR(T^*)$	m^*	δ^*
Without PM	5.55	360.56	–	–
With PM	8.91	268.26	12	1.38

For systems with proportion resilience $\pi = 0.5$, Figure 4 and Table 3 demonstrate the expected cost rate and the optimal maintenance policies for resilient systems with and without PM. Similar results are obtained for systems with resilience, the replacement time increases from 6.17 to 12.23 and $ECR(T^*)$ reduces from 324.04 to 203.42 by 14 PM actions with maintenance level 1.64.

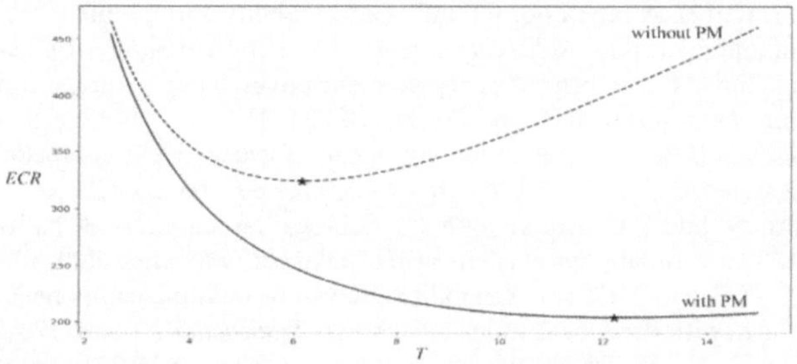

Figure 4. The impact of PM action on the systems with resilience.

Table 3. Optimal maintenance policy with resilience.

	T^*	$ECR(T^*)$	m^*	δ^*
Without PM	6.17	324.04	–	–
With PM	12.23	203.42	14	1.64

Comparing Tables 2 and 3, there are more PM actions performed for systems with proportion resilience, that is, m^* increases from 12 to 14 with higher maintenance level. Furthermore, the replacement time increases from 8.91 to 12.23 and $ECR(T^*)$ reduces from 268.26 to 203.42 by the resilience capability.

5 CONCLUSION

This paper considers systems whose lifetime follows the Weibull distribution to examine the impact of optimal PM and replacement policy on resilient systems. A general expected cost rate model is constructed and the optimal maintenance policy is derived. Some closed-form solutions and properties are obtained for some special cases. Numerical examples indicate that PM action has significantly reduced the expected cost rate for both resilient and non-resilient systems. Furthermore, PM action can extend the operational lifetimes of the system while reducing maintenance costs. Even for resilient systems, PM can further extend the optimal replacement time and reduce the long-term cost rate.

ACKNOWLEDGMENTS

This research is partly supported by a grant (MOST 110-2221-E-011-135) from the Ministry of Science and Technology, Taiwan.

REFERENCES

Badía, F. G., Berrade, M. D., Cha, J. H., and Lee, H. (2018). Optimal replacement policy under a general failure and repair model: Minimal versus worse than old repair. *Reliability Engineering & System Safety*, 180, 362–372.

Haines-Gadd, M., Charnley, F., and Encinas-Oropesa, A. (2021). Self-healing materials: A pathway to immortal products or a risk to circular economy *systems?*. *Journal of Cleaner Production*, 315, 128193.

Holling, C. S. (1973). Resilience and stability of ecological systems. *Annual Review of Ecology and Systematics*, 4(1), 1–23.

Keedy, E., and Feng, Q. (2012). A physics-of-failure based reliability and maintenance modeling framework for stent deployment and operation. *Reliability Engineering & System Safety*, 103, 94–101.

Lim, J. H., Qu, J., and Zuo, M. J. (2016). Age replacement policy based on imperfect repair with random probability. *Reliability Engineering & System Safety*, 149, 24–33.

Liu, H., Yeh, R. H., and Cai, B. (2017). Reliability modeling for dependent competing failure processes of damage self-healing systems. *Computers & Industrial Engineering*, 105, 55–62.

Merryn, H.G., F. Charnley, and E.O. Adriana. (2021). Self-healing materials: A pathway to immortal products or a risk to circular economy systems? *Journal of Cleaner Production*, 128–193.

Song, S., Coit, D. W., and Feng, Q. (2016). Reliability analysis of multiple-component series systems subject to hard and soft failures with dependent shock effects. *IIE Transactions*, 48(8), 720–735.

Tan, P., Wang, H., Xiao, F., Lu, X., Shang, W., Deng, X., ... and Zhou, X. (2022). Solution-processable, soft, self-adhesive, and conductive polymer composites for soft electronics. *Nature Communications*, 13(1), 358.

Wang, H. (2002). A survey of maintenance policies of deteriorating systems. *European Journal of Operational Research*, 139(3), 469–489.

Yang, L., Ye, Z. S., Lee, C. G., Yang, S. F., and Peng, R. (2019). A two-phase preventive maintenance policy considering imperfect repair and postponed replacement. *European Journal of Operational Research*, 274 (3), 966–977.

Yıldırım, G., Keskin, Ö. K., Keskin, S. B., Şahmaran, M., and Lachemi, M. (2015). A review of intrinsic self-healing capability of engineered cementitious composites: Recovery of transport and mechanical properties. *Construction and Building Materials*, 101, 10–21.

Zheng, R., and Makis, V. (2020). Optimal condition-based maintenance with general repair and two dependent failure modes. *Computers & Industrial Engineering*, 141, 106322.

Recent Advances in Reliability and Maintenance Modeling – Okamura, Inoue & Xiao (Eds)
© 2025 the Author(s), ISBN: 978-1-032-79098-5

Mean residual life-based maintenance of lamp systems

Z. Feng
Industrial Engineering Department, CITIC Dicastal Co., Ltd, Qinhuangdao, China

W.Y. Yun
Major in Industrial Data Science and Engineering, Department of Industrial Engineering, Pusan National University, Busan, Korea

ABSTRACT: In this paper, we consider a condition-based preventive replacement policy for two-dimensional lamp systems in which identical lamps cover the system service area. Each lamp located in the service area covers a circular area with the same radius and if the working lamps cannot cover the whole service area, then the system is failed. The failure time of each lamp follows an exponential distribution and can be replaced by a new one. We propose a preventive replacement policy based on the mean residual life and if the mean residual life of the system is less than the given threshold, then we replace all lamps failed. The expected cost rate is used as an optimization criterion and we want to find the optimal threshold to minimize the expected cost rate. The procedures to calculate the expected cost rate and to find the optimal solutions are explained and numerical examples are also studied to analyze the proposed preventive replacement policy.

1 INTRODUCTION

The interesting optimization problem in lamp systems is a coverage problem to determine the positions of a set of lamps (sensors) to cover a service area, where each lamp (sensor) covers a circular area with a given radius. In general, the objective of the optimization is to maximize system reliability, which is the probability that the whole service area can be covered by working sensors. The coverage problem of lamp systems is a topic with wide-ranging implications in sensor placement, system reliability, monitoring, communication systems, and facility layouts. In sensor surveillance and communication coverage scenarios, the reliability concept has emerged as a crucial factor in the design of coverage systems.

A typical application is the telecommunication system network design. In this topic, the optimization problem to determine the locations of base stations presents a key challenge and is related to the location set covering problem (LSCP), which aims to use a minimal number of facilities to serve a set of demands (Toregas *et al.* 1971). The main purpose of LSCP is to minimize the number of facilities needed to cover the entire demands by locating the facilities optimally, but failures in facility operations are not considered. However, facilities can be influenced by various types of interference, leading to mission failures in the system. In autonomous vehicle or telemedicine applications, a dropped connection can cause substantial tangible losses. Sensor network systems are also affected by natural disasters which can disrupt operations and power constraints, which can lead to system shutdowns and reduce overall network reliability. These issues will become more important in designing reliable complex systems.

In sensor networks, numerous sensors collaborate to monitor environments or facilitate telecommunications within a limited communication range. A notable feature of these

DOI: 10.1201/9781003491309-3

systems is the overlapping coverage zones. By locating monitors in overlapping coverage areas, complex systems can be more reliable and robust to the failures of individual monitors. Typically, system failure occurs due to the simultaneous failures of neighboring facilities.

In special cases, the structure of sensor networks resembles two-dimensional consecutive-k-out-of-n: F systems. These systems with $m \times n$ components fail whenever all components within an $r \times s$ submatrix fail. The consecutive (r,s)-out-of-(m,n): F system was introduced by Salvia and Lasher (1990) and has been extensively studied in the reliability engineering literature. Zuo (1993) investigated the optimal arrangement of rectangular and cylindrical connected-(r, s)-out-of-(m, n): F lattice systems and provided necessary conditions for the optimal arrangement, using the sum of disjoint products (SDP) method to calculate system reliability. Yamamoto and Miyakawa (1995) developed the YM algorithm, an efficient recursive algorithm to reduce computing time for (r, s)-out-of-(m, n): F lattice systems. Nakamura and Yamamoto further explored various research issues related to the two-dimensional consecutive-k-out-of-n: F system (Nakamura *et al.* 2019, 2022; Yamamoto *et al.* 2008), focusing on new methods and models to analyze system reliability in various engineering applications.

However, the (r,s)-out-of-(m,n):F system cannot represent facility locations fully in a continuous area. Moreover, we cannot simply find the values of (r,s,m,n) to represent the system state exactly (Feng *et al.* 2023a). The main research topics of lamp problems are how to calculate the system reliability efficiently and optimize the locations of lamps in the service area. Okamura *et al.* (2021) proposed a method to calculate the system reliability using upper/lower bounds in lamp systems. Feng *et al.* (2023a) used Monte Carlo and intersection test methods to check the system state, and the system reliability function could be estimated by simulation. Feng *et al.* (2023b) investigated the location design problem in lamp systems and proposed a redundant heuristic optimal solution, which the performance was checked by the genetic algorithm. Moreover, Feng *et al.* (2024) introduced a Voronoi diagram-based method to calculate the system reliability.

In this paper, we consider a preventive maintenance problem with condition-based preventive replacement in lamp systems. In industrial applications, maintenance plays a crucial role in ensuring system reliability and availability. Effective maintenance strategies can minimize downtime, optimize resource allocation, and enhance overall system performance.

Maintenance strategies typically fall into two main categories: corrective and preventive maintenance. Corrective maintenance involves repairing components after a system failure occurs, while preventive maintenance focuses on scheduled maintenance activities to prevent potential failures and extend component lifespan (Yang 2003). In condition-based preventive maintenance, the system's mean residual life (MRL) serves as a key parameter to describe the system health status. When MRL becomes less than the prespecified threshold, we replace the failed lamps preventively. The prespecified threshold is determined to minimize the expected cost rate and the procedure to find the optimal threshold of MRL is explained. Finally, we investigate some numerical examples to illustrate the method to find the optimal MRL threshold.

2 MODEL ASSUMPTIONS AND DEFINITIONS

In this section, we introduce the assumptions and notations.

Assumptions

A1. The system service area is a 1×1 square.
A2. The number of lamps is fixed.
A3. The locations of lamps are given in the service area.

A4. Each lamp can cover a circular area with the same radius r.
A5. The reliability of all lamps is same and the failure times follows an exponential distribution.
A6. Failures of lamps are independent.
A7. The system is failed if the service area cannot be covered by working lamps and the state of the system is dichotomous: working (1) or failed (0).

Notations & Terms	
n	Number of lamps.
L	Set of lamps, L = $\{l_1, \ldots, l_n\}$, where l_i denote lamp i.
S	System state.
s_i	State of lamp i.
λ	The failure rate of lamps.
θ	Mean failure time of lamps.
r	Coverage radius of each lamp.
E_r	Expected cost rate.
E_c	Expected cost of a renewal cycle.
E_d	Expected duration of a renewal cycle.
K	Failure path counting vector, $K = \{k_1, \ldots, k_n\}$.
k_i	Number of paths where i lamps have failed.
P_i	Probability of i lamps failed in a system.
c_0	Fixed cost of corrective maintenance.
c_1	Fixed cost of preventive maintenance.
c_2	Replacement cost of a lamp.
T_0	Threshold of mean residual life for preventive maintenance.

Suppose that there are n lamps in the system represented as $L = \{l_1, \ldots, l_n\}$. Let s_i represent the state of lamp l_i, where system state vector $S = \{s_1, \ldots, s_n\}$. Each lamp can either be in the "working" (1) or "failed" (0) state. We assume that lamp failures follow an Exponential distribution with failure rate λ and each working lamp can cover a circular area with a radius r. When the entire service area is covered by working lamps, the system is working; otherwise, the system has failed. An efficient Voronoi-based method proposed by Feng and Yun (2024) is used to check the system state.

3 CORRECTIVE MAINTENANCE POLICY

In this section, first we consider the corrective maintenance (CM) policy in which we replace all failed lamps only when a system failure occurs, without preventive replacement. The expected cost rate can be derived from the expected cost for a renewal cycle and the system's mean time to failure ($MTTF_S$), as given in Equation (1).

$$E_r = \frac{E_c}{MTTF_S} \tag{1}$$

To compute $MTTF_S$, we consider the sequence of lamp failures, referred as "paths." A "failure path" indicates a sequence of lamp failures to the system failure. We employ a Depth-first search tree-based method to identify all failure paths (DFP). The algorithm to find DFP is as follows:

Function **Find DFP** (lamp sets, radius)

```
Initialize empty lists 'failure paths sets', and 'current path'.
for each working lamp i remained
    test path = current path + lamp[i]
    Check the system state of the test path by Voronoi checking method (Feng et al. 2024)
    if the system state is the failure state
        Add the test path into failure path sets
    else
        For the remaining lamps, recursively call Find DFP (remaining lamps, radius)
    end
return failure paths sets
end
end function
```

From the DFP, we calculate the failure path counting vector K where k_i represents the number of paths in DFP where i lamps have failed. The probability that the system failure follows the failure path with ith lamp failures (P_i) can be calculated as Equation (3). The expected time to the system failure through the failure path ($MTTF_i$) can be calculated as Equation (4), and then the system MTTF, $MTTF_S$ is calculated in Equation (5). Here, θ is the mean failure time of lamps. Next, the expected cost of a renewal cycle, E_c can be calculated by Equation (6).

$$K = \{k_1, ..., k_n\} \tag{2}$$

$$P_i = \frac{1}{n} \times \frac{1}{(n-1)} \times \ldots \times \frac{1}{(n-i+1)} = \left(\prod_{j=n-i+1}^{n} j \right)^{-1} \tag{3}$$

$$MTTF_i = \frac{\theta}{n} + \frac{\theta}{n-1} + \ldots + \frac{\theta}{n-i+1} = \sum_{j=n-i+1}^{n} \frac{\theta}{j} \tag{4}$$

$$MTTF = \sum_{k_i \in K} k_i \cdot P_i \cdot MTTF_i \tag{5}$$

$$E_c(CM) = \sum_{k_i \in K} k_i \times P_i \times (c_0 + i \times c_2) = \sum_{k_i \in K} k_i \frac{c_0 + i \times c_2}{\prod_{j=n-i+1}^{n} j} \tag{6}$$

4 PREVENTIVE MAINTENANCE POLICY

In this section, we consider a preventive replacement policy based on the MRL in which the lamps failed are replaced preventively if the MRL exceeds to the prespecified threshold (T_0). We want to find the optimal T_0 to minimize the expected cost rate. In order to obtain the expected cost rate for given T_0, we need to obtain the expected cost and duration of a renewal cycle. First, we consider the failure paths to the system failure. For each failure path, calculate MRLs at failure time points of lamps which decrease gradually on the failure path. For a given path S, a sub path means a path that can reach S. for example, S = {1,2}, possible preceding path is only {1} and {1,2,3} and {1,2,4} are following paths.

MRL calculation procedure

Step 1: Use the traversal technique to find the system DFP.
Step 2: For each path in DFP, traverse "preceding path", construct a "working path set" include all preceding paths.
Step 3: Calculate the MRL for each path in the working path set by traversing its "sub paths" using a regressive method.
Step 4: System MRL as the sum of all weighted MRLs of working paths.

Function **Calculate MRL** (working paths, failure paths, n, θ)

Sort the working paths by the size in descending order.
Initialize an empty dictionary 'MRL results' to store MRL for each working path
For each working path
  MRL = **Calculate Path MRL** (path, working paths, failure paths, n, θ)
    MRL results[path] = MRL
  end
return MRL results
end function

Function **Calculate Path MRL** (path, working paths, failure paths, *n*, θ)

Initialize total weighted MRL = 0
Find all sub paths of the given working path
for each sub path
  calculate the weight of current sub path *w*
  if sub path in failure paths:
    sub MRL = 0
  else if sub path in working paths:
    sub MRL = MRL results [sub path]
  else
    i = number of failed lamps in sub path.
    sub MRL = $\theta / (n-i)$
    total weighted MRL = total weighted MRL + ($w \times$ sub MRL)
end
Calculate the expected time for the next lamp failure for the given path
MRL = total weighted MRL + expected time for the next lamp failure
return MRL
end function

For a given replacement threshold, T_0, we check all failure paths and cut the parts on which the MRL is less than T_0. Using failure paths refined, we can calculate the expected duration of a renewal cycle by Equations (3~5). Next, in order to obtain the expected cost of a renewal cycle, we should divide the failure paths into two groups. For failure paths which reaches to the system failure, the expected cost of a renewal cycle is given in Equation (7). For the failure paths in which a part is cut and ends at a working state of the system by preventive replacement, the expected cost of a renewal cycle is given in Equation (8). Finally, the expected cost of a renewal cycle is given in Equation (9), and E_r could be calculated as equation (1).

$$E_c(CM) = c_0 + i \times c_2 \tag{7}$$

$$E_c(PM) = c_1 + i \times c_2 \tag{8}$$

In order to find the optimal T_0 minimizing the expected cost rate, we should find all different mean residual lives on the failure paths and they are candidates of the optimal T_0.

$$E_c = E_c(CMP) + E_c(PMP) = \sum_{k_i \in K(CFP)} k_i \frac{c_0 + i \times c_2}{\prod_{j=n-i+1}^{n} j} + \sum_{k_i \in K(PMP)} k_i \frac{c_1 + i \times c_2}{\prod_{j=n-i+1}^{n} j} \tag{9}$$

5 NUMERICAL ANALYSIS

As a numerical example, we consider a lamp system with 6 lamps evenly distributed and lamp locations are (0.25;0.25), (0.25;0.50), (0.25;0.75), (0.75;0.25), (0.75;0.50), (0.75;0.75) with parameters $r = 0.5$, $\lambda = 0.1$, $\theta = 10$. Let $c_0 = 1$, $c_1 = 0.5$, and $c_2 = 0.2$. First, we consider the corrective replacement policy and obtain the expected cost rate. Failure paths based on the number of failed lamps are given in Table 1. The results show $K = (4,8,8)$, $MTTF_S = 2.5$ same as the result in Feng *et al.* (2023a), $E_c = 1.28$, and $E_r = 0.512$.

Table 1. Failure paths and counting vector.

i	k_i	Failure paths
1	4	[1], [3], [4], [6]
2	8	[2,1], [2,3], [2,4], [2,6], [5,1], [5,3], [5,4], [5,6]
3	8	[2,5,1], [2,5,3], [2,5,4], [2,5,6], [5,2,1], [5,2,3], [5,2,4], [5,2,6]

$$\begin{aligned} MTTF &= \sum_{k_i \in K} k_i \times \frac{1}{\prod_{j=n-i+1}^{n} j} \times \sum_{j=n-i+1}^{n} \frac{\theta}{j} \\ &= \frac{4}{6} \times \frac{10}{6} + \frac{8}{30} \times \left(\frac{10}{6} + \frac{10}{5}\right) + \frac{8}{120} \times \left(\frac{10}{6} + \frac{10}{5} + \frac{10}{4}\right) = 2.5 \end{aligned} \tag{10}$$

$$E_c = \sum_{k_i \in K} k_i \frac{c_0 + i \times c_2}{\prod_{j=n-i+1}^{n} j} = \frac{4}{6} \times (1 + 1 \times 0.2) + \frac{8}{30} \times (1 + 2 \times 0.2) + \frac{8}{120} \times (1 + 3 \times 0.2) = 1.28 \tag{11}$$

$$E_r = \frac{Ec}{MTTF} = \frac{1.28}{2.5} = 0.512 \tag{12}$$

Next we consider a preventive replacement policy based on MRL. By traversing MRL of all working paths, two T_0 intervals are identified: $0 \le T_0 \le 2.5$ (same as CM), or $T_0 > 2.5$. Table 2 shows $MTTF_S$, E_c and E_r. As T_0 increases, PM cost increases, $MTTF_S$ and CM cost decreases. Additionally, E_c decreases, but E_r increases due to reduced $MTTF_S$. Eventually, the interval $0 \le T_0 \le 2.5$ gave a better result.

As another example, we consider a redundantly distributed 9-lamp system in which the minimum number of lamps are used to cover the service area and remaining lamps are

Table 2. PM and CM numerical analysis result.

T_0	PMP	CMP	MTTF	PM cost	CM cost	E_c	E_r
$0 \leq T_0 \leq 2.5$ (corrective)	-	[1], [3], [4], [6], [2,1], [2,3], [2,4], [2,6], [5,1], [5,3], [5,4], [5,6], [2,5,1], [2,5,3], [2,5,4], [2,5,6], [5,2,1], [5,2,3], [5,2,4], [5,2,6]	2.500	0	1.280	1.280	**0.512**
T_0>2.5	[2], [5]	[1], [3], [4], [6]	1.667	0.233	0.800	1.033	0.620

assigned redundantly to same positions (Feng *et al.* 2023b). The parameters used are $\theta = 1$, $c_0 = 10$, $c_2 = 1$, $c_1 = 1,2,5$. For a system with $r = 0.35$, 5 lamps can cover the whole service area and the locations are (0.8381; 0.7166), (0.2558; 0.2028), (0.4730; 0.7446), (0.1349; 0.7028), (0.7558; 0.2166), (0.8381; 0.7166), (0.2558; 0.2028), (0.4730; 0.7446), (0.1349; 0.7028). For a system with $r = 0.30$, 6 lamps can cover the whole service area and the locations are: (0.1485; 0.2594), (0.8228; 0.2406), (0.5287; 0.7572), (0.1772; 0.7594), (0.4713; 0.2428), (0.8515; 0.7406), (0.1485; 0.2594), (0.8228; 0.2406), (0.5287; 0.7572). The expected cost rates in different replacement strategies are compared in Tables 3 and 4.

Table 3. Analysis of expected cost rates for $r = 0.35$.

Strategy	CM	PM (T_0)				
		$0 \leq T_0 \leq 2.500$	$2.500 < T_0 \leq 3.000$	$3.000 < T_0 \leq 3.6667$	$3.667 < T_0 \leq 4.571$	$T_0 > 4.571$
$c_1 = 5$	**2.9141**	**2.9141**	2.9732	3.3125	4.8750	5.9000
$c_1 = 2$	2.9141	2.9141	2.8018	**2.7125**	3.0750	3.5000
$c_1 = 1$	2.9141	2.9141	2.7446	2.5125	**2.4750**	2.7000

Table 4. Analysis of expected cost rates for $r = 0.30$.

Strategy	CM	PM (T_0)			
		$0.000 \leq T_0 \leq 1.667$	$1.667 < T_0 \leq 1.905$	$1.905 < T_0 \leq 2.202$	$T_0 > 2.202$
$c_1 = 5$	**4.6892**	**4.6892**	4.7607	5.1429	6.9000
$c_1 = 2$	4.6892	4.6892	4.6426	**4.6286**	5.1000
$c_1 = 1$	4.6892	4.6892	4.6033	**4.4571**	4.5000

Through numerical examples we can find that PM based on the mean residual life(MRL) can significantly reduce the expected cost rates especially when the PM fixed cost (c_1) is relatively low. These numerical examples illustrate the advantage of the PM strategy.

6 CONCLUSION

This paper considers a preventive replacement policy based on the mean residual life in lamp systems, in which identical lamps cover a system service area. Each lamp located in the service area covers a circular area, and if the working lamps cover the entire service area, then the system is working. The failure of each lamp follows an Exponential distribution and can be replaced by a new one. We propose a procedure to determine the optimal threshold value of the mean residual life. The expected cost rate is used an optimization criterion to evaluate the performance. By numerical examples, we show how to calculate the expected cost rate and that the preventive replacement based on MRL is better than the corrective replacement policy. This research can be extended to maintenance decision making in monitoring systems, production lines, and facility coverage related problems. For promising future research, we could explore more complex maintenance strategies, such as dynamic maintenance scheduling algorithms, and investigate the impact of environmental factors on lamp system reliability.

ACKNOWLEDGEMENT

This work was supported by the National Research Foundation of Korea (NRF) grant funded by the Korea Government (MSIT) (No.2022R1F1A1071057126218206530010 2)

REFERENCES

Feng, Z., Dohi, T. and Yun, W. Y. (2023a). System reliability analysis of a lamp problem by simulation. Proceedings of the Institution of mechanical engineers, *Part O: Journal of Risk and Reliability*, 237(6), 1186–1198.

Feng, Z., Zhang, R. and Yun, W. Y. (2023b). A lamp location problem based on system reliability. *Quality Technology & Quantitative Management*, doi:10.1080/16843703.2023.2283982.

Feng, Z., Okamura, H., Dohi, T. and Yun, W. Y. (2024). Reliability computing methods of probabilistic location set covering problem considering wireless network applications. *IEEE Transactions on Reliability*, 73(1), 290–303.

Nakamura, T., and Yamamoto, H. (2019). Algorithm for solving optimal arrangement problem in Connected-(r, s)-out-of-(m, n): F lattice system. *IEEE Transactions on Reliability*, 69(2), 497–509.

Nakamura, T., Yamamoto, H., and Akiba, T. (2022). Reliability of a toroidal connected-(r, s)-out-of-(m, n): F lattice system. Proceedings of the Institution of Mechanical Engineers, *Part O: Journal of Risk and Reliability*, 236(2), 329–338.

Okamura, H., Zheng, J., Dohi, T. and Yun, W. Y. (2021). A note on computation of system reliability in a lamp problem. *IEICE Technical Report*, 121(47), 52–56.

Salvia, A. A., and Lasher, W. C. (1990). Two-dimensional consecutive-k-out-of-n: F models. *IEEE Transactions on Reliability*, 39(3), 382–385.

Toregas, C., Swain, R., ReVelle, C., and Bergman, L. (1971). The location of emergency service facilities. *Operations Research*, 19(6), 1363–1373.

Yamamoto, H., and Miyakawa, M. (1995). Reliability of a linear connected-(r, s)-out-of-(m, n): F lattice system. *IEEE Transactions on Reliability*, 44(2), 333–336.

Yamamoto, H., Akiba, T., Nagatsuka, H., and Moriyama, Y. (2008). Recursive algorithm for the reliability of a connected-(1, 2)-or-(2, 1)-out-of-(m, n): F lattice system. *European Journal of Operational Research*, 188(3), 854–864.

Yang, S. K. (2023). A condition-based failure-prediction and processing-scheme for preventive maintenance, *IEEE Transactions on Reliability*, 52(3), 373–383.

Zuo, M. (1993). Reliability and design of 2-dimensional consecutive-k-out-of-n systems. *IEEE Transactions on Reliability*, 42(3), 488–490.

Recent Advances in Reliability and Maintenance Modeling – Okamura, Inoue & Xiao (Eds)

Exploring intra-telecom service switching from 4G to 5G: A migration model lens

Avus Hou
Asia Eastern University of Science and Technology, Taipei, Taiwan

ABSTRACT: The fifth-generation mobile network (5G) is widely recognized as one of the biggest market opportunities in the coming years. Despite the commercialization of 5G services by carriers, many consumers are still hesitant to switch from 4G to 5G. To address this emerging issue, this study applies the push-pull migration model to develop a second-order reflective-formative model to investigate users to switch their subscription from 4G to 5G within a telecom carrier. A total of 172 respondents were collected and analyzed using partial least squares technology. Our findings suggest that pushing has a greater effect than pulling, indicating that users are generally satisfied with the incumbent 4G service. However, 5G becomes more attractive when carriers put more effort into promoting its benefits. The results suggest that carriers need to focus more on promoting the benefits of 5G to encourage users to switch.

1 INTRODUCTION

In recent decades, smartphones have catalyzed the development of groundbreaking technologies, revolutionizing service delivery around the globe. The ubiquity of live streaming, mobile gaming, social media, and mobile payments in smart cities is largely due to the widespread adoption of fourth-generation (4G) mobile networks. As of 2022, global mobile network subscribers numbered 8.2 billion (Ericsson 2021). However, as 4G networks approach their capacity, the more advanced fifth-generation (5G) networks are being deployed worldwide, with numerous countries initiating commercial 5G services since 2020 (Massaro & Kim 2021).

The shift from 4G to 5G is vital due to the numerous benefits that 5G technology provides. However, many users still maintain 4G subscriptions for reasons such as cost, contractual obligations, and concerns about 5G infrastructure availability (Imoize *et al.* 2020). User hesitancy and carriers' delayed investments during the initial construction phase have hindered the adoption of 5G, making it challenging to achieve economies of scale (Massaro & Kim 2021). Intra-telecom switching refers to customer shift from an incumbent 4G service to a new 5G service within the same telecom provider. This switching typically involves signing a new contract, which outlines the updated terms and conditions associated with the upgraded network. Intra-telecom switching is an essential aspect of the ongoing shift from 4G to 5G, as it allows customers to access the latest technology and improved services without changing their service provider. While the majority of research has focused on inter-providers switching, there remains a significant gap in understanding the complexities of intra-telecom switching services (Fitchard *et al.* 2019; Moozakis 2022; Parcu *et al.* 2021). This gap emphasizes the importance of exploring the ongoing switching from 4G to 5G, which is the central aim of our study.

The process of switching service providers resembles the experiences of immigrants leaving their birthplace for a new destination. Hou & Shiau 2020 employed the migration

DOI: 10.1201/9781003491309-4

Push-Pull model (PP model) as a theoretical framework to examine users' switching behavior in social media. The PP model has been extensively applied to various service-switching contexts, including airline, and social media switching (Hou and Shiau 2020; Jung *et al.* 2017, 2021). Consequently, the PP model is well-suited for investigating user switching behavior from 4G to 5G networks but less applying in inter-telecom service switching. We propose the following research questions:

Research Question 1: What factors influence users' decisions to switching from 4G to 5G services?
Research Question 2: How can the identified factors be categorized as either pushing or pulling in the context of 5G switching?

This study aims to explore the factors that influence users' decision to switch from 4G to 5G services in a same carrier. In particular, we focus on users who are currently subscribed to a telecommunications service provider and are considering migration from 4G to 5G services. The push effects that contribute to users moving away from 4G include planned obsolescence and low network speed. On the other hand, the pull effects that drive users' move towards 5G are the freshness, peer influence, and carrier promotions. A total of 172 respondents participated in the study, and their responses were analyzed using Structural Equation Modeling (SEM). The results demonstrate that pushing has more effects on switching intention than pulling. We further identify the constructs different from the physical world migration and look forward to providing more insights for the telecom service research.

The structure of this study is as follows: Section II provides an overview of 5G network characteristics and discusses the theoretical background of migration; Section III introduces the research hypotheses; Section IV details the methodology employed and presents the results of the study; finally, Section V offers a discussion of the findings and draws conclusions.

2 LITERATURE REVIEW

2.1 *5G mobile network*

The 5G is driven by two primary factors: telecom technology innovations and 4G limitations. On one hand, the evolution from 4G to 5G offers numerous benefits, including lower battery consumption, reduced cell tower costs, and improved performance (Parcu *et al.* 2021). On the other hand, 4G struggles to accommodate the increasing number of devices and the resulting surge in mobile data demands. Compared to 4G distributed antennas, 5G technology incorporates more antennas in a single cell tower. As a key feature of 5G networks, these additional antennas enable large-scale multi-in-multi-out (MIMO) capabilities, increasing spectrum efficiency and reducing the likelihood of spectrum bottlenecks, thereby providing high-speed bandwidth networks. Furthermore, compared to 4G, 5G infrastructure allows telecom carriers to optimize resources more effectively and at lower costs (Lakshitha *et al.* 2018).

Three key characteristics of 5G are lightning-fast speeds, low latency, and massive connectivity. In theory, 5G boasts speeds that are up to 20 times faster than 4G, potentially reaching gigabit-per-second levels, which would significantly enhance data transfer rates and overall network performance. Data transfers on 5G networks are remarkably fast, taking less than 2 milliseconds to complete (Zikria *et al.* 2018). This results in ultra-low latency, even surpassing the speed of human visual processing, which allows for near-instantaneous communication and data exchange. As a result of the ultra-low latency offered by 5G networks, human response times could become the limiting factor in certain 5G usage contexts, as our natural reaction times may not be able to keep up with the rapid pace of data exchange. In contrast, machine-to-machine communication can fully leverage the speed and efficiency of 5G networks, enabling seamless interactions and real-time data sharing among connected devices. The exceptional combination of high speed and low latency offered by

5G networks creates an optimal environment for virtual reality (VR) and augmented reality (AR) applications (Moozakis 2022). These cutting-edge technologies rely on rapid data transfer and minimal lag to deliver immersive, interactive experiences, making 5G the ideal foundation for their widespread adoption and future growth.

The network capacity of 5G is theoretically 1,000 times greater than that of 4G, enabling it to accommodate an enormous number of low-power consumption devices (Zikria *et al.* 2018). This enhanced capacity is particularly well-suited for applications in the Internet of Things (IoT), where countless sensors and devices work together to collect, process, and share data across various networks and systems. With the ability to handle connections for hundreds or even thousands of devices simultaneously, 5G has the potential to revolutionize various aspects of modern life, including cities, factories, farms, schools, and homes (O'Connell *et al.* 2020). As a result, numerous industries, such as agriculture, manufacturing, and logistics, stand to benefit from the widespread adoption of 5G networks. By providing the necessary infrastructure, this new generation of telecommunications systems can facilitate significant advancements and transformations across society (Massaro & Kim 2021).

2.2 *The push-pull model*

In 1885, Ravenstein published "Laws of migration". He observed that human migration is influenced by negative factors in the original place and positive factors to attract people to new destinations (Moon, 1995). More, he has proved that the great body of our migrants only proceed a short distance. Following Ravenstein's ideas, Lee (1966) proposed the push-pull migration model in demographic study to better demonstrate Ravenstein's arguments. This model regards migration as the consequence of the interaction between the push effect at the original place and the pull effect at the distinct places. To date, the push-pull model is one of the most common models used to interpret human migration (Hou & Shiau 2020).

Push effects are the negative factors that compel people to leave their original location, such as a lack of job opportunities, difficulties in finding a spouse, or unfavorable climate. In contrast, pull effects are the positive aspects of a destination that attract people, including better developmental opportunities like higher-paying jobs, better living quality, or high-quality school districts (Hou & Shiau 2020).

The push-pull (PP) model has been applied to both online and offline relationship marketing disciplines. Hou and Shiau (2020) demonstrated the PP model to social media usage and found that it effectively explained consumers' switching behavior from Facebook to Instagram. Among the various predictors, some are associated with utility functions (e.g., dissatisfaction with customer service, low system quality, low information quality, attractiveness of alternatives), while others are related to social functions (e.g., critical mass, peer influence). These factors have different effects on customer responses and can either encourage or inhibit switching intentions. They suggested that future researchers utilize the robust PP model to better understand consumer switching behaviors in other contexts.

In the context of 5G intra-switching, the push-pull model can explain why individuals motivated to switch from their current 4G to 5G service if we see the switching as national migration. For instance, a user may feel "pushed" away from their current 4G network due to slow speeds, poor coverage, or other issues that make the existing network less appealing. On the other hand, they may feel "pulled" towards a 5G network because of its faster speeds, lower latency, and other benefits. The push-pull model can be helpful in identifying the specific factors that push users away from their current 4G network or pull them towards a 5G network within same provider.

2.3 *Telecom switching vs. migration*

The term "migration," as defined by Boyle and Halfacree (1998), refers to the movement of people between two locations for a certain period. Migration can be broadly categorized into two main types: national migration and international migration (Hou *et al.* 2011). National migration, also known as internal migration, involves the movement of people within a country, typically relocating from one region or city to another. Factors driving this migration can include employment opportunities, education, and quality of life. Conversely, international migration entails individuals moving from one country to another, influenced by factors such as economic, social, political, or environmental conditions. This type of migration often results in a more diverse and interconnected global society (Hou and Shaiu 2020).

The concept of migration can be analogously applied to telecom switching. Internal telecom service switching, for instance, refers to a customer migrating from an incumbent 4G service to a new 5G service in the same telecom provider. This switch could be driven by factors such as network performance, enhanced features, or better pricing plans (Massaro & Kim 2021). Just like in the context of national migration, the decision to switch telecom service within same provider involves a complex interplay of individual and environmental factors. These may include the user's personal preferences, their perceptions of the quality and value of the service provided by the provider, the availability of alternative services, and the influence of social networks or marketing efforts (Chuang 2011). This migration may be motivated by a range of factors, such as a desire for better network coverage, more affordable pricing plans, or access to newer and more advanced technologies. This switching typically involves signing a new contract, which outlines the updated terms and conditions associated with the upgraded network. Intra-telecom switching is an essential aspect of the ongoing switching from 4G to 5G, as it allows customers to access the latest technology and improved services without changing their service provider, and the telecom carriers, who can retain their customer base while upgrading their network infrastructure.

3 HYPOTHESES

This research examines the user's switching from 4G to 5G within the same carrier, comparing it to national migration patterns. The decision to migrate is influenced by a combination of push and pull effects. Figure 1 presents the research model which operates as a second-order formative model. The second-order constructs include both push and pull with no items. The push effect consists of two: planned obsolescence and slow network speed. The pull effect consists of three: freshness, peer influence, and carrier promotion. This section details our choice of these constructs and their positions in the PP model.

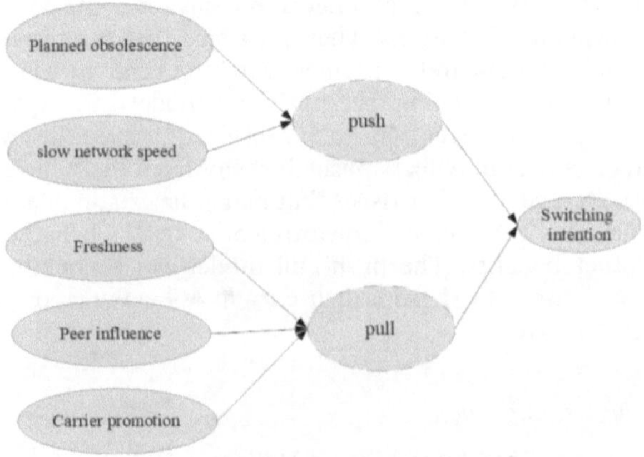

Figure 1. Research model.

3.1 *Push effect*

Planned obsolescence is a phenomenon that refers to users' smartphones beginning to exhibit poor battery life, slow processor performance, or are no longer supported by Apps up-to-date. According to (Longmuss & Poppe 2017), customers purchase new smartphones because their current handsets are too old to support the Apps and mobile operating systems updates. New smartphones come equipped with advanced hardware, new network functions, and the latest mobile operating systems. In many cases, users opt to purchase a new smartphone with 5G capability, regardless of whether they currently have access to 5G services (Kuppelwieser *et al.* 2019). This study suggests that planned obsolescence may a driver of this trend, as consumers are encouraged to upgrade to newer devices to keep up with the latest technology.

Slow network speed can have various adverse effects on individuals and organizations, such as poor user experience, impact on productivity, and limited innovation. It can lead to frustrating user experiences, such as buffering while streaming videos, slow loading times for web pages, and lag during online gaming. As a result, users may feel compelled to seek out alternative providers that offer better network speeds and reliability (Dey *et al.* 2020). Moreover, Slow network speeds can negatively affect productivity of users who rely on their mobile devices for business purposes. When they experience delays in accessing emails, using cloud-based services, or downloading files, they may upgrade for a faster network to improve their experience. Slow 4G networks associate with a lack of investment in infrastructure by the provider (Chuang 2011). As the result, slow 4G network speeds can indeed have a push effect, causing people to leave or switch to other carriers.

H1: The higher push effects lead to a higher intention to switch to 5G.

3.2 *Pull effects*

Freshness refers to the extent to which a new product or service offers novelty and interest to users. It is characterized by the introduction of new and innovative features that differentiate it from existing offerings in the market (Bansal *et al.* 2005). Fresh products or services often generate consumer curiosity and enthusiasm, which can lead to early adoption by technology enthusiasts (Hou *et al.* 2011). Research in demographics has shown that the freshness of a new place to live is a key factor for people considering moving to a new destination, even if they have limited knowledge of the area. Given this, this study infers that the freshness of 5G is a construct in shaping its pulling effect.

Peer influence refers to the extent to which individuals are influenced by the opinions and actions of their peers when making decisions (Hou & Shiau 2020). It is a construct in shaping consumer behavior and can have a significant impact on the adoption of new technology services (Elhai *et al.* 2017). When individuals perceive that their peers are adopting 5G technology, they may be more likely to adopt it themselves, because they may perceive 5G adoption as a social norm and feel pressure to conform to the behaviors of their peers (Liao *et al.* 2021). In addition, peer influence can be reinforced by social media and other online platforms, where individuals can see and share information about their experiences with 5G technology. Positive reviews and comments from peers can create a powerful pulling effect of 5G technology that encourages users to switch to 5G.

Carriers promotion can be an effective way to shape the pull effect of 5G mobile networks switching within carriers (Hosseini *et al.* 2020). Promotions can take various forms, such as discounts, special offers, and marketing campaigns highlighting the advantages of 5G technology over 4G or other competing technologies (Massaro & Kim 2021). For example, carriers may offer free or discounted 5G devices to users who switch to 5G, or provide extra data allowances for new 5G users. In addition to pricing, carriers can also leverage other marketing strategies to create pull effect for 5G. This can include targeted advertising campaigns that reach out to specific user groups, such as early adopters or tech-savvy users, as well as content marketing efforts that highlight the unique features and benefits of 5G technology (Shah *et al.*

2018). When carriers actively promote the benefits of 5G technology to current users, they can create a pull effect that encourages users to switch to 5G mobile networks.

According to previous discussions, we suggest that freshness, peer influence, and carrier promotion can all contribute to the pulling of 5G. Therefore, the following hypothesis is proposed:

H2: The higher pull effects lead to a higher intention to switch to 5G.

4 METHODOLOGY

4.1 *Instrument development*

According to the hypotheses presented in last section, a survey instrument was developed based on prior relevant research. Questionnaire items were modified slightly to fit our 5G switching context. There are six constructs with a total of 19 items. All items were measured using either 5-point Likert scale or 5-point semantic differential scale, except for the two sections of the questionnaire used to collect users' experience and demographic backgrounds. The questionnaire items are listed in Appendix A.

We considered the following three dimensions to formulate push effects: The scale used for planned obsoleted (PO) was referring to green study with four items (Kuppelwieser *et al.* 2019). Slow network sPeed (SP) was adopted from Chuang's scale (2011). The pull effects were measured with freshness, peer influence, and carrier's promotion. The freshness (FR) was measured using the scale described by Hou *et al.* (2011). Peer influence (PI) was adopted from Hou and Shiau's (2020) scales with 3 items. The carrier PRomotion (PR) was measured using the three-item scale proposed by (Chuang 2011). Finally, switching intentions (SI) was measured using Hou and Shiau's (2020) method.

4.2 *Data collection*

Given that our study focused on investigating why users intend to switch from 4G to 5G within incumbent carriers, it was important to be able to probe perceptions of those who had experience on using 4G and knowing about 5G. Our survey approach was developed as follows. First, we placed messages on the most popular online survey sites in Taiwan, Myservey.com (www.mysurvey.com) to recruit volunteers with experience using 4G to join this survey. Then researchers pick out respondents randomly from the qualified user pool and send them an invitation message with the URL to the online questionnaire included. Thus the problems with sample frame and non-random volunteer samples were solved. A gift certificate with a value of US$3 was then sent to each respondent who filled out the questionnaire as an incentive to boost the response rate and to insure the quality of the data.

The survey yielded 207 responses. After removing those with unanswered items, we ended up with 172 usable responses. Approximately 51% of the respondents were male. The majority of respondents were between 21 and 40 years old (58%), and 23% were more than 40 years old.

5 RESULTS

5.1 *Measurement model*

All scales were assessed in terms of reliability, convergent validity, and discriminant validity. We used Cronbach's alpha to verify the reliability of the survey instrument. According to Bagozzi & Yi's suggestions (1988), all of the constructs obtained a Cronbach's alpha value greater than 0.6 and composite reliability score greater than 0.7, indicating the internal consistency of the measurement. Descriptive statistics for the constructs are given in Table 1.

Table 1. Reliability.

Construct	alpha	CR	AVE
PO	0.72	0.80	0.59
FR	0.86	0.87	0.79
SI	0.74	0.75	0.80
PI	0.82	0.83	0.74
PR	0.89	0.89	0.82
SP	0.81	0.85	0.73

Discriminant validity highlights the degree to which two constructs' measure-ments are empirically distinct (Williams & Clippinger 2002). As Table 2 shows, all constructs share more variance with their indicators than with the other constructs, which reinforces the discriminant validity of our model.

Table 2. Discriminant validity.

Construct	PO	FR	SI	PI	PO	SP
PO	1.00					
FR	0.32	1.00				
SW	0.63	0.59	1.00			
PI	0.49	0.44	0.79	1.00		
PR	0.16	0.17	0.34	0.54	1.00	
SP	0.21	0.20	0.36	0.32	0.24	1.00

5.2 *Structural model*

The model was tested by using Partial Least Squares (PLS), a latent structural equations modeling technique with a component-based approach to estimation. Because of this, minimal demands were made on sample size and residual distributions (Chin 1998). Furthermore, PLS is suited for predictive applications and theory building (Gefan *et al.* 2000). The SmartPLS 4.0 package was used to perform the PLS assessment (Ringle *et al.* 2014).

Figure 2 shows the result of the PLS assessment. The predictive validity was evaluated by examining the R square and the path coefficient. The results indicate that our research model

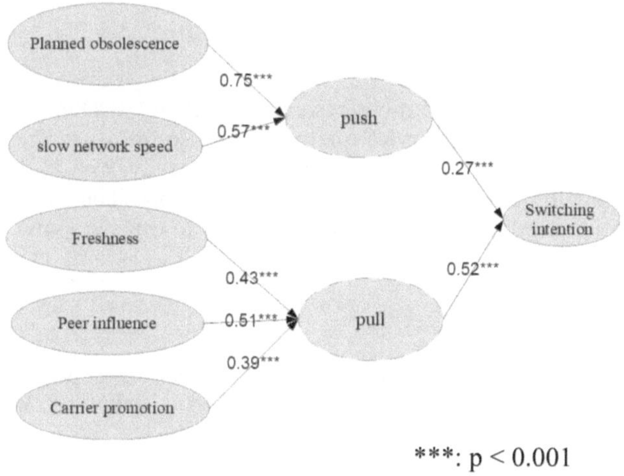

Figure 2. Results.

explains a 46.2 percent variance, which is acceptable (Hair *et al.* 2013). It can be seen that all hypotheses are supported. Pushing shows a significant negative influence on switching intentions while pulling does the opposite. Among pushing factors, low speed, and user support formulate the push effect. Among pulling, the attractiveness of alternative, peer influence, and critical mass are significant, meaning they all constitute the pull effects.

6 DISCUSSION

The study utilizes PP model original from demography to investigate users' switching intentions from 4G to 5G. Empirical results show that factors embodying pull and push effect were all able to trigger switching intentions. The discovery sides with the view of social media that PP model serve as the solid theoretical framework in service switching context (Hou & Shiau 2020). By viewing intra-telecom switching as a form of users' national migration, researchers and practitioners can gain insights into the factors that motivate users to switch providers and design strategies to retain customers or attract new ones.

According to theories of demographic migration, push effects are negative factors generated by original place that causes people to leave. The study hypothesized planned obstacle and slow network speed formulate push effects. The findings show that pushing is significant associate with switching intentions, meaning that whether users care about network speed. Therefore, fulfilling the needs in network speed becomes important issues for carriers. Fast network speed can be generated by increased cell towers, redundant mechanisms, good infrastructures in backbone networks.

The findings indicate that the pull effects was the most significant drivers of users' intentions to switch to 5G. Freshness, as a pull effect, suggests that users are attracted to the latest and most advanced technology. 5G networks are relatively new, and their increased speed and connectivity capabilities make them more attractive to users than the older 4G networks. Peer influence, another pull effect, indicates that users may be influenced by the opinions or experiences of their friends or family members who have already switched to 5G. This influence can create social pressure that drives users to switch to 5G networks.

Carrier promotion suggests that the marketing efforts of telecommunication companies can play a significant role in promoting the benefits of 5G networks and influencing users to switch. Telecommunication companies can leverage these effects by focusing their marketing efforts on promoting the benefits of 5G networks, emphasizing their freshness, and leveraging social networks to promote 5G among friends and family members.

7 LIMITATION

The study only examines a limited number of variables related to switching intention. More, other variables, such as pricing or service quality, may also influence the decision to switch telecommunications services and were not included in the study. The study's sample size and representativeness may limit the generalizability of the findings. The study may not represent the broader population, and the findings may not apply to other populations or contexts.

REFERENCES

Bansal, H. S., Taylor, S. F., and James, Y. S. (2005). "Migrating" to new service providers: Toward a unifying framework of consumers' switching behaviors. *Journal of the Academy of Marketing Science*. 33, pp. 96–115.

Chuang, Y.-F. (2011). Pull-and-suck effects in Taiwan mobile phone subscribers switching intentions. *Telecommunications Policy*, 35(2), 128–140.

Dey, B. L., Al-Karaghouli, W., Minov, S., Babu, M. M., Ayios, A., Mahammad, S. S., and Binsardi, B. (2020). The role of speed on customer satisfaction and switching intention: A study of the UK mobile telecom market. *Information Systems Management*, 37(1), 2–15.

Elhai, J. D., Dvorak, R. D., Levine, J. C., and Hall, B. J. (2017). Problematic smartphone use: A conceptual overview and systematic review of relations with anxiety and depression psychopathology. *Journal of Affective Disorders*. 207, pp. 251–259.

Hosseini, S. H., Zadeh, F. H., Shafiee, M. M., and Hajipour, E. (2020). The effect of price promotions on impulse buying: The mediating role of service innovation in fast moving consumer goods. *International Journal of Business Information Systems*. 33, pp. 320–336.

Hou, A., Chern, C.-C., Chen, H.-G., Chen, Y.-C., and (2011). 'Migrating to a New Virtual World': Exploring MMORPG Switching Through Human Migration Theory. 27, pp. 1892–1903.

Hou, A., and Shiau, W.-L. (2020). Understanding Facebook to Instagram migration: a push-pull migration model perspective. *Information Technology & People*. 33, 272–295.

Kuppelwieser, V. G., Klaus, P., Manthiou, A., and Boujena, O. (2019). Consumer responses to planned obsolescence. *Journal of Retailing and Consumer Services*, 47, 157–165.

Lee, E. S. (1966). A Theory of Migration. *Demography*. 3: 47–57.

Longmuss, J., and Poppe, E. (2017). Planned obsolescence: who are those planners? *Product Lifetimes and The Environment*: 217–221.

Massaro, M., and Kim, S. (2021). Why is South Korea at the forefront of 5G? Insights from technology systems theory. *Telecommunications Policy*. 102290.

Moon, B. (1995). Paradigms in migration research: Exploring 'moorings' as a schema. *Progress in Human Geography*. Vol. 19:504–524.

Moozakis, C. (2022). The Pros and Cons of 5G Networks.

O'Connell, E., Moore, D., and Newe, T. (2020). Challenges Associated with Implementing 5G in Manufacturing. *Telecom* 1:48–67.

Shah, M. A. R., Husnain, M., and Zubairshah, A. (2018). Factors Affecting Brand Switching Behavior in Telecommunication Industry of Pakistan: A Qualitative Investigation. *American Journal of Industrial and Business Management*. 8:359–372.

Zikria, Y. B., Kim, S. W., Afzal, M. K., Wang, H., and Rehmani, M. H. (2018). 5G mobile services and scenarios: Challenges and solutions. *Sustainability*. 10:1–9.

Recent Advances in Reliability and Maintenance Modeling – Okamura, Inoue & Xiao (Eds)

Describing behavior of safety-related systems with proof-testing and its applications

S. Inoue
Kansai University, Suita, Japan

S. Yamada
Tottori University, Tottori, Japan

ABSTRACT: We describe the behavior of E/E/PE safety-related systems based on a continuous-time Markov chain with proof-testing which is known as maintenance activities especially for undetected dangerous faults. The undetected dangerous failure / fault cannot be detected by self-diagnostic system generally implemented in the E/E/PE safety-related systems. Therefore, major concern in the operation phase is when to conduct the proof-testing to maintain designed safety level. Analyzing our continuous-time Markov chain gives us useful discussions on optimal proof-testing intervals, and resolution of IEC 61508-based discontinuous safety assessment depending on the operation mode of the E/E/PE safety-related systems. We addresses these discussions and show numerical examples for them by following some definitions in IEC 61508, which is the international basic standard for the functional safety of the E/E/PE safety-related systems.

1 INTRODUCTION

With the development of AI technology and computer performance, the needs for computer systems are diversifying in our daily lives, including in the industrial world. For example, the number of situations in which computer systems are responsible for safety functions, such as safety support systems in automobiles, has been rapidly increasing in recent years. Airbag systems and pre-crash systems in automobiles are concrete examples of such systems. These systems are called safety-related systems which are installed in addition to the main system that performs intended functions, and ensure the safety of the overall system from the functional aspect. Generally, such safety ensuring concept is called functional safety. Discussions on the safety assessment methods for the safety-related systems that play a role in functional safety have been getting activated internationally since the early 2000s, when the safety-related systems with embedded software systems, which is called electrical / electronic / programmable electronic (E/E/PE) safety-related systems, began to become popular. For example, IEC 61508 (IEC 61508 2010) which is the international basic standard for the functional safety of the E/E/PE safety-related systems was released in 2000. This standard defines the whole safety life cycle of the E/E/PE safety-related systems, which consists of 16 phases from the conceptualization phase to the end-of-life phase, and also defines the safety integrity level (SIL) for indicating the safety levels and assessing the safety of the E/E/PE safety-related systems. Nowadays, safety evaluation based on this international basic standard is widely required internationally. This international basic standard has also led to discussion in academic fields on methods of assigning safety integrity levels, taken into account the internal structure and operational requirements of the safety-related systems that are not assumed in the standard (Kato & Sato 2000; Misumi & Sato 1999; Sato 2014; Yoshimura & Sato 2008).

DOI: 10.1201/9781003491309-5

On the other hand, maintenance activities in the operation phase of the E/E/PE safety-related systems are indispensable to keeping the safety features built in during the design phase. In addition, the maintenance activities, especially those called proof-testing require suspending the entire system, inspecting the entire safety-related system in detail, and repairing defects if any are found, which incurs costs due to the suspension of operations, including the cost of maintenance activities themselves. Therefore, it is desirable to implement the proof-testing activities to the extent that they are reasonably feasible for potential hazardous risks.

Furthermore, in the safety evaluation of the E/E/PE safety-related systems based on SIL, safety integrity levels are assigned according to different safety assessment measures (or so-called target failure measures) depending on the frequency of safety demands to the E/E/PE safety-related system. For example, the SIL is allocated based on the average probability of a dangerous failure on demand (PFD_{avg}) for the low-demand operation mode, and based on the probability of a dangerous failure per hour (PFH) for the high-demand or continuous operation mode. Therefore, in the case that we conduct the safety assessment following IEC 61508 standard, the different and appropriated safety assessment measures are required to apply depending on the frequency of the safety demands, that leads discontinuous safety assessment. There is also the problem of evaluation accuracy around the boundary value of these safety demands operation modes. Table 1 shows the safety integrity level defined in IEC 61508.

Following the above background, we address the above-mentioned issues of interest surrounding the E/E/PE safety-related systems. Specifically, we describe the behavior of E/E/PE safety-related systems based on a continuous-time Markov chain with proof-tests, and derive useful measures for safety evaluation from the results of Markov analysis. Using the results, we also discuss how to derive the optimal proof-testing intervals that minimizes the loss cost based on the hazardous risk and the cost of the proof test.

Table 1. Safety integrity level of functional safety (IEC 61508).

SIL	Low Demand Mode	High Demand or Continuous Mode
4	$10^{-5} \leq PFD_{avg} < 10^{-4}$	$10^{-9} \leq PFH\ (1/hour) < 10^{-8}$
3	$10^{-4} \leq PFD_{avg} < 10^{-3}$	$10^{-8} \leq PFH\ (1/hour) < 10^{-7}$
2	$10^{-3} \leq PFD_{avg} < 10^{-2}$	$10^{-7} \leq PFH\ (1/hour) < 10^{-6}$
1	$10^{-2} \leq PFD_{avg} < 10^{-1}$	$10^{-6} \leq PFH\ (1/hour) < 10^{-5}$

2 DESCRIBING BEHAVIOR OF SAFETY-RELATED SYSTEMS

In describing probabilistic behavior of the E/E/PE safety-related systems, it should be noted that a dangerous fault for the E/E/PE safety-related systems can be classified into two types, such as a dangerous detected fault (DD fault) and a dangerous undetected fault (DU fault) (Rausand 2014; Sato 2014). The DD fault is usually diagnosed and detected at a very high frequency by the self-diagnostic systems installed in the E/E/PE safety-related systems. However, the DU fault is not able to be detected by the diagnostic detections and reveals only when a hazardous event occurs or the proof-testing is completely conducted. Then, we describe the behavior focusing on the effect of the DU failure occurrence.

Let us denote the following each possible state for the E/E/PE safety-related systems during the operation phase.

- State 0 : the safety-related system is ready to work as intended without any failures,
- State 1 : the safety-related system is working as intended without any failures due to safety demand issued from main systems,

- State 2 : the safety-related system is in the DU fault, but a hazardous event is not occurred because a safety-demand has not been issued yet,
- State 3 : the whole system is in the hazardous state,
- State 4 : the whole system is not working due to the proof-testing.

Further, we introduce the following state transition rates:

- λ_d : the safety demand rate (1/hour),
- λ_{DU} : the DU failure occurrence rate (1/hour),
- λ_D : the dangerous failure occurrence rate (1/hour),
- λ_P : the proof-testing rate (1/hour),
- μ_d : the safety demand completion rate (1/hour),
- μ_M : the recovery rate of the whole system (1/hour),
- μ_P : the proof-testing completion rate (1/hour),

respectively. It is worth mentioning that there is the relationship between λ_{DU} and λ_D with the diagnostic rate (DC) which represents the fraction of the DD failure rate for the dangerous failure rate, such as $\lambda_{DU} = (1 - DC)\lambda_D$. Based on the definitions for the possible states and the transition rates mentioned above, our Markov transition diagram expressing the stochastic behavior of the E/E/PE safety-related system with the proof-testing can be developed as shown in Figure 1.

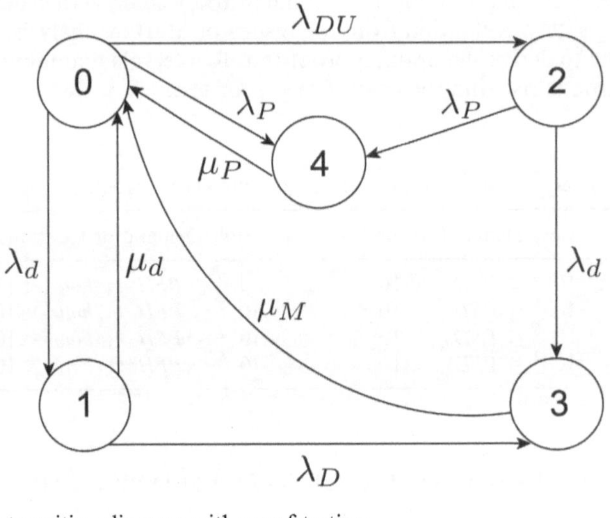

Figure 1. Markov transition diagram with proof-testing.

From Chapman-Kolmogorov forward equations (Osaki 1992; Ross 2009; Trivedi and Bobbio 2017) based on the Markov transition diagram shown in Figure 1, we obtain

$$
\begin{aligned}
(\lambda_d + \lambda_{DU} + \lambda_P)p_0 &= \mu_d p_1 + \mu_M p_2 + \mu_P p_4, \\
\lambda_d p_0 &= (\mu_d + \lambda_D)p_1, \\
\lambda_{DU} p_0 &= (\lambda_P + \lambda_d)p_2, \\
\lambda_D p_1 + \lambda_d p_2 &= \mu_M p_3, \\
\lambda_P(p_0 + p_2) &= \mu_P p_4,
\end{aligned}
\tag{1}
$$

by considering the limiting probability assuming the long-time operation. In Equation (1), p_i $(i = 0, 1, 2, 3, 4)$ represents the limiting probabilities for each state considering the long-time operation. We note that $\sum_{i=0}^{4} p_i = 1$. Solving the system of the equations in Equation (1), the limiting probabilities can be obtained as

$$\begin{aligned}
p_0 &= J^{-1}, \\
p_1 &= \left(\frac{\lambda_d}{\mu_d + \lambda_D}\right) p_0, \\
p_2 &= \left(\frac{\lambda_{DU}}{\lambda_P + \lambda_d}\right) p_0, \\
p_3 &= \left(\frac{\lambda_d}{\mu_d + \lambda_D} + \frac{\lambda_{DU}}{\lambda_P + \lambda_d}\right) \frac{\lambda_D p_0}{\mu_M}, \\
p_4 &= \left(1 + \frac{\lambda_{DU}}{\lambda_P + \lambda_d}\right) \frac{\lambda_P p_0}{\mu_P},
\end{aligned} \tag{2}$$

where J in Equation (2) is

$$J = 1 + \left(\frac{\lambda_d}{\mu_d + \lambda_D}\right)\left(1 + \frac{\lambda_D}{\mu_M}\right) + \left(\frac{\lambda_{DU}}{\lambda_P + \lambda_d}\right)\left(1 + \frac{\lambda_D}{\mu_M} + \frac{\lambda_P}{\mu_P}\right) + \frac{\lambda_P}{\mu_P}. \tag{3}$$

3 SAFETY RELATED ASSESSMENT MEASURES

From the results of the Markov analysis, useful assessment measures for the safety or availability of the E/E/PE safety-related systems can be derived. Before deriving these measures, we get simplified the mathematical structure of J in Equation (3) by considering possible situations in the operation phase of the E/E/PE safety-related systems. First of all, we consider that the whole system is replaced with new one immediately when the safety-related system moves to the State 3 (hazardous state). That is, we assume μ_M takes a very large value and then $\lambda_D \ll \mu_M$ in Equation (3). Furthermore, the time-duration needed to complete the proof-testing is assumed to be much shorter than the proof-testing interval. That is, we also assume $\lambda_P \ll \mu_P$. Following these assumptions, J in Equation (3) is simplified as

$$J_s = 1 + \frac{\lambda_d}{\mu_d + \lambda_D} + \frac{\lambda_{DU}}{\lambda_P + \lambda_d}. \tag{4}$$

Based on the above situation, we first derive the frequency of the hazardous event occurrence, $h_{EO}(\lambda_d)$ (1/hour), as

$$\begin{aligned}
h_{EO}(\lambda_d) &= p_1\lambda_D + p_2\lambda_d \\
&= \left(\frac{\lambda_D\lambda_d}{\mu_d + \lambda_D} + \frac{\lambda_{DU}\lambda_d}{\lambda_P + \lambda_d}\right)\frac{1}{J_s}.
\end{aligned} \tag{5}$$

It should be noted that this safety assessment measures in Equation (5) is depending on the safety demand rate λ_d. Actually, IEC 61508 recommends assessing the safety of the E/E/PE safety-related systems by using different assessment measures depending on the frequency of the safety-demand. That is, IEC 61508 defined to use PFD_{avg} as the safety assessment measure for the case that the E/E/PE safety-related systems receives the safety demands less than once per year and PFH for the case the systems receives the demands greater than equal once a year. However, in Equation (5), the safety can be estimated by using $h_{EO}(\lambda_{sd})$ that is not depending on the frequency of the safety demands,

As an assessment measure related to the unavailability of the E/E/PE safety-related systems due the proof-testing, we derive the frequency of the proof-testing, $u_{PT}(\lambda_d)$, as

$$\begin{aligned} u_{PT}(\lambda_d) &= \lambda_P(p_0 + p_2) \\ &= \left(1 + \frac{\lambda_{DU}}{\lambda_P + \lambda_d}\right)\frac{\lambda_P}{J_s}. \end{aligned}$$

This measure is also depending on the safety demand rate λ_d and represents suspending rate of the whole system due to the proof-testing.

On the other hand, treating $h_{EO}(\cdot)$ and $u_{PT}(\cdot)$ as the functions depending on the proof-testing rate λ_P under the given safety-demand rate λ_d, we can formulate the instantaneous cost with the hazardous risk and the proof-testing cost as

$$Q(\lambda_P) = c_h h_{EO}(\lambda_P) + c_m u_{PT}(\lambda_P), \tag{6}$$

where c_h and c_m are cost parameters representing the cost per hazardous event occurrence and per proof-testing, respectively.

4 OPTIMAL CYCLES FOR PROOF-TESTING

Economic proof-testing intervals with the hazardous risk and the cost for the proof-testing can be derived from Equation (6). The optimal solution of $\lambda_P = \lambda_P^*$ minimizing Equation (6) satisfies

$$A(\lambda_P) \equiv K_1\left(\lambda_P + \frac{K_2}{K_1}\right)^2 + K_3 - \frac{K_2^2}{K_1} = \frac{c_h}{c_m}, \tag{7}$$

which is obtained by differentiating Equation (6) with respect to λ_P and making this equal to zero. In Equation (7),

$$K_1 = \frac{\lambda_D + \lambda_d + \mu_d}{\lambda_{DU}\lambda_d(\lambda_d + \mu_d)} \quad (> 0),$$

$$K_2 = \frac{\lambda_d(\lambda_D + \lambda_d + \mu_d) + \lambda_{DU}(\lambda_D + \mu_d)}{\lambda_{DU}\lambda_d(\lambda_d + \mu_d)} \quad (> 0),$$

$$K_3 = \frac{(\lambda_{DU} + \lambda_D + \lambda_d + \mu_d)\lambda_d^2 + (\lambda_D + \mu_d)(2\lambda_{DU}\lambda_d + \lambda_{DU}^2)}{\lambda_{DU}\lambda_d(\lambda_d + \mu_d)} \quad (> 0),$$

respectively. From the above equations, we see that $K_2/K_1 > 0$. Then, the function $A(\lambda_P)$ is a monotonically increasing function in the range of $\lambda_P > 0$ and $A(0) = K_3$. Therefore, if $K_3 < c_h/c_m$, there is a unique solution $\lambda_P = \lambda_P^*$ satisfying Equation (7). Furthermore, the expected proof-testing interval can be obtained as the reciprocal of λ_P^* due to the relationship between the proof-testing rate and the expected proof-testing interval. On the other hand, if $K_3 \geq c_h/c_m$, there is no solution of λ_P satisfying Equation (7) in the range of $\lambda_P > 0$. That means $Q(\lambda_P)$ in Equation (6) is a non-decreasing function of λ_P. In that situation, the instantaneous cost with the hazardous risk and the proof-testing cost in Equation (6) increases as the proof-testing interval becomes short. This implies that the hazardous risk is comparatively small against to the proof-testing cost. Therefore, it is better to consider $\lambda_P^* \to 0$ or $1/\lambda_P^* \to \infty$ that means the optimal policy is not to conduct proof-testing until the hazardous event occurrence whose risk is very small from the economic viewpoint.

5 NUMERICAL EXAMPLES

We show numerical examples for the frequency of the hazardous event occurrence $h_{EO}(\lambda_d)$ in Equation (5) and the estimation of the optimal proof-testing intervals by following the discussion in the previous section.

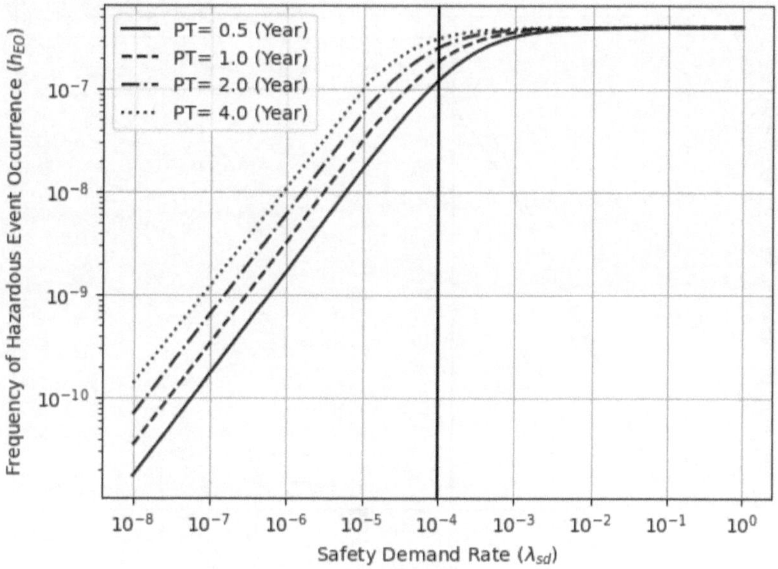

Figure 2. Estimated frequency of the hazardous event occurrence, $\widehat{h_{EO}}(\lambda_{sd})$.

Figure 2 shows the estimated frequency of the hazardous event occurrence $\widehat{h_{EO}}(\lambda_d)$ depending on the safety demand rate λ_d under that the proof-testing intervals have been set as 0.5, 1.0, 2.0, and 4.0 (year), respectively. Just for the reference, these proof-testing intervals can be given as $\lambda_P = 2.2831 \times 10^{-4}$, 1.1416×10^{-4}, 5.7078×10^{-5}, and 2.8539×10^{-5} (1/hour), respectively. Furthermore, in Figure 2, we also set $\lambda_D = 1.0 \times 10^{-6}$, $DC = 0.6$, and $\mu_d = 1.0 \times 10^{2}$, respectively. In this situation, the DU failure rate can be obtained as

$$\lambda_{DU} = \lambda_D(1 - DC) = 1.0 \times 10^{-6} \times (1 - 0.6) = 4.0 \times 10^{-7}(1/hour). \tag{8}$$

It is worth mentioning that IEC 61508 requires that PFD_{avg} have to be applied as the safety assessment measures if the safety demand rate λ_d is less than 1.0×10^{-4} (1/hour), i.e., the E/E/PE safety-related systems is operated as the low-demand operation mode in which the system receives the safety demand less than 1.0×10^{-4}. On the other hand, PFH is required for assessing safety of the E/E/PE safety-related systems requested to work as the high-demand or continuous operation mode in which the safety demands are issued more than 1.0×10^{-4} (1/hour).

Anyway, the PFD_{avg} is essentially equivalent to the unavailability of the E/E/PE safety-related systems and the PFH is the dangerous hazard rate of the systems themselves. Accordingly, we need to assess the safety of the E/E/PE safety-related systems by using different safety assessment measures, such as PFD_{avg} and PFH, depending on the operation mode of the E/E/PE safety-related systems if we follow the regulation of the IEC 61508. In other words, the estimated safety assessment results behave discontinuously after $\lambda_{sd} = 1.0 \times 10^{-4}$ which is the threshold value of the operation mode. However, by using the frequency of hazardous event occurrence derived as Equation (5), the safety of the E/E/PE safety-related systems can be continuously evaluated according to the frequency of the safety-demands, regardless of the value of the frequency of the safety-demand rate or the operation mode of the E/E/PE safety-related systems. From Figure 2, we can say that the frequency of the hazardous event occurrence decreasing in the low-demand mode region and converges to $\lambda_{DU} = 4.0 \times 10^{-7}$ in the high-demand or continuous operation mode.

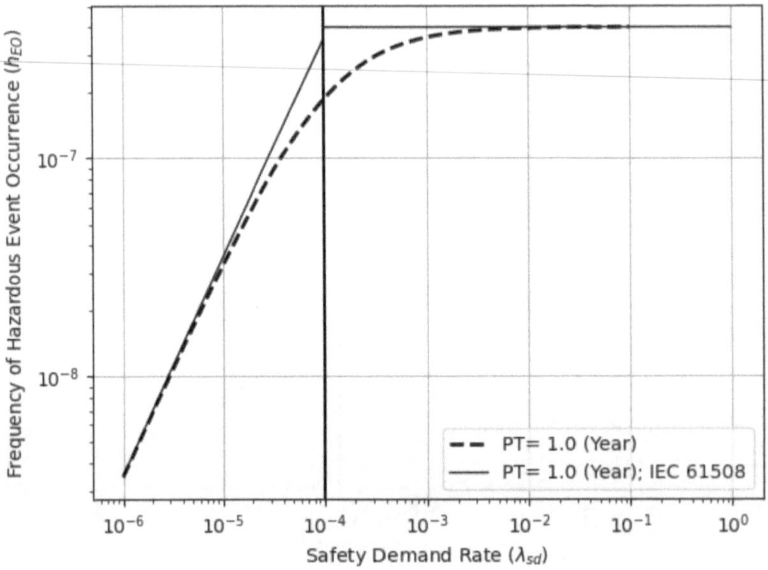

Figure 3. Comparison of the estimated behaviors of the frequency of hazardous event occurrence with the regulation of IEC 61508.

Figure 3 compares the estimated behavior of the frequency of hazardous event occurrence in Equation (5) with IEC 61508-based estimation. By following the notion of PFD_{avg} and PFH in IEC 61508, we note that the the frequency of the hazardous event occurrence estimated based on PFD_{avg} is given as

$$h_{EO,PFD}(\lambda_d) = \lambda_d \times PFD_{avg}, \tag{9}$$

for the range of the low-demand operation mode. And for the range of the high-demand or continuous operation mode, the frequency estimated based on PFH is also given as

$$h_{EO,PFH}(\lambda_d) = PFH, \tag{10}$$

From Figure 3, the estimated behavior of the frequency of hazardous event occurrence in Equation (5) is almost same as the behavior of $h_{EO,PFD}(\lambda_d)$ which is based on the PFD_{avg} in the range where the value of λ_{sd} is less than 10^{-5}. Furthermore, the behavior of $h_{EO}(\lambda_d)$ is also almost same as the behavior $h_{EO,PFH}(\lambda_d)$ in the range where the value of λ_d is greater than 10^{-3}. However, in the range where the value of λ_d takes from 10^{-5} to 10^{-3}, the frequency of the hazardous event occurrence in Equation (5) differs significantly from the values of $h_{EO,PFD}(\lambda_d)$ and $h_{EO,PFH}(\lambda_d)$ estimated according to the regulation of IEC 61508. This is because the safety assessment measures specified in IEC 61508 changes from PFD_{avg} to PFH after the safety demand rate $\lambda_{sd} = 1.0 \times 10^{-4}$. Therefore, we can say that the frequency of the hazardous event occurrence based on IEC 61508 around $\lambda_{sd} = 1.0 \times 10^{-4}$ is not well estimated. On the other hand, the proposed method in Equation (5) is able to express the behavior of the frequency of hazardous events continuously regardless of the value of the frequency of safety demands or the operation mode of the E/E/PE safety-related systems.

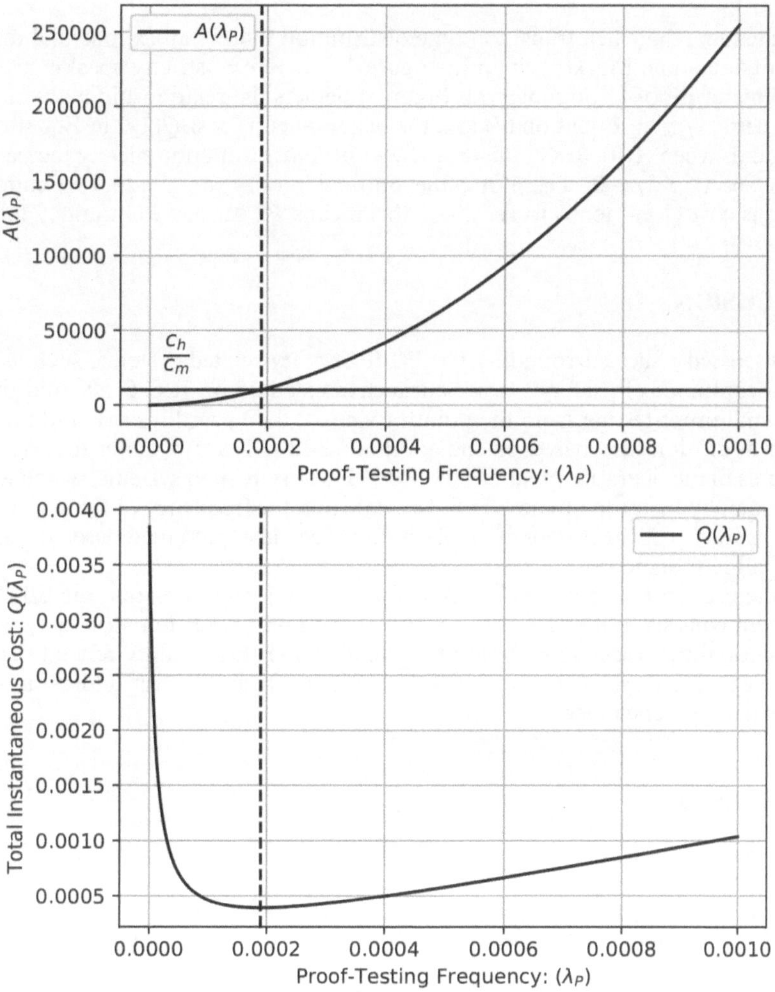

Figure 4. Estimated optimal proof-testing rate for λ_D= 1.0×10^{-6}, DC= 0.6, μ_{sd}= 1.0×10^{2}, and λ_{sd}= 1.0×10^{-5}.

For the calculation of optimal proof-testing, we set $\lambda_D = 1.0 \times 10^{-6}$, $DC = 0.6$, $\mu_d = 1.0 \times 10^{2}$, which are the same values mentioned before, and we also set $\lambda_d = 1.0 \times 10^{-5}$ which represents that the E/E/PE safety-related system is received safety demands once a decade. Further, we consider situation where the cost ratio is set as $c_h/c_m = 1.0 \times 10^{4}$. Under these situations, we see that

$$A(0) = K_3 = 27.0400 < c_h/c_m = 1.0 \times 10^{4}. \tag{11}$$

Therefore, there is a unique solution of $\lambda_P = \lambda_P^*$ minimizing the instantaneous cost with the hazardous risk and the proof-testing cost. Solving Equation (7) numerically for λ_P, λ_P^* is calculated as 1.896×10^{-4} (1/hours). Then, the optimal proof-testing interval PT^* can be estimated as

$$PT^* = \frac{1}{\lambda_{PT}^*} = 0.6021(years), \tag{12}$$

because we assume that each transition time distribution follows an exponential distribution in our continuous-time Markov chain in Figure 1. Therefore, strictly speaking, PT^* is the expected value of proof-testing interval. Figure 4 depicts the relationship between the function $A(\lambda_P)$ with c_h/c_m in Equation (7) and the instantaneous cost $Q(\lambda_P)$ in Equation (6). On the other hand, when $A(0) = K_3 > c_h/c_m$, $A(\lambda_P)$ becomes a monotonically increasing function with respect to $\lambda_P > 0$. Therefore, the optimal proof-testing rate λ_P^* minimizing the instantaneous cost $Q(\lambda_P)$ tends to zero, i.e., this means PT^* becomes infinitely long.

6 CONCLUSION

We have discussed issues surrounding the E/E/PE safety-related systems, such as the resolution of discontinuous in safety assessment measures based on IEC 61508 and the estimation of optimal proof-testing from the point of view of the hazardous risk and maintenance cost. These discussions are based on the continuous-time Markov chain that describes the possible states in the operation phase of the E/E/PE safety-related systems, which include the proof-test state, by focusing on the DU failure rate and the frequency of the safety demands. Numerical examples of the proposed methods have been also presented based on the realistic situations and IEC standards.

The approaches discussed here is based on the implicit assumption that the E/E/PE safety-related system consists of a single-channel (not a multiple protection structure). In our further researches, these discussions should be extended to E/E/PE safety-related systems with more complex structure, such as *k*-out-of-*n* system structure, for addressing extended approaches from practical aspects.

REFERENCES

IEC 61508. (2010). Functional Safety of Electrical/Electronic/Programmable Electronic Safety-related Systems, Edition 2.0.

Kato, E. and Sato, Y. (2000). Safety integrity levels model for IEC 61508—examination of modes of operation—. *IEICE Transactions on Fundamentals of Electronics, Communications and Computer Sciences*, E83-A (5): pp. 863–865.

Misumi, Y. and Sato, Y. (1999). Estimation of average hazardous-event-frequency for allocation of safety-integrity levels. *Reliability Engineering and System Safety*, 66(2): 135–144.

Osaki, S. (1992). *Applied Stochastic System Modeling*. Springer-Verlag.

Rausand, M. (2014). *Reliability of Safety-Critical Systems: Theory and Application*, John Wiley & Sons.

Ross, S.M. (2009). *Introduction to Probability Models* (10th Edition). Academic Press.

Sato, Y. (2014). *Fundamentals of Functional Safety (in Japanese)*. Japanese Standards Association.

Trivedi, K. S. and Bobbio, A. (2017). *Reliability and Availability Engineering — Modeling, Analysis, and Applications —*, Cambridge University Press.

Yoshimura, I. and Sato, Y. (2008). Safety achieved by the safe failure fraction (SFF) in IEC 61508. *IEEE Transactions on Reliability*, 57(4): 662–669.

Recent Advances in Reliability and Maintenance Modeling – Okamura, Inoue & Xiao (Eds)
 ISBN: 978-1-032-79098-5

Maintenance policy for deteriorating k-out-of-n systems subject to random shocks

M. Kasuya & L. Jin
University of Electro-Communications, Chofu, Tokyo, Japan

ABSTRACT: We investigate a k-out-of-n system in which each unit deteriorates according to a gamma process and is susceptible to various random external shocks. Upon the arrival of a shock, the deterioration states of the relevant units experience a jump. There is an interdependence between the deterioration state of individual units and the magnitude of the system's jump. At equally spaced time intervals, the deterioration state and jump magnitude of each unit are fully observed. At the beginning of each time interval, the decision-maker determines maintenance actions (either to do nothing or to replace) for each unit based on the deterioration state and cumulative jump magnitude caused by random shocks, aiming to minimize the total expected maintenance cost over an infinite horizon. We optimize the condition-based policy using a Markov decision process and show that the total expected discounted cost monotonically increases with both the deterioration state and the jump magnitude.

1 INTRODUCTION

Although efficient maintenance policies for multi-unit systems are required to operate real-world systems efficiently, the inherent complexities of multi-unit systems make this challenging (Kobbacy *et al.* 2008; Keizer *et al.* 2017). Two difficulties hinder the development of efficient maintenance policies for such systems. The first difficulty is due to the curse of dimensionality (Bellman 1957). In multi-unit systems, the state and action spaces Fincrease with the number of units considered, exponentially increasing the number of candidates for the optimal policy. Consequently, searching for an optimal policy becomes computationally infeasible within practical timeframes. The second difficulty arises from the interdependencies among units, such as economic dependencies, structural dependencies, and stochastic dependencies. Neglecting these interdependencies may jeopardize the system's reliability. Thus, properly considering these interdependencies is essential for effectively operating multi-unit systems.

Research on maintenance policies for multi-unit systems can be categorized into two types: numerical research and analytical research. In numerical research, the optimization problem of maintenance decision-making is formulated, and then the optimal policy is derived through computation. Oakley *et al.* (2019) focused on stochastic dependence, where failure of a unit increases the load of other units, making the units more prone to deterioration. They also numerically demonstrated that the proposed policy is more cost-effective than the conventional policy. In numerical research, the main objective is that the proposed policy leads to improvements such as cost reductions and shorter computation time than the conventional policy. Therefore, optimality is not investigated to evaluate how much room for improvement there is in the numerically obtained optimal policy. On the other hand, in analytical research, the main objective is to analyze the structure of the objective function and clarify the properties of the optimal policy. Sun *et al.* (2018) proved

DOI: 10.1201/9781003491309-6

that the optimal policy for k-out-of-n systems without stochastic dependence has a structural property. Zhang *et al.* (2023) considered stochastic dependence, where the failure of one unit induces the failure of others, and then demonstrated that the optimal policy also exhibits structural properties. If it is known in advance that the optimal policy satisfies certain properties, it is only necessary to search for candidates that fulfill these properties. Consequently, the number of candidates is reduced, enabling the efficient search for the optimal policy. Furthermore, the properties obtained analytically facilitate the evaluation of the potential for improvement of the numerically obtained optimal policy. For example, if the approximate solution obtained numerically does not satisfy the properties of the optimal policy, it indicates that there is room for improvement to satisfy these properties.

Many systems become disordered due to not only system deterioration but also random external shocks. For example, over-charging and over-discharging of the batteries in electric vehicles, which reduce their capacity, are considered random external shocks (Zhao *et al.* 2019). Lightning strikes on the blades of wind turbines are random external shocks that induce or accelerate cracking (Katsaprakakis *et al.* 2021). Research on maintenance policies that consider random external shocks has been extensively undertaken. Rafiee *et al.* (2015) devised a maintenance policy for single-unit systems that considers extreme shock, run shock, and δ-shock. Dui *et al.* (2023) devised a policy for multi-unit systems that determines maintenance priorities on the basis of the reliability of each unit subjected to random external shocks. In the research cited above, the deterioration of the target system and the magnitude of random external shocks affecting the system were assumed to be independent. However, in real situations, not only does the deterioration rate change with each shock, but the current deterioration state affects the magnitude of the shock (Yousefi *et al.* 2020). Kurt & Maillart (2009) considered that the deterioration rate depends on the amount of damage from shocks and demonstrated that their policy for single-unit systems has a structural property. Wang *et al.* (2023) considered that the deterioration state and the amount of damage from shocks are interdependent and demonstrated that their policy for single-unit systems has the structural property.

If the maintenance target system is a single-unit system, typically only the shocks affecting the target unit are considered. However, in multi-unit systems, not all units are equally affected by shocks. Each unit experiences different effects due to various types of shocks, and the correspondence between each unit and a shock varies (Dong *et al.* 2021). For instance, deterioration occurs in a micro-electromechanical system (MEMS) oscillator composed of multiple MEMS resonators. MEMS resonators may be impacted by diverse random external shocks, such as thermal and vibration shocks. If a shock occurs in MEMS oscillators, the deterioration state and deterioration rate of the resonator where the shock directly arrives are affected. Concurrently, the deterioration rate of the resonators where the shock does not arrive may increase (Shen *et al.* 2018). Therefore, maintenance policies should consider the variability in the effects of random external shocks on different units.

This research considers a multi-unit system consisting of identical units as our maintenance target. We assume that the system is subject to various types of random external shocks, causing the deterioration state of the corresponding units to jump. We consider stochastic dependence among units induced by random external shocks. Additionally, we incorporate the interdependence between the deterioration state of units and the magnitude of shocks: the more significant the cumulative shock magnitude in the system, the faster each unit deteriorates, and conversely, the more a unit deteriorates, the higher the magnitude of shocks it experiences. Our main contributions are outlined below:

- We propose an optimal maintenance policy for a k-out-of-n system that considers the interdependence between unit deterioration and random external shocks. This policy has the potential to be applied to many real-world applications and can enable more efficient operation of the system.
- We derived a structural property of the optimal policy. This property can serve as a key insight in the development of algorithms for efficiently searching for the optimal policy.

In Section 2, we describe the model of the maintenance target system and formulate the optimal decision-making problem for CBM by using a MDP. In Section 3, we prove that the optimal policy has a structural property. In Section 4, we present a case study of the three-bladed rotor system in offshore wind turbines and discuss the significance of the property. Finally, we conclude our research in 5.

2 MODEL

2.1 *System deterioration*

We consider a k-out-of-n: G system with n identical units. The system works if and only if at least k of the n units are operational. In other words, when more than $n-k$ units fail, the system becomes out of order and cannot meet system requirements. In this research, each unit suffers from both internal deterioration and external shocks. Let $S_i(t)$ be the deterioration state of unit i at time t, and $S_i(0)=0$. We assume that M types of external shocks occur randomly in the system. $\mathbf{U}_m$ represents the subset of units directly impacted by type m shock. The occurrence of each type of shock is stationary and independent. The time interval between two successive occurrences of type $m \in \{1, ..., M\}$ shocks follows an exponential distribution, $g_m(t)$. If type m shock occurs at time t, unit $i \in \mathbf{U}_m \subseteq \{1, ..., n\}$ is directly affected, causing the deterioration state $S_i(t)$ to increase by $\Delta Z_i(t)$, which represents a sudden incremental jump in the deterioration state of unit i. We assume that $\Delta Z_i(t)$ is a random variable and following a probability distribution $f_i^m(\Delta Z|S(t))$. Let $Z_i(t)$ be the cumulative sudden incremental jump in the deterioration state of unit i at time t. $Z_i(t)$ is calculated by accumulating ΔZ_i that has occurred in unit i up to time t. When $S_i(t)$ reaches or exceeds a predetermined threshold L, unit i fails.

Let $\mathbf{S}(t)=(S_1(t), ..., S_n(t))$ and $\mathbf{Z}(t)=(Z_1(t), ..., Z_n(t))$ be a system deterioration state and a system cumulative jump, respectively. We assume that each unit deteriorates following a gamma process, a commonly used model in reliability engineering to characterize phenomena of monotonically increasing deterioration, such as crack growth, erosion, and creep of concrete. In this research, we assume that unit deterioration is independent of the deterioration state of other units, but depends on the cumulative jump of other units. Given $S_i(t)=s_i$ and $\mathbf{Z}(t)=\mathbf{z}=(z_1, ..., z_n)$, the transition probability from s to s' after Δt is expressed as follows.

$$p_i(s'|s, \mathbf{z}, \Delta t) = \frac{(s'-s)^{\alpha_i \Delta t - 1}}{\Gamma(\alpha_i \Delta t)\beta_i(\mathbf{z})^{\alpha_i \Delta t}} e^{-\frac{s'-s}{\beta_i(\mathbf{z})}}{}_{\{s'-s>0\}}, \tag{1}$$

where α_i is the shape parameter, and $\beta_i(\mathbf{z})$ is the scale parameter depending on $\mathbf{z}$. ${}_{\{\cdot\}}$ is an indicator function.

The system deterioration state $\mathbf{s}$ and cumulative jump $\mathbf{z}$ are inspected perfectly at discrete inspection epochs $t=k\tau(k=1,2,...)$, where τ is the fixed inspection interval. Let $q(\mathbf{s}', \mathbf{z}'|\mathbf{s}, \mathbf{z}, \tau)$ be the conditional transition probability that $\mathbf{s}=(s_1, s_2, ..., s_n)$ and $\mathbf{z}=(z_1, z_2, ..., z_n)$ transition to $\mathbf{s}'=(s_1', s_2', ..., s_n')$ and $\mathbf{z}'=(z_1', z_2', \ldots, z_n')$, respectively, after τ. We express $q(\mathbf{s}', \mathbf{z}'|\mathbf{s}, \mathbf{z}, \tau)$ as a recursive function as follows

$$\begin{aligned} &q(\mathbf{s}', \mathbf{z}'|\mathbf{s}, \mathbf{z}, \tau) \\ &= \sum_{m=1}^{M} \int_{z_n}^{z_n'} \cdots \int_{z_1}^{z_1'} \int_{s_n+y_n-z_n}^{s_n'} \cdots \int_{s_1+y_1-z_n}^{s_1'} \int_0^{\tau} q(\mathbf{s}', \mathbf{z}'|\mathbf{x}, \mathbf{y}, \tau-u) g_m(u) \prod_{l=1, l\neq m}^{M} [1-G_l(\tau)] \\ &\quad \times \prod_{i=1}^{n} \{\cdot_{\{i\in U_m\}} \times f_i^m(y_i - z_i|x_i) p(x_i|s_i, \mathbf{z}, u) + {}_{\{i\notin U_m, y_i=z_i\}} \times p(x_i|s_i, \mathbf{z}, u).\} du dx_1 \cdots dx_n dy_1 \cdots dy_n \\ &\quad + {}_{\{\mathbf{z}'=\mathbf{z}\}} \prod_{m=1}^{M} [1-G_m(\tau)] \prod_{i=1}^{n} p(s_i'|s_i, \mathbf{z}, \tau), \end{aligned} \tag{2}$$

where $1 - G_m(\tau)$ is the probability that no shock arrives until τ. Note that $Z_i(t) \leq S_i(t)$ is always satisfied. This is because $Z_i(t)$ represents only the increased deterioration state caused by external shocks, whereas $S_i(t)$ represents the total deterioration including both the effects of external shocks and internal deterioration.

2.2 *Actions and costs in condition-based maintenance*

In this research, one of two maintenance actions, namely do nothing or replacement, is selected for each unit based on system information $(\mathbf{s}, \mathbf{z})$, where $\mathbf{s}$ is the system deterioration state and $\mathbf{z}$ represents the system cumulative jump. Let $\mathbf{r} = (r_1, r_2, ..., r_n)$ denote the maintenance action for the entire system, indicating whether each unit is replaced or not. Here, $r_i = 0$ indicates that unit i is replaced, and $r_i = 1$ indicates that unit i undergoes no action. The action space of $\mathbf{r}$ is defined as follows,

$$\mathcal{A} = \{(r_1, r_2, ..., r_n) | r_i \in \{0, 1, i \in \{1, 2, ..., n\}\}.$$

When do nothing is selected for unit i, unit i is operated for one time period. Conversely, when unit i is replaced, the operating unit is replaced with an as-good-as-new unit. In this situation, replacement cost is incurred depending on s_i. If $s_i < L$, a preventive replacement cost c_p is incurred. On the other hand, a corrective replacement cost c_f is incurred if $s_i \geq L$. We set the following assumption regarding c_p and c_f.

Assumption 1. $0 < c_p < c_f$.
Assumption 1 indicates that the corrective replacement cost is higher than preventive replacement cost in all units to capture the harm caused by failure. A setup cost c_s is incurred when one or more units are replaced. Maintenance time is assumed to be negligible; thus, the unit is immediately operated after replacement completion. Furthermore, we assume that the system cannot be self-detected but can be identified through inspection. If the system fails during operation until the next inspection reveals the failure, a downtime cost c_d is incurred per unit time.

2.3 *Optimal maintenance policy*

An optimal maintenance decision-making problem for CBM is formulated using a discrete-time Markov decision process. In this research, an optimal maintenance policy is defined as a sequence of maintenance actions that minimize the total expected discounted cost based on the system deterioration state and the system cumulative jump. In other words, one of our objectives is to minimize the total expected discounted cost given system information $(\mathbf{s}, \mathbf{z})$.

We define $V(\mathbf{s}, \mathbf{z})$ as the total expected discounted cost and express it as follows

$$V(\mathbf{s}, \mathbf{z}) = \min_{\mathbf{r}}\{V_{\mathbf{r}}(\mathbf{s}, \mathbf{z})\}, \tag{3}$$

where

$$V_{\mathbf{r}}(\mathbf{s}, \mathbf{z}) = \sum_{i=1}^{n} c_{p\{r_i=0, s_i<L\}} + \sum_{i=1}^{n} c_{f\{r_i=0, s_i \geq L\}} + W(\mathbf{s} \circ \mathbf{r}, \mathbf{z} \circ \mathbf{r}) + c_s\left\{\sum_{i=1}^{n} r_i \neq n\right\}$$
$$+ e^{-\gamma\tau} \int_{z_n r_n}^{\infty} \cdots \int_{z_1 r_1}^{\infty} \int_{s_n r_n}^{\infty} \cdots \int_{s_1 r_1}^{\infty} q(\mathbf{s}', \mathbf{z}' | \mathbf{s} \circ \mathbf{r}, \mathbf{z} \circ \mathbf{r}, \tau) V(\mathbf{s}', \mathbf{z}') ds'_1 \cdots ds'_n dz'_1 \cdots dz'_n$$

and $e^{-\gamma\tau}(\gamma > 0)$ is the discounted factor. In Eq. (3), $(\mathbf{s} \circ \mathbf{r}, \mathbf{z} \circ \mathbf{r}) = ((s_1 r_1, ..., s_n r_n), (z_1 r_1, ..., z_n r_n))$ denotes the system information after maintenance action $\mathbf{r}$. $W(\mathbf{s}, \mathbf{z})$ is the expected downtime cost. If T_f is the time elapsed from the last inspection to system failure, $W(\mathbf{s}, \mathbf{z})$ is expressed as

$$\begin{aligned} W(\mathbf{s}, \mathbf{z}) &= \int_0^\tau \frac{c_d(e^{-\gamma t} - e^{-\gamma\tau})}{\gamma} f_{T_f}(t|\mathbf{s}, \mathbf{z})dt \\ &= \int_0^\infty \frac{c_{d\{t \leq \tau\}}(e^{-\gamma t} - e^{-\gamma\tau})}{\gamma} f_{T_f}(t|\mathbf{s}, \mathbf{z})dt. \end{aligned} \tag{4}$$

3 STRUCTURAL PROPERTIES OF OPTIMAL MAINTENANCE POLICY

In an MDP, the number of candidate maintenance policies grows exponentially as the number of units and state dimensions increase. Therefore, when dealing with high dimensions, it may become impractical to search for all policies and identify the optimal one within a reasonable timeframe. In this research, we investigate the properties of the optimal policy to address this dimensionality issue. Knowing these properties in advance may help alleviate the curse of dimensionality, as only policies satisfying the properties need to be considered as candidates. To this end, we first introduce definitions and lemmas to discuss the structural property of the optimal policy.

Definition 1. *For n-dimentional vectors* $\mathbf{x} = (\mathrm{x}_1, ..., \mathrm{x}_\mathrm{n})$ *and* $\mathbf{y} = (\mathrm{y}_1, ..., \mathrm{y}_\mathrm{n})$, *if* $\mathrm{x}_\mathrm{i} \leq \mathrm{y}_\mathrm{i}$ *for any* $\mathrm{i} \in \{1, ..., \mathrm{n}\}$, $\mathbf{x}$ *is smaller than* $\mathbf{y}$ *in a product order, written* $\mathbf{x} \leq_c \mathbf{y}$.

Definition 2. *Stochastic Increasing (SI) (Shaked and Shanthikumar 2007)*

For random variables X and Y, if $\Pr\{X > t\} \leq \Pr\{Y > t\}$ *for all t, X is stochastically smaller than Y in the sense of* SI, *written* $X \preceq_{\mathrm{SI}} Y$.

For random vectors $\mathbf{X} = (X_1, ..., X_n)$ and $\mathbf{Y} = (X_1, ..., X_n)$ that take on values in $\mathbb{R}_+^n = [0, \infty)^n$, if $\Pr\{\mathbf{X} \in U\} \leq \Pr\{\mathbf{Y} \in U\}$ for all upper sets $U \subseteq \mathbb{R}_+^n$, $\mathbf{X}$ is stochastically smaller than $\mathbf{Y}$ in the sense of SI, written $\mathbf{X} \preceq_{\mathrm{SI}} \mathbf{Y}$.

Lemma 1. *Shaked and Shanthikumar (2007)*

If $\mathbf{X} \preceq_{\mathrm{SI}} \mathbf{Y}$ *is satisfied, then* $E[f(\mathbf{X})] \leq E[f(\mathbf{Y})]$ *for any non-decreasing function* $f(\cdot)$, *and* $E[f(\mathbf{X})] \geq E[f(\mathbf{Y})]$ *for any non-increasing function* $f(\cdot)$

Lemma 2. *Shaked and Shanthikumar (2007)*

For random vectors $\mathbf{X} = (X_1, ..., X_n)$ *and* $\mathbf{Y} = (Y_1, ..., Y_n)$ *that take on values in* $\mathbb{R}_+^n = [0, \infty)^n$, *if* $X_1 \preceq_{\mathrm{SI}} Y_1$, $\langle X_2 | X_1 = x_1 \rangle \preceq_{\mathrm{SI}} \langle Y_2 | Y_1 = y_1 \rangle$ *whenever* $x_1 \leq y_1$, *and in general, for* $i = 2, 3, ..., n$, $\langle X_i | X_1 = x_1, ..., X_{i-1} = x_{i-1}] \rangle \preceq_{\mathrm{SI}} \langle Y_i | Y_1 = y_1, ..., Y_{i-1} = y_{i-1} \rangle$ *whenever* $x_j \leq y_j$, $j = 1, ..., i-1$, *then* $\mathbf{X} \preceq_{\mathrm{SI}} \mathbf{Y}$.

We provide Lemma 3 about the transition of system information.

Lemma 3. *If* $\mathbf{s}^- = (s_1^-, ..., s_n^-) \leq_c \mathbf{s}^+ = (s_1^+, ..., s_n^+)$ *and* $\mathbf{z}^- = (z_1^-, ..., z_n^-) \leq_c \mathbf{z}^+ = (z_1^+, ..., z_n^+)$ *are satisfied, then* $\langle \mathbf{s}', \mathbf{z}' | \mathbf{s}^-, \mathbf{z}^-, \tau \rangle \preceq_{\mathrm{SI}} \langle \mathbf{s}', \mathbf{z}' | \mathbf{s}^+, \mathbf{z}^+, \tau \rangle$ *for any* τ.

In the following, we set two assumptions to consider the interdependence between deterioration and external shocks.

Assumption 2. If $\mathbf{z}^- \leq_c \mathbf{z}^+$, then $\beta_i(\mathbf{z}^-) \leq \beta_i(\mathbf{z}^+)$ for any $i \in \{1, 2, ..., n\}$.

Assumption 3. If $s^- \leq s^+$, then $f_i^m(\Delta z|s^-) \preceq_{SI} f_i^m(\Delta z|s^+)$ for any $i \in \{1, 2, ..., n\}$.

Assumption 2 indicates that all units are more susceptible to deteriorates the system is subjected to random external shocks. Assumption 3 indicates that all units are more susceptible to external shocks as they deteriorate.

We provide Theorem 1 about the expected downtime cost.

Theorem 1. *The expected downtime cost* $\mathrm{W}(\mathbf{s}, \mathbf{z})$ *is non-decreasing in* $\mathbf{s}$ *and* $\mathbf{z}$ *ordered by a product order.*

We provide Theorem 3 about the total expected discounted cost.

Theorem 2. *The total expected discounted cost* $\mathrm{V}(\mathbf{s}, \mathbf{z})$ *is non-decreasing in* $\mathbf{s}$ *and* $\mathbf{z}$ *ordered by a product order.*

Theorem 3. *We consider any two actions* $\mathbf{r} = (\mathrm{r}_1, ..., \mathrm{r}_n)$ *and* $\tilde{\mathbf{r}} = (\tilde{\mathrm{r}}_1, ..., \tilde{\mathrm{r}}_n)$ *that satisfy* $\mathrm{r}_i = 0$ and $\tilde{\mathrm{r}}_i = 1$, *respectively. For any* $\mathbf{s}$ *and* $\mathbf{z}$*, the maximum number of times that* $\mathrm{V}_{\mathbf{r}}(\mathbf{s}, \mathbf{z})$ *and* $\mathrm{V}_{\tilde{\mathbf{r}}}(\mathbf{s}, \mathbf{z})$ *change their order is three in* s_i*, and the maximum number of times that* $\mathrm{V}_{\mathbf{r}}(\mathbf{s}, \mathbf{z})$ *and* $\mathrm{V}_{\tilde{\mathbf{r}}}(\mathbf{s}, \mathbf{z})$ *change their order is one in* z_i.

4 CASE STUDY OF THREE-BLADED ROTOR SYSTEM OF OFFSHORE WIND TURBINES

4.1 *Parmeter settings*

As interest in climate change and environmental issues grows, the importance of clean energy usage has become prominent. Offshore wind energy is anticipated to play a significant role in future energy markets, not only due to its environmental benefits such as pollution-free operation and absence of harmful emissions but also because of its capacity for stable power generation compared to onshore wind (Chen and Kim 2022). Operation and maintenance constitute approximately one-fourth of the investment in offshore wind turbines (Ren *et al.* 2021), making the improvement of maintenance efficiency a crucial strategy for cost reduction (Irawan *et al.* 2017).

Based on the above, the proposed CBM policy is applied to a three-bladed rotor system of offshore wind turbines, which is considered a 3-out-of-3 system. The combined effects of sustained tensile stress and a corrosive environment induce stress corrosion cracks on the blades of the offshore wind turbine (Hao *et al.* 2018). Since the cracks may cause serious risks such as environmental pollution and economic losses, they are considered targets of condition monitoring (Raza and Ulansky 2019). We characterize the propagation of stress corrosion cracks on a three-bladed rotor system as a gamma process. In this section, each gamma process has the same parameters. Referring to Shafiee *et al.* (2015), we set $\alpha_i = 0.542$, $\beta_i(\mathbf{z}) = 0.872$ with $\mathbf{z} = (0, ..., 0)$ for all $i \in \{1, 2, 3\}$. Furthermore, if the length of the crack is over 20 cm, the unit is regarded as a failure, which means $L = 20$, $i \in \{1, 2, 3\}$. Because the external shocks may accelerate the deterioration of the entire system, we calculate $\beta_i(\mathbf{z}) = 0.872(1 + (2z_i + \sum_{j=1, j \neq i}^{3} z_j)/16)$ given $\mathbf{z}$ for all i.

Offshore wind turbines are not only subject to internal deterioration but also to various random external shocks such as sea waves, wind gusts, lightning, hailstones and icing. In this research, lightning, and hailstones are considered external shocks. Lightning is considered the shock that easily affect blades and create favorable conditions, such as delamination, debonding, and shell detachment, for crack initiation and propagation (Katsaprakakis *et al.* 2017). In addition, lightning strikes any one of the three blades, and only the struck blade is affected (Katsaprakakis *et al.* 2017). Therefore, we assume that there are 3 types of external shocks caused by lightning: $\mathbf{U}_1 = \{1\}$, $\mathbf{U}_2 = \{2\}$ and $\mathbf{U}_3 = \{3\}$. We also assume that the jump in the deterioration state caused by each lightning follows an exponential distribution with an expected value $\kappa(s_i) = \max(0.01, 0.2 - 0.0075s_i)$. Lightning strikes affect approximately 4–8% of wind turbines in Europe annually (Rachidi *et al.* 2008). Therefore, we assume that the lightning strikes each blade with a probability of 1/600 on average per month and define $g_m(x) = \frac{1}{600}e^{-\frac{1}{600}x}$, $m \in \{1, 2, 3\}$. External shocks from hailstones may cause delamination, resulting in cracks, cause reduced lift and increased drag (Katsaprakakis *et al.* 2017). We assume that there is 1 type of shock caused by hailstones and define $\mathbf{U}_4 = \{1, 2, 3\}$. We also assume that the jump of the deterioration state caused by hailstones follows an exponential distribution with an expected value $h_i(s_i) = \max(0.1, 0.4 - 0.015s_i)$. We assume that the arrival of the external shock by hailstones with a probability of 1/200 on average per month and define $g_4(x) = \frac{1}{200}e^{-\frac{1}{200}x}$. Referring to Besnard and Bertling (2010), we set $c_d = 72$, $c_s = 40$, $c_p = 130$ and $c_f = 400$. Assuming $\tau = 2$, two months is considered as the inspection interval. We set the discounted factor for each month as $e^{-0.01\tau}$.

4.2 *Optimal maintenance policy for three-bladed rotor system of offshore wind turbines*

An optimal maintenance policy for a three-bladed rotor system of offshore wind turbines is obtained using a value iteration algorithm. We discretize continuous deterioration states and continuous jumps in amount into discrete intervals to facilitate easier computation. If the stress corrosion cracks size (cm) of unit i is in the interval $[0.0, 5.0)$, then the deterioration level is $s_i = 0$. For the intervals $[5.0, 10.0)$, $[10.0, 15.0)$, $[15.0, 20.0)$, $[20.0, \infty)$, the deterioration levels are $s_i = 1, 2, 3$ and 4, respectively. Similarly, if the cumulative jump (cm) of unit i falls within the interval $[0.0, 5.0)$, then the cumulative jump level is $z_i = 0$, and for the intervals $[5.0, 10.0)$, $[10.0, 15.0)$, $[15.0, 20.0)$, $[20.0, \infty)$, the cumulative jump levels are $z_i = 1, 2, 3$ and 4, respectively. Since $L = 20$, the unit fails when the deterioration level s_i reaches 4.

Figures 1 and 2 show the corresponding optimal action of the three-bladed rotor system for each $\mathbf{s}$ when $\mathbf{z} = (0, 0, 0)$ and $\mathbf{z} = (1, 1, 1)$, respectively. In Figures 1 and 2, each figure shows the optimal maintenance policy given s_1 and $\mathbf{z}$. The color in Figures 1 and 2 represents the optimal number of units to be replaced, turning black as the number of units increases. For example, the action $\mathbf{r} = (1, 0, 0)$, which replaces units 2 and 3, is optimal given $\mathbf{s} = (0, 4, 4)$, $\mathbf{z} = (0, 0, 0)$ in Figure 1.

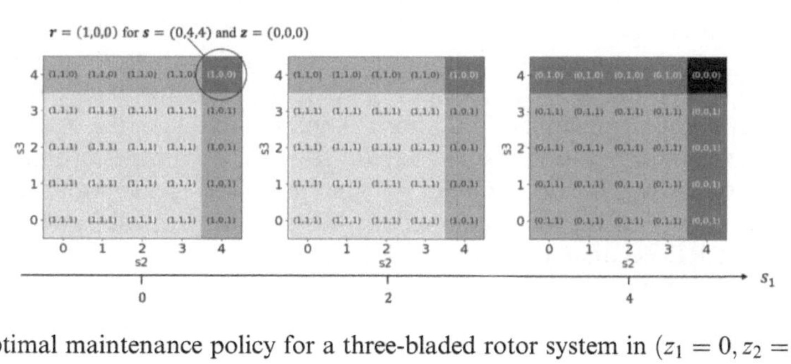

Figure 1. Optimal maintenance policy for a three-bladed rotor system in $(z_1 = 0, z_2 = 0, z_3 = 0)$.

From Figure 1, it is optimal to correctively replace only units that transition to $s_i = 4$. On the other hand, in Figure 2, preventive maintenance appears as the optimal action somewhere. For instance, the optimal action is $\mathbf{r} = (0, 1, 1)$ given $\mathbf{s} = (3, 1, 1)$ and $\mathbf{z} = (1, 1, 1)$ in Figure 2. This indicates that the optimal policy becomes conservative as $\mathbf{z}$ becomes larger.

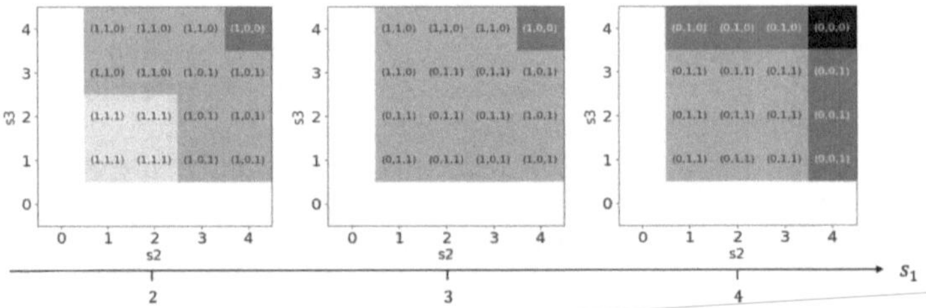

Figure 2. Optimal maintenance policy for a three-bladed rotor system in $(z_1 = 1, z_2 = 1, z_3 = 1)$.

Furthermore, Figures 1 and 2 show that the optimal policy exhibits the property indicated by Theorem 3.

We compare the proposed policy with other policies to investigate its effectiveness. Under the assumption that unit deterioration depends on the cumulative jump of the deterioration state in the system and that the jump due to external shocks also depends on the deterioration state, we consider the following three policies.

4.2.1 *Considering only dependence of the jump (DJ policy)*

Under this policy, only the dependence of the cumulative jump on unit deterioration is considered. In other words, $\beta_i(\mathbf{z})$ is calculated based on $\mathbf{z}$, but $f_i^m(\Delta z_i|s_i)$ is assumed to be stationary with s_i. In this policy, $f_i^m(\Delta z_i|s_i) = f_i^m(\Delta z_i|0)$ is assumed for all s_i.

4.2.2 *Considering only dependence of the deterioration state (DS policy)*

Under this policy, only dependence of the deterioration state on the jump due to external shocks is considered. In other words, $f_i^m(\Delta z|s)$ is calculated based on s, but $\beta_i(\mathbf{z})$ is assumed to be stationary with $\mathbf{z}$. In this policy, $\beta_i(\mathbf{z}) = \beta_i(\mathbf{z} = (0, ..., 0))$ is assumed for all $\mathbf{z}$.

4.2.3 *Considering no dependencies (ND policy)*

Under this policy, unit deterioration and the jump due to external shock are assumed to be independent of each other. In other words, $\beta_i(\mathbf{z})$ is assumed to be stationary with $\mathbf{z}$, and $f_i^m(\Delta z|s)$ is assumed to be stationary with s_i. In this policy, $\beta_i(\mathbf{z}) = \beta_i(\mathbf{z} = (0, ..., 0))$ is assumed for all $\mathbf{z}$, and $f_i^m(\Delta z|s) = f_i^m(\Delta z|0)$ is assumed for all s.

We use the average of total discounted costs as an indicator when comparing policies and investigating the effectiveness of the proposed policy. We outline the method for calculating the average of total discounted costs. At the beginning, we define an episode as a sequential decision-making process with an initial state $(\mathbf{s} = (0, ..., 0), \mathbf{z} = (0, ..., 0))$ for a finite period that is long enough to be considered infinite in accordance with the maintenance policy. Here, let the total discounted cost be the sum of costs incurred for each period within one episode. Subsequently, we repeat this simulation 2000 times and take the average of the total discounted cots based on simulation results. Table 1 shows the averages of total maintenance costs and the cost reduction rate (CRR) of the proposed policy and other policies. Specifically, if the average of the total maintenance costs of the proposed policy is v_1 and that of DJ policy is v_2, CRR is calculated as $(v_2 - v_1)/v_2$.

Table 1. Comparison of averages of total discounted costs incurred when following optimal policies for three-bladed rotor system.

Proposed policy		DJ policy		DS policy		ND policy	
cost	CRR (%)	cost	CRR (%)	cost	CRR (%)	cost	CRR (%)
7742.00	-	7755.55	0.2	7832.99	1.2	7843.48	1.3

From Table 1, it is evident that the proposed policy, considering the interdependence of unit deterioration and the jump due to external shock, is the most effective. Additionally, the table demonstrates that it is more effective to consider that the unit deterioration depends on the jump than that the jump depends on the deterioration state.

5 CONCLUSION

We focused on a k-out-of-n: G system with n identical units, where each unit deteriorates independently following a gamma process and is subject to various types of random external shocks. We explored condition-based maintenance for such a system, considering the interdependence between unit deterioration and the jump caused by random external shocks. Using a Markov decision process, we proposed an optimal condition-based maintenance policy to minimize the total expected discounted cost. Additionally, we analytically demonstrated that the proposed policy exhibits a structural property. Furthermore, we presented a case study involving a three-bladed rotor system of offshore wind turbines and illustrated that the policy also possesses a structural property.

In our research, we assumed that the deterioration of each unit is independent of the deterioration states of other units. However, in reality, the failure or deterioration state of one unit may influence the deterioration of other units. Therefore, investigating and incorporating such interdependencies into our model is one direction for future research.

REFERENCES

Bellman, R. E. (1957). *Dynamic Programming*. Princeton University Press, Princeton, NJ.

Chen, J. and Kim, M. H. (2022). Review of recent offshore wind turbine research and optimization methodologies in their design. *Journal of Marine Science and Engineering* 10(1): 28.

Dong, W., Liu, S., Bae, S. J. and Cao, Y. (2021). Reliability modelling for multi-component systems subject to stochastic deterioration and generalized cumulative shock damages. *Reliability Engineering & System Safety* 205: 107260.

Dui, H., Zhang, H. and Wu, S. (2023). Optimisation of maintenance policies for a deteriorating multi-component system under external shocks. *Reliability Engineering & System Safety*, 238: 109415.

Hao, W., Liu, Z., Wu, W., Li, X., Du, C. and Zhang, D. (2018). Electrochemical characterization and stress corrosion cracking of E690 high strength steel in wet-dry cyclic marine environments. *Materials Science and Engineering: A* 710: 318–328.

Irawan, C. A., Ouelhadj, D., Jones, D., Stalhane, M. and Sperstad, I. B. (2017). Optimisation of maintenance routing and scheduling for offshore wind farms. *European Journal of Operational Research* 256(1): 76–89.

Katsaprakakis, D. A., Papadakis, N. and Ntintakis, I. (2021). A comprehensive analysis of wind turbine blade damage. *Energies* 14(18): 5974.

Keizer, M. C. O., Flapper, S. D. P. and Teunter, R. H. (2017). Condition-based maintenance policies for systems with multiple dependent components: A review. *European Journal of Operational Research* 261(2): 405–420.

Kurt, M. and Maillart, L. M. (2009). Structured replacement policies for a Markov-modulated shock model. *Operations Research Letters*, 37(4): 280–284.

Kobbacy, K. A., Murthy, D. P., Nicolai, R. P. and Dekker, R. (2008). *Optimal Maintenance of Multi-component Systems: A Review*: 263–286. Springer London.

Oakley, J. L., Wilson, K. J. and Philipson, P. (2022). A condition-based maintenance policy for continuously monitored multi-component systems with economic and stochastic dependence. *Reliability Engineering & System Safety* 222: 108321.

Rafiee, K., Feng, Q. and Coit, D. W. (2015). Condition-based maintenance for repairable deteriorating systems subject to a generalized mixed shock model. *IEEE Transactions on Reliability*, 64(4): 1164–1174.

Ren, Z., Verma, A. S., Li, Y., Teuwen, J. J. E. and Jiang, Z. (2021). Offshore wind turbine operations and maintenance: A state-of-the-art review. *Renewable and Sustainable Energy Reviews* 144: 110886.

Shafiee, M., Finkelstein, M. and Berenguer, C. (2015). An opportunistic condition-based maintenance policy for offshore wind turbine blades subjected to degradation and environmental shocks. *Reliability Engineering & System Safety* 142: 463–471.

Shen, J., Elwany, A. and Cui, L. (2018). Reliability analysis for multi-component systems with degradation interaction and categorized shocks. *Applied Mathematical Modelling* 56: 487–500.

Sun, Q., Ye, Z. and Chen, N. (2017). Optimal inspection and replacement policies for multi-unit systems subject to deterioration. *IEEE Transactions on Reliability* 67(1): 401–413.

Wang, Z., Jin, L. and Yamamoto, W. (2023). Optimal maintenance policy for deteriorating systems subject to external shock damage under a semi-Markov decision process. *Total Quality Science* 9(2): 83–97.

Yousefi, N., Coit, D. W. and Zhu, X. (2020). Dynamic maintenance policy for systems with repairable components subject to mutually dependent competing failure processes. *Computers & Industrial Engineering* 143: 106398.

Zhang, N., Tian, S., Cai, K. and Zhang, J., 2023. Condition-based maintenance assessment for a deteriorating system considering stochastic failure dependence, *IISE Transactions* 55(7): 687–697.

Zhao, X., Sun, L., Wang, M. and Wang, X. (2019). A shock model for multi-component system considering the cumulative effect of severely damaged components. *Computers and Industrial Engineering* 137: 106027.

Recent Advances in Reliability and Maintenance Modeling – Okamura, Inoue & Xiao (Eds)
© 2025 the Author(s), ISBN: 978-1-032-79098-5

Optimal CBM policy for a degrading system with two weighted components

Shinji Kawano
Graduate School of Science and Engineering, Hosei University, Tokyo, Japan

Nobuyuki Tamura
Department of Industrial and Systems Engineering, Hosei University, Tokyo, Japan

ABSTRACT: We consider a system in which the degradation of two components follows a Wiener process with positive drift, and a failure occurs when the sum of these weighted degradation amounts, which is called the degradation level, exceeds a threshold value given in advance. The deterioration level is periodically observed, and if preventive maintenance is necessary, we choose whether to replace one or two components. Also, in the case that one component is replaced, we decide which is more appropriate. We investigate various properties of the optimal maintenance policy by modeling the system via a Markov decision process (MDP). Then, we theoretically show some structural results on the optimal maintenance policy which are consistent with the relevant studies in this area. Also, we provide the significance that the weight is introduced through numerical analysis.

Keywords: Weighted system, Markov decision process, Wiener process, IG distribution, Control limit policy, Monotone property

1 INTRODUCTION

We know that it is very important to avoid failures any systems such as productive facilities and equipments for infrastructure or reduce negative impacts on the failures from the viewpoints of availability and economic aspects. Thus, there are many papers for addressing maintenance problems on items mathematically. Recently, compared with time-based maintenance, condition-based maintenance is more available with advances of monitoring techniques. For example, Jardine *et al.* (2006) summarized the recent research and development in diagnostics and prognostics of mechanical system implementing CBM for data processing. Also, Olde Kaizer *et al.* (2017) and Alaswad & Xiang (2017) reviewed stochastic models focused on condition-based maintenance.

When Markov processes have been used for expressing deterioration of a system, many papers have considered that the system consists of a single-unit. However, since redundant structures are usually employed to avert sudden failures in realistic situations, we need to consider maintenance problems for multi-unit systems like series and parallel systems. Although Sun *et al.* (2018), Zhang and Si [11], Zhang *et al.* (2020) addressed the studies on the CBM policy for k-out-of N systems, we pay attention to a weighted k-out-of N system as its generalization. While Khorshidi *et al.* (2016), Atashgar & Abdollahzadeh (2017), and Hamdan (2021) studied maintenance problems on weighted systems whose structure are different, they supposed that the degradations were discrete in time. However, in the case that temperature, humidity, electrical current, and voltage are measured as a degradation of a system, it may be more convenient or appropriate to express the degradation as a continuous stochastic process for decision making based on the mathematical model since some researchers have treated these values as the set of real numbers in the past (see e.g. Lawless (2003) and Meeker and Escobar (1988)).

In this paper, we propose a weighted system which includes two components and each components degrade according to a Wiener process. We express a maintenance problem for the weighted system as a Markov decision process and investigate properties of the optimal maintenance policy minimizing a total expected discounted cost for unbounded horizon. Furthermore, numerical examples are provided for quantitative analysis.

The remainder of the paper is summarized as follows. We explain the structure of a weighted system in Section 2. Section 3 describes a maintenance policy for the weighted system and a total expected cost for unbounded horizon is derived via the theory of Markov decision process in Section 4. In Section 5, we provide some properties of the optimal maintenance policy including the control-limit policy. We numerically analyze the optimal maintenance policy and discuss its characteristics in Section 5. And finally, Section 6 concludes this paper.

2 WEIGHTED SYSTEM

Different from k-out-of N systems, the mechanism of failure for weighted systems in the previous studies are different. We consider a weighted system consisting of two components that each degradation follows a Wiener process. We let $Y_i(t)$ denote the degradation at time t of component i ($i = 1, 2$) and suppose

$$Y_i(t) = \mu_i t + \sigma_i B_i(t), \tag{1}$$

where $\{B_i(t); t \geq 0\}$ are standard Brownian motions, μ_i and σ_i are the positive drift parameter and the diffusion parameter, respectively. Whenever $Y_i(t)$ reaches a fixed threshold L_i, component i fails. For positive weights w_i, we assume that $Y_S(t)$ is the degradation level of the system at time t and it is expressed as a linear combination of the two components as follows:

$$Y_S(t) = w_1 Y_1(t) + w_2 Y_2(t). \tag{2}$$

If $Y_S(t) \geq \kappa$ then the system fails even though $Y_i(t) \leq L_i$, where κ is constant and given in advance. Thus, the weight means the significance of the component. Hereafter, we call κ the failure level to distinguish the threshold of the components. Since it is natural to consider that the system certainly fails whenever the two components reach the threshold, we suppose

$$0 < \kappa \leq w_1 L_1 + w_2 L_2. \tag{3}$$

Also, Equation (3) means that the system fails even if the two components do not fail. Thus, deciding the weights w_i and κ appropriately, the weighted system can express series and parallel systems. Although a similar idea has been proposed by Wu and Castro [9], they did not suppose a condition-based maintenance and the failure of each component, different from our study.

For the preparation of mathematical analysis, we transform $Y_S(t)$ into

$$\begin{aligned} Y_S(t) = w_1 Y_1(t) + w_2 Y_2(t) &= (w_1\mu_1 + w_2\mu_2)t + (w_1\sigma_1 B_1(t) + w_2\sigma_2 B_2(t)) \\ &= \mu_S t + \sigma_S B_S(t), \end{aligned} \tag{4}$$

where

$$\mu_S = w_1\mu_1 + w_2\mu_2,$$

$$B_S(t) = \left(\frac{w_1\sigma_1}{\sigma_s}\right) B_1(t) + \left(\frac{w_2\sigma_2}{\sigma_S}\right) B_2(t),$$

$$\sigma_S^2 = (w_1\sigma_1)^2 + (w_2\sigma_2)^2.$$

Then $\{Y_S(t); t \geq 0\}$ is the Wiener process with the drift parameter μ_S and the diffusion parameter σ_S since $B_S(t)$ follows a standard Brownian motion.

3 MAINTENANCE POLICY

We suppose that the weighted system constructed in Section 2 is maintained under the following rules:

(i) The weighted system is periodically observed at each inspection epoch, $0, \tau, \cdots, k\tau, \cdots$, and the true degradation levels of the two components are always detected.

(ii) Based on the degradation levels for the two components after inspection, we select one of the following actions: NULL, PM1-i $(i = 1, 2)$, PM2.

- When action NULL is selected, the system is operated until the next inspection without any maintenance. Then, we incur a downtime cost c_d per unit time when it fails between two inspection epochs.
- When action PM1-i $(i = 1, 2)$ is selected, component i is replaced by a new one. Then, we incur a setup cost c_s and a preventive replacement cost $c_p(w_i)$.
- When action PM-2 is selected, the two components are replaced by new ones. Then, we incur the setup cost and the preventive replacement costs for the two components.

(iii) Whenever the degradation level of the system $Y_S(t)$ exceeds the failure level κ at inspection epoch, action CM is selected and the two components are replaced by new ones. Then we incur corrective replacement costs $c_c(w_1)$ and $c_c(w_2)$.

For simplicity, we suppose that PM1 indicates that PM-1 or PM-2 is selected and we call $Y_S(t)$ the system-state.

Then, the costs satisfy some assumptions to reflect intuitively appropriate situations.

Assumption 1. $c_p(w)$ and $c_c(w)$ are increasing in w, respectively.

Assumption 1 indicates that the costs for preventive and corrective replacements increase as the weight increases.

4 FORMULATION

To analyze the optimal maintenance policy mathematically, we model the problem as a Markov decision process. Thus, we introduce a discount factor $e^{-r\tau}$ at time t for a fixed positive constant r. And $V(y_1, y_2)$ means the total expected discounted cost for unbounded horizon when an optimal action is selected at degradation level (y_1, y_2), where y_i $(i = 1, 2)$ is the degradation level of component i at each inspection epoch. Also, C_1, C_2, C_3 denote the expected cost when action NULL, PM1, and PM2 are selected at degradation level (y_1, y_2), respectively.

Firstly, we derive the expected downtime cost during two inspections for C_1. We let $T_f^{(m)}$ and $D_T(t)$ denote the time from the m-th inspection to the failure of the system and the downtime cost from the time epoch that the system fails at time t to the current inspection epoch, respectively. Then, we have

$$D_T\left(T_f^{(m)}\right) = \int_{T_f^{(m)}}^{\tau} c_d e^{-rt} dt = -\frac{c_d\left\{\exp(-r\tau) - \exp(-rT_f^{(m)})\right\}}{r}.$$

$T_f^{(m)}$ is the random variable which follows the Inverse Gaussian distribution since $T_f^{(m)}$ is derived as the first passage time of the Wiener process. When the random variable T follows the Inverse Gaussian distribution, its PDF $f_T(t; a, b)$ with parameters (a, b) is given by

$$f_T(t; a, b) = \left(\frac{b}{2\pi t^3}\right)^{1/2} \exp\left(-\frac{b(t-a)^2}{2a^2 t}\right), \quad t > 0, \tag{5}$$

where

$$a = (\kappa - y_S)/\mu_S, \quad b = ((\kappa - y_S)/\sigma_S)^2, \quad y_S = w_1 y_1 + w_2 y_2.$$

When $Dc(y_S)$ denotes the downtime cost for one period given that the system-state is observed, we have

$$\begin{aligned} Dc(y_S) &= E\left[D_T\left(T_f^{(m)}\right) | Y_S^{(m)} = y_S\right] \\ &= \int_0^{\tau} - \frac{c_d\{\exp(-r\tau) - \exp(-rt)\}}{r} f_T(t)dt, \end{aligned} \tag{6}$$

where $Y_S^{(m)}$ is the system-state at the m-th inspection epoch.

Thus, we have

$$C_1 = Dc(y_S) + e^{-r\tau} U(y_1, y_2), \tag{7}$$

where

$$U(y_1, y_2) = E\left[V\left(Y_1^{(m+1)}, Y_2^{(m+1)}\right) | Y_1^{(m)} = y_1, Y_2^{(m)} = y_2\right],$$

and $Y_i^{(m+1)}$ is the degradation level of component i at the m-th inspection epoch. Although it is not necessary to introduce the number of inspections for the degradation, we use the notation as $Y_i^{(m)}$ to clarify a state-transition of each component based on the Markov process.

Similarly, we have

$$C_2 = \min\left[C_p(w_1) + c_s + V(0, y_2), C_p(w_2) + c_s + V(y_1, 0)\right], \tag{8}$$

$$C_3 = C_p(w_1) + C_p(w_2) + c_s + V(0, 0). \tag{9}$$

Hence, the total expected discounted cost for unbounded horizon when the degradation level of the components is y_i is expressed as

$$V(y_1, y_2) = \begin{cases} C_c(w_1) + C_c(w_2) + V(0,0) & (y_S \geq \kappa) \\ \min[C_1, C_2, C_3] & (otherwise) \end{cases}.$$

Since a new system is installed at the beginning, the optimal maintenance policy is obtained by minimizing $V(0, 0)$.

5 PROPERTIES OF THE OPTIMAL MAINTENANCE POLICY

This section investigates the structure of the optimal maintenance policy that minimizes $V(0, 0)$. Before the discussion, we provide two basic results.

Lemma 1. *$Dc(y_S)$ is increasing in y_S.*

Since $y_S = w_1 y_1 + w_2 y_2$, $Dc(y_S)$ is also increasing in y_1 (y_2) for any y_2 (y_1). This lemma indicates that as the degradation proceeds, the downtime cost increases. And by using Lemma 1, the properties of the Wiener process, and the assumptions on the costs, we have Theorem 1.

Theorem 1. *$V(y_1, y_2)$ is increasing in y_1 for any y_2, and is increasing in y_2 for any y_1.*

Many previous studies in this area also obtained the similar result to Theorem 1. This result is available for deriving other results on the optimal maintenance policy.

In the next, we provide some results on the structure of the optimal maintenance policy. Now we suppose that $D(y_1, y_2)$ is an optimal decision at degradation level (y_1, y_2). Then, the structure of the optimal maintenance policy is provided as Theorem 2.

Theorem 2. *If $L_1 = L_2$, $w_1 = w_2$, $\mu_1 = \mu_2$ and $\sigma_1 = \sigma_2$, then the optimal maintenance policy has the following form:*

(1) For any a and b, $V(a, b) = V(b, a)$ holds. And for $y_1 = y_2$, $V(y_1, 0) = V(0, y_2)$ holds.
(2) There exist a real number $\delta_2 \leq L_1$ and a control limit $g_2(y_1)$ such that $D(y_1, y_2) = PM2$ when $y_2 > g_2(y_1)$ for any $y_1 \geq \delta_2$. Also, $g_2(y_1)$ is non-increasing in y_1.
(3) Suppose that there exists a degradation level $(y_1, g_1(y_1))$ such that $C_1 = C_2 \leq C_3$. Then, $D(y_1, y_2) = NULL$ for $y_2 < g_1(y_1)$ and $D(y_1, y_2) = PM1$ for $y_2 \geq g_1(y_1)$.
(4) For any y_1, $g_2(y_1) \geq g_1(y_1)$ holds.

In Theorem 3, g_1 means the control limit for PM1 and g_2 means the control limit for PM2. Then, we find that the control limits are symmetric against the line $y_1 = y_2$ and the four results in Theorem 2 hold in the case that y_1 and y_2 are exchanged. Thus, hereafter, we also call the optimal maintenance policy the two-dimensional control limit policy.

In the next, we pay attention to the case that $w_1 \neq w_2$ under $L_1 = L_2$.

Theorem 3. *If $L_1 = L_2$, $w_1 \neq w_2$, $\mu_1 \neq \mu_2$* and *$\sigma_1 \neq \sigma_2$, then the optimal maintenance policy has the following form:*

(1) Suppose

$$y_2^* = \min[y_2 : C_2 \leq C_1]$$

for any y_1. If

$$C_p(w_1) + c_s + V(0, y_2) > C_p(w_2) + c_s + V(y_1, 0)$$

and $D(y_1, y_2^*) = PM1$ then $D(y_1, \tilde{y}_2) \neq NULL$ for $\tilde{y}_2 > y_2^*$.
(2) Suppose

$$y_1^* = \min[y_1 : C_2 \leq C_1]$$

for any y_2. If

$$C_p(w_1) + c_s + V(0, y_2) \leq C_p(w_2) + c_s + V(y_1, 0)$$

and $D(y_1^*, y_2) = PM1$ then $D(\tilde{y}_1, y_2) \neq NULL$ for $\tilde{y}_1 > y_1^*$.
(3) If $D(y_1', y_2) = PM2$ for any y_2, then $D(y_1'', y_2) = PM2$ for $y_1' < y_1''$. Also, this result holds if we exchange y_1 and y_2.

Theorem 3 indicates that the optimal maintenance policy is not symmetric different from Theorem 2 while the two-dimensional control limit policy holds even if $w_1 \neq w_2$.

6 NUMERICAL ANALYSIS

In this section, we quantitatively investigate the behavior of the optimal maintenance policy through some examples. For numerical analysis, we discretize the deterioration levels of the components as

$$S_{c_i} = \{0, \Delta, 2\Delta, \cdots, L_i, L_1 + \Delta\}, \quad S_s = \{w_1 a + w_2 b \,|\, a \in S_{c_1}, b \in S_{c_2}\}.$$

S_{c_i} is the state space of component i and S_s is the state space of the weighted system for the Markov decision process, respectively. And the parameters for the weighted system are set as Table 1.

Table 1. Parameters for the weighted system.

w_1	4.0
w_2	2.5
κ	50
L_1	10
L_2	10
μ_1	3.2
μ_2	3.2
σ_1	0.7
σ_2	0.7

Table 1 indicates that the two components are identical while the weights are different. Also, for the costs, we suppose that

$$C_p(w) = c_p f_p(w), C_c(w) = c_c f_c(w), f_p(w) = 1.2^w, f_c(w) = 1.3^w,$$

and set the parameters as Table 2.

Table 2. Parameters for the costs.

c_p	3
c_c	17
c_d	3
c_s	3
r	0.01

These settings give us the following results,

$$C_p(w_1) + C_p(w_2) + c_s \approx 13.953, \;\; C_c(w_1) + C_c(w_2) + c_s \approx 81.311,$$

which mean that the cost of CM is bigger than that of PM. Many previous studies on optimal maintenance problems have introduced the similar condition on costs for PM and CM from the realistic point of view. Then, the optimal maintenance policy is obtained as Figure 1.

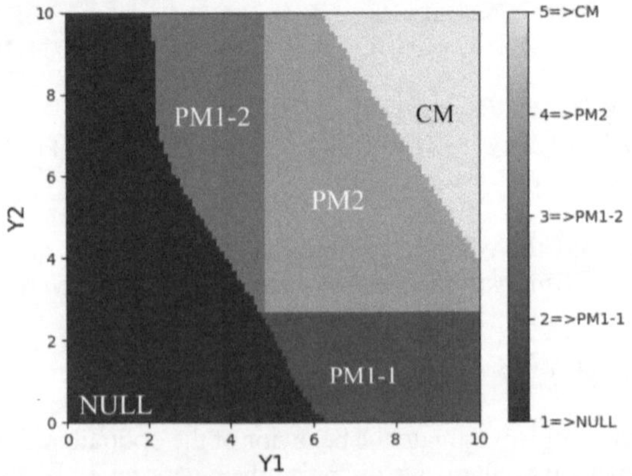

Figure 1. Example of an optimal maintenance policy when $w_1 \neq w_2$.

In Figure 1, the horizontal axis is the degradation level of component 1 and the vertical axis is the degradation level of component 2, respectively. Figure 1 illustrates that the two-dimensional control limit policy holds as shown in Theorem 3. Also, we find that the smallest control limit for PM1-2 is less than that for PM1-1 because Table 1 supposes that $w_2 < w_1$, $\mu_1 = \mu_2$ and $\sigma_1 = \sigma_2$.

Hereafter, the subsequent examples set the weights and the drifts only since the other parameters are unchanged. In the next, we suppose that $w_1 = w_2 = 3.0$ and $\mu_1 = \mu_2 = 3.2$ and the system fails whenever the system-state reaches 80 % of the weighted sum of the failure thresholds for the two components, that is, the failure level of the weighted system is expressed as

$$\kappa = 0.8(w_1 L_1 + w_2 L_2). \tag{10}$$

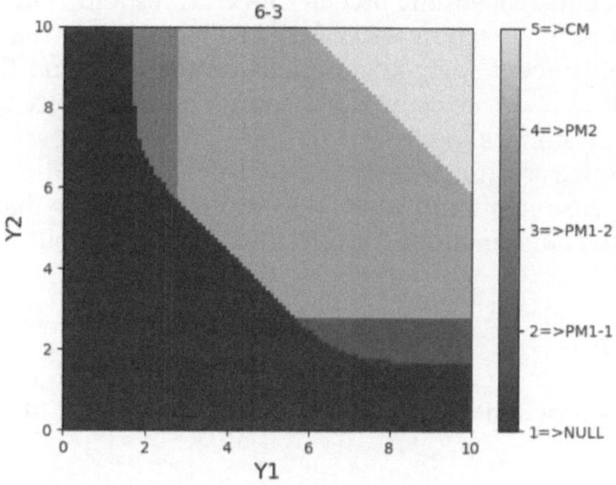

Figure 2. Example of an optimal maintenance policy when $w_1 = w_2$.

Figure 2 illustrates that the two-dimensional control limit policy holds and it is symmetric as shown in Theoren 2 since the costs and the parameters for degradation satisfy the assumptions for Theorem 2. Also, Figure 2 shows that the regions for PM-1 is smaller and the regions for PM-2 is larger compared with the case in Figure 1.

These two optimal maintenance policies include all actions of NULL, PM1-1, PM1-2, PM2, and CM. However, the number of actions in the optimal maintenance policy differ depending on the costs and the parameters for degradation. When $w_1 = 1$, $w_2 = 1.5$, $\mu_1 = 1.2$, and $\mu_2 = 1.8$ under Equation (10), the optimal policy is given by Figure 3.

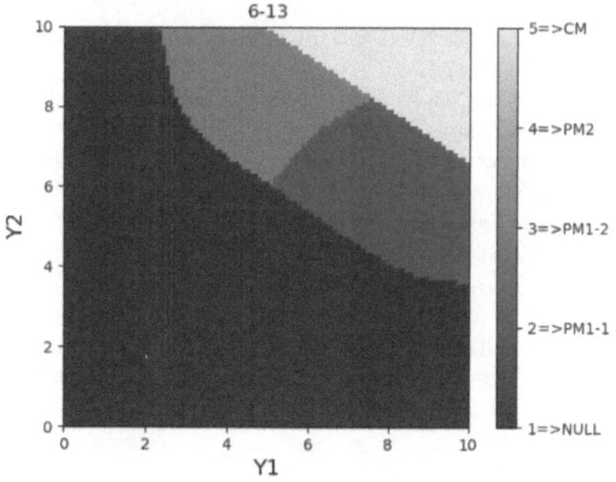

Figure 3. Example of an optimal maintenance policy when four actions are selected.

Figure 3 illustrates that PM2, which means the two components are simultaneously replaced as preventive maintenance action, is not selected in the optimal maintenance policy. This result gives us an interpretation that it is optimal to replace one component only until the system fails.

7 CONCLUSION

We have constructed the weighted system consisting of the two components whose degradations follow the Wiener processes, and the system fails the weighted sum of the degradation of the two components reaches the predetermined threshold. The system is periodically observed and then one optimal action is selected. We have shown that the two-dimensional control limit policy holds under some assumptions on the costs and the parameters for the degradation. Furthermore, we have provided some considerations on the optimal maintenance policy through numerical analysis.

Future work is to construct and analyze a weighted system that includes more than two components as an extension of our study. However, we guess that the analysis is not straightforward since the combinations for selecting an action should be so complicated.

ACKNOWLEDGMENT

This work was supported by JSPS KAKENHI Grant Number JP20K04989.

REFERENCES

Alaswad, S. and Xiang, Y. (2017). A review on condition-based maintenance optimization. *Reliability Engineering and System Safety*, 157, 54–63.

Atashgar, K. and Abdollahzadeh, H. (2017). A joint reliability and imperfect opportunistic maintenance optimization for a multi-state weighted k-out-of-n system considering economic dependence and periodic inspection. *Quality and Reliability Engineering International*, 33(8), 1685–1707.

Hamdan, K., Tavangar, M. and Asadi, M. (2021). Optimal preventive maintenance for repairable weighted k-out-of-n systems. *Reliability Engineering and System Safety*, 205, 1–9.

Jardine, A.K.S., Lin D. and Banjevic, D. (2006). A review on machinery diagnostics and prognostics implementing condition-based maintenance. *Mechanical Systems and Signal Processing*, 20(1), 1483–1510.

Khorshidi, H. Gunawan, I. and Ibrahim, Y. (2016). A dynamic unreliability assessment and optimal maintenance strategies for multistate weighted k-out-of-n: F systems. *Applied Stochastic Models in Business and Industry*, 32(4), 485–493.

Lawless, J. F. (2003). *Statistical Models and Methods for Lifetime Data* (Second Edition). John Wiley & Sons, New Jersey.

Meeker, W.O. and Escobar, L.A. (1988). *Statistical Methods for Reliability Data*. John Wiley & Sons, New York.

Olde Keizer, M.C.A. Flapper, S.D.P. and Teunter, R.H. (2017). Condition-based maintenance policies for systems with multiple dependent components. *European Journal of Operational Research*, 261(2), 405–420.

Wu, S. and Castro, I.T. (2020). Maintenance policy for a system with a weighted linear combination of degradation processes. *European Journal of Operational Research*, 280(1), 124–133.

Sun, Q., Ye, Z. and Chen, N. (2018). Optimal inspection and replacement policies for multi-unit systems subject to degradation. *IEEE Transactions on Reliability*, 67(1), 401–413.

Zhang, N. and Si, W. (2020). Deep reinforcement learning for condition-based maintenance planning of multi-component systems under dependent competing risks. *Reliability Engineering and System Safety*, 203, 1–10.

Zhang, N., Tian, S., Cai, K. and Zhang, J. (2023). Condition-based maintenance assessment for a deteriorating system considering stochastic failure dependence. *IISE Transactions*, 55(7), 687–697.

Recent Advances in Reliability and Maintenance Modeling – Okamura, Inoue & Xiao (Eds)

Modified reliability model of high-performance and flexible protocol with hybrid ARQ

M. Kimura
Faculty of Business Administration, Aichi University, Aichi, Japan

M. Imaizumi
Faculty of Home Economics, Aichi Gakusen University, Aichi, Japan

T. Nakagawa
Department of Business Administration, Aichi Institute of Technology, Aichi, Japan

ABSTRACT: This paper considers a modified reliability model of High-performance and Flexible Protocol with Hybrid ARQ. We have already discussed some reliability models of a window flow control scheme based on High-performance and Flexible Protocol. That is, when a server transmits data packets to a client, packet loss has occurred multiple times, and then the server transmits the data packets at longer interval in order to reduce network congestion. These models presumed that when packet loss has occurred multiple times, the data transmission interrupts and the retransmission of all data restarts from the beginning, and then the server transmits the data packets at longer interval in order to mitigate congestion. In this paper, we consider a stochastic model that when packet loss has occurred, only error packets are retransmitted at longer interval in order to mitigate congestion. That is, when the transmission fails, only error packets are retransmitted and a packet transmission interval is extended. Moreover, when the retransmission has failed again, the server checks it again and the connection is made newly. We derive the mean time until packet transmissions succeed, and discuss an optimal policy analytically.

1 INTRODUCTION

Enhancing the reliability of a communication system operating in a radio environment and transmission video streams is crucial. For instance, in cases of network congestion, packets may be lost within a router. The window flow control schemes have been employed to address this issue (Jacobson 1988; Stevens 1977). In other words, a server transmits the amount of packets equivalent to the window size and in the event of network congestion, the half of window size is transmitted. However, it has been noted that there is an issue regarding the degradation of throughput. To address this concern, High-performance and Flexible Protocol (HpFP) has been proposed (Murata *et al.* 2016). It is the method that in the event of network congestion, packet transmission intervals are prolonged. We considered the reliability models of a window flow control scheme with HpFP (Kimura *et al.* 2020, 2021). These models presumed that when packet loss has occurred multiple times, the data transmission interrupts and the retransmission of all data restarts from the beginning.

This paper considers a stochastic model of a window flow control scheme that when packet loss has occurred multiple times, only error packets are retransmitted at longer interval in order to mitigate congestion. Moreover, we adopt Hybrid ARQ in code error

DOI: 10.1201/9781003491309-8

correction schemes (Zhao and Valenti 2005). That is, a server transmits error correction packets concurrently with data packets and a client restores the loss packets by them. If a client can not recover all lost packets, it requests the remaining loss packets from a server and the server retransmits them at longer transmission interval. When the retransmission has failed again, the server checks it again and the connection is made newly. We derive the mean time until packet transmissions succeed, and discuss an optimal policy to maximize throughput analytically.

2 MODEL AND ANALYSIS

We consider a communication system where a web server transmits data packets to a client. When the client transmits the request for data packets to the server, it establishes a connection between them. Then, the server transmits $n(n = 1, 2, \cdots)$ data packets and $m(m = 1, 2, \cdots; m \leq n)$ redundant packets. If the number of failed packets is within the limit of m, the client restores all error packets using the redundant packets. If the quantity of failed packets surpasses m, the transmission interval of packets is prolonged to mitigate congestion.

The stochastic model is formulated as follows (Kimura *et al.* 2021):

(1) The client transmits the request for data packets to the server, and it establishes a connection between them according to a general distribution $A(t)$ with mean a.
(2) The server transmits n data packets, which corresponds to a window size, and m redundant packets of FEC to the client successively after the connection is established.
(3) The time of transmission interval of one packet has a general distribution $B_1(t)$ with mean b_1.
(4) If the number of failed packets is within m, the client restores all error packets using the redundant packets. The time to restore an error packet has a general distribution $C(t)$ with mean c, and the probability that a packet error occurs is p_1 $(0 < p_1 < 1)$.
(5) When all n packets have been received correctly, the client returns ACK to the server and the transmission succeeds.
(6) If the number of failed packets exceeds m, m error packets are restored and NAK are returned by the client. The time to receive ACK or NAK has a general distribution $D(t)$ with mean d.
    (i) The server retransmits the quantity exceeding the limit of m error packets with extended transmission intervals. That is, the time of transmission interval of one packet has a general distribution $B_2(t)$ with mean $b_2 (b_2 > b_1)$. The probability that a packet error occurs is $p_2 (p_2 \leq p_1)$.
    (ii) When the quantity exceeding the limit of m error packets has been received correctly, the transmission succeeds.
    (iii) If the retransmission fails again, the connection is made newly after constant time μ, where $G(t) \equiv 0$ for $t < \mu$ and 1 for $t \geq \mu$.

Under the above assumptions, the following states of the system is defined:

State 0: Client transmits the request for data and it establishes a connection at the time with $A(t)$.
State 1: Server transmits $n + m$ packets at the time with $B_1(t)$.
State 2: When the number of failed packets exceeds m, server restores m error packets and retransmits the quantity exceeding the limit of m at the time with $B_2(t)$.
State F: When server fails again, the connection is reestablished after time μ.
State S: The transmission of n packets succeeds.

The system states defined above form a Markov renewal process, where S is an absorbing state (Nakagawa 2011). A transition diagram between system states is shown in Figure 1.

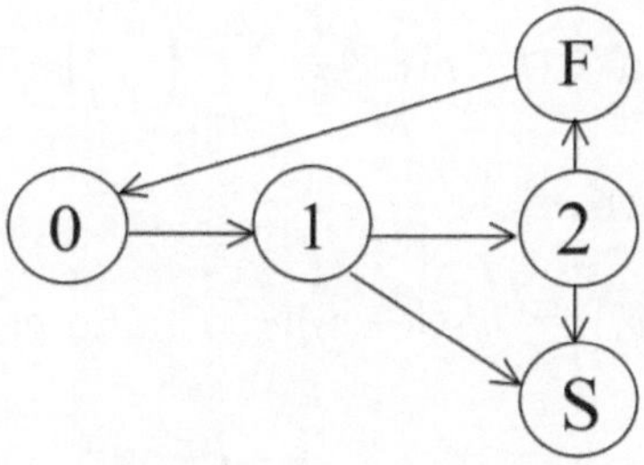

Figure 1. Transition diagram among system states.

We derive transition probabilities $Q_{i,j}(t)$ from State $i(i=1)$ to State $j(j=2,S)$ are given by the following equations:

$$Q_{0,1}(t) \equiv A(t),$$

$$Q_{1,S}(t) \equiv \sum_{j=0}^{m} \binom{n}{j} p_1^j (1-p_1)^{n-j} B_1^{(n+m)}(t) * D(t) * C^{(j)}(t),$$

$$Q_{1,2}(t) \equiv \sum_{j=m+1}^{n} \binom{n}{j} p_1^j (1-p_1)^{n-j} B_1^{(n+m)}(t) * D(t) * C^{(m)}(t),$$

where $\Phi^{(i)}(t)(i=1,2,\cdots)$ is the i-fold Stieltjes convolution of $\Phi(t)$, i.e., $\Phi^{(i)}(t) \equiv \int_0^t \Phi^{(i-1)}(t-w)d\Phi(w)(i=1,2,\cdots), \Phi^{(0)}(t) \equiv 1$ for $t \geq 0$, and the asterisk mark denotes the Stieltjes convolution with each other, i.e., $\Phi_1(t) * \Phi_2(t) \equiv \int_0^t \Phi_1(t-w)d\Phi_2(w)$ (Nakagawa 2011).

On the other hand, using $Q_{1,S}(t)$ and $Q_{1,2}(t)$, transition probabilities $H_{1,S}(t)$ and $H_{1,F}(t)$ are given by following equations:

$$\begin{aligned} H_{1,S}(t) &\equiv Q_{1,2}(t) * Q_{2,S}(t) \\ &= \sum_{j=m+1}^{n} \binom{n}{j} p_1^j (1-p_1)^{n-j} B_1^{(n+m)}(t) * D(t) * C^{(m)}(t) * (1-p_2)^j B_2^{(j)}(t) * D(t), \end{aligned}$$

$$\begin{aligned} H_{1,F}(t) &\equiv Q_{1,2}(t) * Q_{2,F}(t) \\ &= \sum_{j=m+1}^{n} \binom{n}{j} p_1^j (1-p_1)^{n-j} B_1^{(n+m)}(t) * D(t) * C^{(m)}(t) * [1-(1-p_2)^j] B_2^{(j)}(t) * D(t). \end{aligned}$$

Laplace Stieljes (LS) transforms $q_{1,S}(s)$ and $h_{i,j}(s)$ of $Q_{1,S}(t)$ and $H_{i,j}(t)(i=1; j=S,F)$ are

$$q_{1,S}(s) = \sum_{j=0}^{m} \binom{n}{j} [p_1 c(s)]^j (1-p_1)^{n-j} b_1(s)^{n+m} d(s),$$

$$\begin{aligned} h_{1,S}(s) &= \sum_{j=m+1}^{n} \binom{n}{j} [p_1(1-p_2)b_2(s)]^j (1-p_1)^{n-j} b_1(s)^{n+m} c(s)^m d(s)^2, \\ &= \left\{ [p_1(1-p_2)b_2(s) + (1-p_1)]^n - \sum_{j=0}^{m} \binom{n}{j} [p_1(1-p_2)b_2(s)]^j (1-p_1)^{n-j} \right\} \\ &\quad \times b_1(s)^{n+m} c(s)^m d(s)^2, \end{aligned}$$

$$h_{1,F}(s) = \left\{\left[\sum_{j=m+1}^{n}\binom{n}{j}[p_1 b_2(s)]^j(1-p_1)^{n-j} - \sum_{j=m+1}^{n}\binom{n}{j}[p_1(1-p_2)b_2(s)]^j(1-p_1)^{n-j}\right]\right\}$$

$$\times\, b_1(s)^{n+m}c(s)^m d(s)^2,$$

$$= \left\{[p_1 b_2(s) + (1-p_1)]^n - [p_1(1-p_2)b_2(s) + (1-p_1)]^n\right.$$

$$\left. - \left[\sum_{j=0}^{m}\binom{n}{j}[p_1 b_2(s)]^j(1-p_1)^{n-j} - \sum_{j=0}^{m}\binom{n}{j}[p_1(1-p_2)b_2(s)]^j(1-p_1)^{n-j}\right]\right\}$$

$$\times\, b_1(s)^{n+m}c(s)^m d(s)^2,$$

where $\phi(s) \equiv \int_0^\infty \mathrm{e}^{-st}\mathrm{d}\Phi(t)$ for a general function $\Phi(t)$, i.e., $\int_0^\infty \mathrm{e}^{-st}d[\Phi_1(t) * \Phi_2(t)] = \phi_1(s)\phi_2(s)$ and $q_{1,S}(0) + h_{1,S}(0) + h_{1,F}(0) = 1$.

First, we derive the mean time $\ell_{0,S}(n)$ until the transmissions of n packets succeeds. Letting $h_{0,S}(s)$ be LS transform of the first-passage distribution $H_{0,S}(t)$ from State 0 to State S,

$$H_{0,S}(t) = \sum_{i=1}^{\infty}[A(t) * H_{1,F}(t) * G(t)]^{(i-1)} * A(t) * [Q_{1,S}(t) + H_{1,S}(t)],$$

and its LS transform is

$$h_{0,S}(s) = \frac{a(s)[q_{1,S}(s) + h_{1,S}(s)]}{1 - a(s)h_{1,F}(s)g(s)}.$$

Hence, the mean time $\ell_{0,S}(n)$ is

$$\ell_{0,S}(n) \equiv \int_0^\infty t dH_{0,S}(t) = \lim_{s\to 0} -\frac{dh_{0,S}(s)}{ds}$$

$$= \frac{\begin{array}{c} a + n(b_1 + b_2 p_1) + m(b_1 + c) + 2d + (c - b_2)\sum_{j=0}^{m}\binom{n}{j} j p_1^j (1-p_1)^{n-j} \\ +(mc + d)\sum_{j=0}^{m}\binom{n}{j} p_1^j (1-p_1)^{n-j} + \mu \end{array}}{(1 - p_1 p_2)^n + \sum_{j=0}^{m}\binom{n}{j} p_1^j (1-p_1)^{n-j} - \sum_{j=0}^{m}\binom{n}{j}[p_1(1-p_2)]^j(1-p_1)^{n-j}} - \mu. \quad (1)$$

3 OPTIMAL POLICY

We discuss an optimal policy which maximizes the transmission of all packets per unit of time until it succeeds. Then, we define throughput $E(n)$, which represents the rate of n packets to their mean transmission times,

$$E(n) \equiv \frac{n}{\ell_{0,S}(n)}. \quad (2)$$

The mean times are given by, respectively

$$\ell_{0,S}(n) = Z(n) - \mu,$$

where

$$Z(n) \equiv \frac{\begin{array}{c} a + n(b_1 + b_2 p_1) + m(b_1 + c) + 2d + (c - b_2)\sum_{j=0}^{m}\binom{n}{j} j p_1^j (1-p_1)^{n-j} \\ +(mc + d)\sum_{j=0}^{m}\binom{n}{j} p_1^j (1-p_1)^{n-j} + \mu \end{array}}{(1 - p_1 p_2)^n + \sum_{j=0}^{m}\binom{n}{j} p_1^j (1-p_1)^{n-j} - \sum_{j=0}^{m}\binom{n}{j}[p_1(1-p_2)]^j(1-p_1)^{n-j}}.$$

We discuss optimal n^* to maximize $E(n)$. We put formally that $A(n) \equiv 1/E(n)$ and seek n^* to minimize $A(n)$. From the inequality $A(n+1) - A(n) \geq 0$,

$$nZ(n+1) - (n+1)Z(n) + \mu \geq 0. \tag{3}$$

Denoting the left side of (3) by $L(n)$, we have

$$L(n+1) - L(n) = (n+1)[Y(n+1) - Y(n)],$$

where

$$Y(n) \equiv Z(n+1) - Z(n).$$

In particular, it can be easily shown that

$$Z(1) = a + b_1 + 3d + b_2 p_1 + m(b_1 + 2c) + (c - b_2)p_1 + \mu,$$

$$\lim_{n\to\infty} Z(n) = \infty.$$

Therefore, there exists a finite n^* to minimize $A(n)$.

In particular, when $p_2 \leq p_1 = 1$,

$$Y(n+1) - Y(n) = \frac{[a + (m+1)b_1 + mc + b_2 + 2d + \mu]p_2{}^2 + (b_1 + b_2)p_2(1 + np_2)}{(1 - p_2)^{n+2}} > 0.$$

$L(n)$ is strictly increasing in n from $L(1)$ to ∞, we have the following optimal policy:

(i) If $L(1) < 0$, then there exists a finite and unique $n^*(> 1)$ which satisfies (3),
(ii) If $L(1) \geq 0$ then $n^* = 1$.

In particular case, when $p_1 > p_2 = 0$,

$$\begin{aligned} A(n) &= \frac{Z(n)}{n} \\ &= \frac{1}{n}\{a + (n+m)b_1 + mc + nb_2 p_1 + 2d \\ &\quad + (c - b_2)\textstyle\sum_{j=0}^{m}\binom{n}{j} jp_1^j(1-p_1)^{n-j} + (mc + d)\textstyle\sum_{j=0}^{m}\binom{n}{j} p_1^j(1-p_1)^{n-j} + \mu\} \end{aligned},$$

$$A(1) = a + (m+1)b_1 + (2m + p_1)c + 3d + \mu,$$
$$\lim_{n\to\infty} A(n) = b_1 + b_2 p_1.$$

Therefore, we can compute $A(n)$ numerically and obtain optimal n^* to minimize $A(n)$.

4 NUMERICAL EXAMPLES

We compute numerically the optimal window size n^*. Suppose that the mean time b_1 until transmitting one packet is a unit time. It is assumed that the mean time required for data packets is $a/b_1 = 10$, the mean time for the server to transmit all packets to receive ACK or NAK is $d/b_1 = 2, 4$, the mean time until transmitting one packet which is longer than the regular time is $b_2/b_1 = 2, 4$, the number of redundant packets of FEC is $m = 16, 32$, the mean time to restore one packet loss is $c/b_1 = 2$, the mean time for the server to interrupt n

retransmission to restart again is $\mu/b_1 = 30$ and the probability that loss packets occur is $p_1 = 0.05, 0.1$ and $p_2 = 0.01 \sim 0.1$.

Table 1. Optimal window size n^* to maximize $E(n)$ with hybrid ARQ when $a/b_1 = 10$, $c/b_1 = 2$ and $\mu/b_1 = 30$.

p_1	p_2	m	d/b_1	b_2/b_1	n^*	$\ell_{0,S}(n^*)$	$E(n^*)$
0.05	0.01	16	2	2	306	451.0	0.7684
				4	274	414.5	0.6610
			4	2	311	464.1	0.6702
				4	278	426.3	0.6520
		32	2	2	521	766.2	0.6800
				4	504	747.3	0.6745
			4	2	523	775.1	0.6747
				4	505	754.7	0.6691
	0.05	16	2	2	226	357.2	0.6328
				4	221	351.5	0.6288
			4	2	228	366.5	0.6222
				4	223	360.7	0.6182
		32	2	2	461	697.3	0.6612
				4	456	691.4	0.6595
			4	2	463	704.9	0.6555
				4	457	699.0	0.6538
0.1	0.05	16	2	2	121	254.4	0.4756
				4	117	249.2	0.4694
			4	2	121	260.6	0.4643
				4	118	257.5	0.4582
		32	2	2	241	479.5	0.5026
				4	237	474.2	0.4998
			4	2	241	485.6	0.4962
				4	238	482.3	0.4935
	0.1	16	2	2	114	245.1	0.4651
				4	111	240.8	0.4609
			4	2	114	251.3	0.4536
				4	112	249.1	0.4495
		32	2	2	237	474.3	0.4997
				4	234	470.5	0.4973
			4	2	237	480.5	0.4933
				4	234	476.6	0.4910

Table 2 presents the optimal window size n^* to maximize the throughput, the mean time $\ell_{0,S}(n^*)$ and the throughput $E(n^*)$ with Hybrid ARQ. For example, when $p_1 = 0.05, p_2 = 0.01, d/b_1 = 2, b_2/b_1 = 2$, the optimal number of packets is $n^* = 306$, the mean time is $\ell_{0,S}(n^*) = 451.0$ and the throughput is $E(n^*) = 0.7684$. This indicates that n^* and $\ell_{0,S}(n^*)$ decrease with p_1, p_2 and b_2/b_1 and increase with d/b_1. But $E(n^*)$ decrease with $p_1, p_2, d/b_1$ and b_2/b_1.

Table 1 presents the optimal window size n^* to maximize the throughput, the mean time $\ell_{0,S}(n^*)$ and the throughput $E(n^*)$ with Selective Repeat ARQ (Kimura *et al.* 2023). These indicate that n^*, $\ell_{0,S}(n^*)$ and $E(n^*)$ similarly show the above trends of Table 1. Moreover, by comparison with Table 1, n^* and $\ell_{0,S}(n^*)$ of Table 2 are smaller than those of Table 1. But $E(n^*)$ is smaller that that of Table 1 when p_1, p_2 and m are large. That is, $E(n^*)$ of Hybrid ARQ is much better than those of Selective Repeat ARQ when p_1, p_2 and m are large.

Table 2. Optimal window size n^* to maximize $E(n)$ with selective repeat ARQ when $a/b_1 = 10$ and $\mu/b_1 = 30$ (Kimura *et al.* 2023).

p_1	p_2	d/b_1	b_2/b_1	n^*	$\ell_1(n^*)$	$E(n^*)$
0.05	0.01	2	2	152	197.9	0.7681
			4	146	205.8	0.7094
		4	2	172	228.5	0.7527
			4	165	237.2	0.6957
	0.05	2	2	63	102.6	0.6143
			4	61	106.4	0.5731
		4	2	71	120.5	0.5892
			4	68	123.5	0.5506
0.1	0.05	2	2	41	84.43	0.4856
			4	38	87.55	0.4340
		4	2	46	99.92	0.4604
			4	43	104.2	0.4128
	0.1	2	2	26	67.49	0.3852
			4	25	71.38	0.3502
		4	2	29	80.57	0.3600
			4	28	85.26	0.3284

5 CONCLUSION

We have considered a modified reliability model of High-performance and Flexible Protocol with Hybrid ARQ. That is, the server transmits the window size of n data packets and m redundant packets. If the number of failed packets is within m, the client restores all error packets using the redundant packets. If the number of failed packets exceeds m, m error packets are restored and the quantity exceeding the limit of m are retransmitted at longer transmission interval in order to alleviate congestion. Moreover, when the retransmission has failed again, the server checks it again and the connection is made newly. We have derived the mean time until packet transmissions succeed, and discussed an optimal window size to maximize the amount of packets per unit of time until the transmission succeeds. From numerical examples, we have shown that the optimal window size increase with the number of redundant packets of FEC and the time of receiving ACK or NAK, conversely, and decrease with the probability that loss packets occur and the one packet interval time. Moreover, the optimal throughput increase with the number of redundant packets of FEC, and decrease the other parameters. On the other hand, by comparison with Hybrid ARQ and Selective Repeat ARQ, optimal throughput of Hybrid ARQ is much better than those of Selective Repeat ARQ when the probability that loss packets occur and the number of redundant packets of FEC are large.

In environments prone to instability, one would anticipate additional research into methods for correcting code errors and controlling network congestion.

ACKNOWLEDGMENT

This research was partially supported by the joint research project "Modeling and Analyzing Information Security Measures for Server Systems" of the Institute of Managerial Research, Aichi University.

REFERENCES

Jacobson, V. (1980). Congestion avoidance and control. *Computer Communication Review*, 18(4): 314–329.

Kimura, M., Imaizumi, M. and Araki, T. (2021). Reliability of congestion control based on packet transmission interval with hybrid ARQ. *International Journal of Reliability, Quality and Safety Engineering*, 28 (6): 2140004-1–2140004-17, (17pages), DOI: 10.1142/S0218539321400040.

Kimura, M., Imaizumi, M. and Nakagawa, T. (2020). Modelling of congestion control based on packet transmission interval with hybrid ARQ. *Proceedings of the 4th International Conference on Mathematical Techniques in Engineering Applications*, 91.

Kimura, M., Imaizumi, M. and Nakagawa, T. 2023. Reliability consideration of modified reliability model with high-performance and flexible protocol. *Proceedings of the 2023 IEICE Society Conference*, 19.

Murata, T., Mizuhara, T., Takagi, A., Fukushima, K., Yamamoto, K., Nagaya, Y., Muranaga, K. and Kimura, E. (2016). HpFP : A new protocol for LFNs with packet-loss based on UDP : A basic concept and detailed design of the protocol. *IEICE Technical Report*, 115(484): IN2015-109, 7–12.

Nakagawa, T. 2011. *Stochastic Processes with Applications to Reliability Theory*, Springer, London.

Stevens, W. 1997. TCP Slow Start, Congestion Avoidance, Fast Retransmit and Fast Recovery Algorithms. RFC2001.

Zhao, B. and Valenti, M. C. 2005. Practical relay networks: A generalization of hybrid-ARQ. *IEEE J Selected Areas Commun*, 23(1): 7–18.

Recent Advances in Reliability and Maintenance Modeling – Okamura, Inoue & Xiao (Eds)
© 2025 the Author(s), ISBN: 978-1-032-79098-5

Integration model of maintenance and spare provisioning by considering spare parts deterioration and leadtime delay

Nani Kurniati, Zukhrof Romadhon & Rindi Kusumawardani
Department of Industrial and Systems Engineering, Institut Teknologi Sepuluh Nopmeber, Surabaya, Indonesia

ABSTRACT: Holding spare parts on hand is crucial for addressing maintenance demands, especially when the maintenance needs replacement. Managing inventory parameters efficiently is essential for balancing inventory costs and part availability. This study proposes an optimization model for managing spare parts by considering deterioration and lead times. This study utilizes the periodic inventory policy (R, S). The maintenance cost, ordering cost, holding cost, and shortage cost equations are modified by incorporating the lead time function and spare parts deterioration function. Delay lead time directly increases total costs, while the speed of spare parts deterioration reduces them. The total cost can be minimized by balancing maintenance intervals and maximum inventory. The algorithm was developed to solve the model; moreover, numerical examples verify the deduction of the proposed model.

1 INTRODUCTION

Maintenance is one of the crucial activities in the manufacturing process that requires proper management. Maintenance mainly involves corrective and preventive replacement. Adequate inventory of spare parts in the company's warehouse is essential for successful maintenance. During operations, if spare parts are on hand, faulty or underperforming component is instantly replaced upon failure (viz. CR) to ensure continued optimal functioning. While all operating component will be replaced at scheduled interval (viz. PR). In addition, downtime of component may take longer due to shortage of spares representing loss of system functioning. Company should maintain inventory of spare parts, taking into account both the corrective and preventive replacement purposes.

An effective preventive maintenance can substantially lessen the system failure and replacement costs. Preventive maintenance should not be performed too often to avoid the expense surpassing the advantages. The goal of preventive maintenance is to reduce long-term expenses. Appropriate maintenance schedules i.e. block replacement and age replacement policies, have been implemented for the systems. This paper focuses on block replacement policies in the context of uncertain lifetimes. Block replacement is typically employed when dealing with numerous similar units in a system, as per (Ke and Yao 2016). Nakagawa (1982) suggests a block replacement policy modification where a unit gets replaced both at failure and during its schedule time within (*0*, *To*).

In a periodic review inventory system, the ordering interval is constant whereas the order quantity is determined by the maximum demand during that period. This study follows a periodic review process with stages ranging from R to S, meaning in every R time unit (the review period) an order is placed to maintain the inventory level as S (Zahedi-Hosseini *et al.* 2017). For convenience purpose in this research, the maintenance interval and the stock review interval are considered identical (R = T = S).

DOI: 10.1201/9781003491309-9

The literature extensively covers the joint optimization of maintenance replacement policy and spares inventory. Most are focused on the optimization of age replacement policy, e.g., Amstrong and Atkins (1996, 1998), Kabir and Farrash (1996, 1997), Kabir and Al-Olayan (1996), Liao and Yuan (1998), Park and Park (1986), Shue and Griffith (2001), Shue and Chien (2004). Meanwhile, Chelbi and Ait-Kaidi (2001) and Brezavscek and Hudoklin (2003) proposed an analytical model for joint optimization of block replacement policy with constant lead time. A further model proposed by Huang *et al.* (2008) by considering the random lead time of the spare parts inventory, with the same policy. Moreover, Jiang *et al.* (2015) proposed a joint model for maintenance and spare part inventory decision making that incorporates spare part deterioration.

Problems are inevitable for companies due to the actual lead time not being constant, thus necessitating further investigation into delay mitigating policy is important. Moreover, inadequate inventory conditions result in spare parts deteriorating may lead to further inventory reduction and increased risk of shortages. The impact of both spare parts deterioration and leadtime is taken into account for maintenance and inventory management. This research proposes a model to determine maintenance costs by integrating block replacement maintenance policy (interval), periodic review inventory policies, consideration on spare parts deterioration and leadtime delays.

2 MODEL DEVELOPMENT

Model development is carried out on a multi-unit system by considering spare part deterioration and random lead time. The research model was created to obtain an optimal period for maintenance and reviewing inventory, and the maximum inventory level, to reduce downtime due to non available spare parts. Delay lead time in delivery, causing a delivery lag time (δ) cause spare part unavailability. With random lead times, the interval time between PR (T_{cycle}) that has been planned will change and vary, with variations in *Tcycle*, the total cost will also change.

Asumptions are same as those in (Huang *et al.* 2008) and (Jiang *et al.* 2015):

- Holding costs during one preventive replacement cycle are directly proportional to spare parts inventory levels, the greater the spare parts inventory level, the greater holding cost.
- Shortage costs during one preventive replacement cycle are directly proportional to the level of spare parts shortages, so that the higher the level of spare parts shortages, the greater the shortage costs.
- During the cycle, the decrease in inventory level is linier and stationary.
- Corrective and preventive replacement times are negligible and all failed components can be detected and replaced.
- When n is large enough, the distribution of X_n can be approximated by a normal distribution with the mean $E[X_n] = n$, the variance (rather than standard deviation) $Var[X_n] = n$, and the PDF is $gX_n(x_n)$, where $E[X] = M(T)$ is the component renewal function.
- spare part deterioration follows weibull distribution, and random lead time follows uniform distribution.

In this model development, there are two scenarios considered affect the holding costs and shortage costs:

(1) The new spares ordered arrive after the pre-arranged PR interval, but the original inventory is enough to supply the components for CR in one cycle. The ordered spares arrive after T, but before T_1. See Figure 1.

(2) The new spares ordered arrive after the pre-arranged PR interval, but the original inventory is not enough to supply the components for CR in one cycle. The ordered spares arrive after T and T_1. See Figure 2.

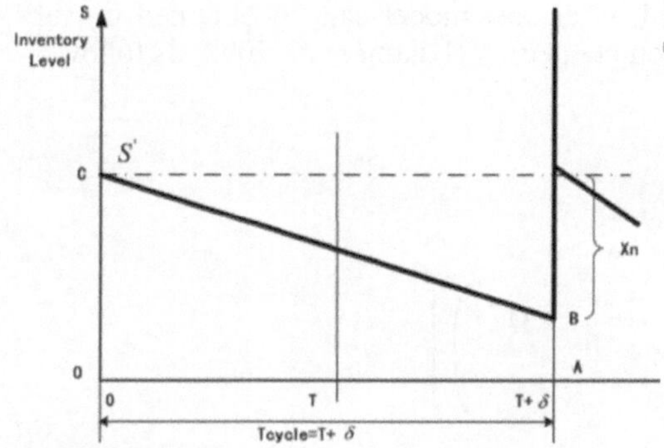

Figure 1. Scenario 1.

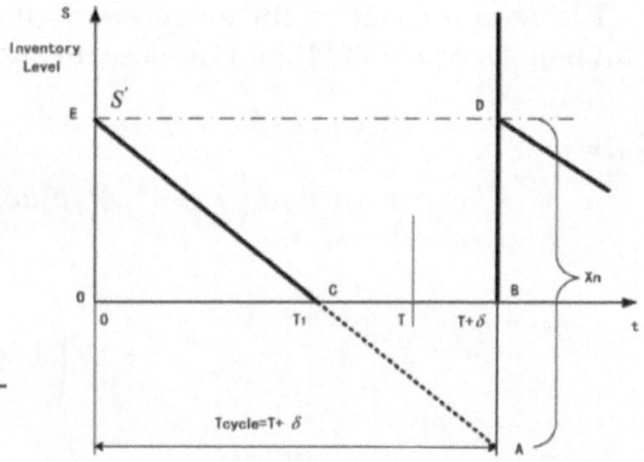

Figure 2. Scenario 2.

Delivery lag time (δ) can affect the planned preventive replacement time interval (T) which causes the value of T_{cycle}. The actual value of lead time (τ_0) may vary in value, it will affect the costs and maintenance actions due to shortage of spare parts. In (Huang 2008) T_{cycle} can be obtained as follow:

$$E(T_{cycle}) = T + \int_0^{\infty} \delta\xi(\delta)d\delta \tag{1}$$

The number of spare parts ordered during one cycle is the accumulation of spare part demand both for preventive and corrective replacement during one cycle, respectively, as follow:

$$n + E[X_n] \tag{2}$$

where n is the number of spare parts for preventive replacemenr and $E[X_n]$ is the number of spare parts for corrective replacement. The $E[X_n]$ define as:

$$\begin{aligned} E[X_n] &= n\,.\,M\left[E\left(T_{cycle}\right)\right] \\ E[X_n] &= n\,.\,M[T + \int_0^{\infty} \delta\xi(\delta)d\delta] \end{aligned} \tag{3}$$

The spare part deterioration level can be formulated as follows:

$$E(Y) = Sa - Sa\,.\,e^{-\int_0^T g(\tau)d\tau} = Sa(1 - e^{-\int_0^T g(\tau)d\tau}) \tag{4}$$

Where Sa is the expected inventory level just after PR.
By involving equation (3) and (4), the equation of total cost can be obtained as follow:

$$CT(S_{\max}, T) = \frac{Cm + Co + Ch + Cs}{T + \int_0^{\infty} \delta\xi(\delta)d\delta} \tag{5}$$

The revised Sa by considering leadtime delay and spare part deterioration can be formulated as follow:

$$Sa = S_{\max} - n - n\left[M\left(T + \int_0^{\infty} \delta\xi(\delta)d\delta\right) - M\left(T + \int_0^{\infty} \delta\xi(\delta)d\delta - \tau o\right) - Sa\left(e^{-\int_0^{T-\tau} g(\tau)d\tau} - e^{-\int_0^T g(\tau)d\tau}\right)\right. \tag{6}$$

The revised equation for maintenance and order cost model can be obtained by substituting the equation (4) into the cost equation created by (Huang *et al.* 2008) as follows:

$$
\begin{aligned}
Cm &= p\,.\,n + c\,.\,E(X_n) \\
&= p\,.\,n + c.n.M\left(T + \int_0^{\infty} \delta\xi(\delta)d\delta\right)
\end{aligned}
\tag{7}
$$

$$
\begin{aligned}
Co &= K + s\,.\left[n + E(X_n) + Sa\left(1 - e^{-\int_0^T g(\tau)d\tau}\right)\right] \\
&= K + s\,.\left[n + M\left(T + \int_0^{\infty} \delta\xi(\delta)d\delta\right) + Sa\left(1 - e^{-\int_0^T g(\tau)d\tau}\right)\right]
\end{aligned}
\tag{8}
$$

Similar way of subtitution applied in developing the revised equation for holding cost and shortage cost. Using the graph of scenario, the positive area represents holding cost and the negative area represents shortage cost. The holding and shortage cost for each scenario, respectively, is obtain as follows:

2.1 *Holding cost*

Scenario 1. When $0 \leq (X_n + Y) \leq S_a e^{-\int_0^T g(t)}$ and $\delta > 0$, the positive area is:

$$L_{H3} = L_{OABC} = \frac{1}{2}(S_a + S_a - X_n)(T + \delta) = \left(S_a - \frac{X_n}{2}\right)(T + \delta) \tag{9}$$

Scenario 2. When $(X_n + Y) \geq S_a e^{-\int_0^T g(t)}$ and $\delta > 0$, the positive area is:

$$L_{H4} = L_{OAE} = \frac{1}{2}\left(S_a \quad \frac{S_a}{X_n + Y}\right)(T + \delta) = \frac{S_a^{\,2}}{2(X_n + Y)}(T + \delta) \tag{10}$$

The total holding cost is sum of all the positive areas each scenario and multiplying it by the failure distribution for *CR* and the leadtime delay distribution, as follows:

$$Ch = h\,.\left(\int_0^{\tau_0 S_a} \int_0^{e^{-0Tg(t)}} L_{H1} + \int_{0\ S_a}^{\tau_0} \int_{e^{-0Tg(t)}}^{X_n + Y} L_{H2}\,.\,(g(X_n)\,.\,\xi(\delta)) \quad dX_n \quad d\delta\right) \tag{11}$$

2.2 *Shortage cost*

Scenario 3. When $(X_n + Y) \geq S_a e^{-\int_0^T g(t)}$ and $\delta > 0$ so the positive area can be obtained with:

$$
\begin{aligned}
L_{S4} = L_{ABC} &= \frac{1}{2}(X_n + Y - S_a)\left(T + \delta - \frac{S_a}{X_n + Y}(T + \delta)\right) \\
&= \frac{1}{2(X_n + Y)}(X_n + Y - S_a)^2(T + \delta)
\end{aligned}
\tag{12}
$$

The total shortage cost is sum of all the negative areas each scenario and multiplying it by the component failure distribution for *CR* and the leadtime delay distribution, as follows:

$$C_{sh} = z \,.\left(\left(\int_{-\tau_0}^{0} \int_{S_a \cdot \frac{T}{T+\delta} \cdot e^{-0Tg(t)}}^{X_n+Y} L_{S2} \int_{0}^{\tau_0} \int_{S_a e^{-0Tg(t)}}^{X_n+Y} L_{S4}\right) . \left(g(X_n) \,.\, \xi(\delta)\right) dX_n \, d\delta\right) \tag{13}$$

The derivatives of the remaining terms in the r.h.s. of C_T with respect to Sa as follows:

$$Q(S_a) = \frac{\partial C_T}{\partial S_a}$$

$$Q(S_a) = h \,.\left\{ \int_0^{\infty} \left[\int_0^{S_a e^{-0Tg(t)}} 1 + \frac{\left(1 - e^{-\int_0^T g(t)}\right)}{2} (T+\delta) g(X_n) - \int_0^{S_a e^{-0Tg(t)}} \frac{(T+\delta) {S_a}^2 \left(1 - e^{-\int_0^T g(t)}\right)}{2\left(X_n + S_a\left(1 - e^{-\int_0^T g(t)}\right)\right)^2} + \frac{S_a (T+\delta)}{\left(Xn + S_a\left(1 - e^{-\int_0^T g(t)}\right)\right)} g(X_n) \right] \zeta(\delta) \right\}$$

$$-z\left\{ \int_0^{\infty} \int_{\infty}^{S_a\, e^{-0Tg(t)}} \frac{(T+Z)\left(\left(X_n + S_a\left(1 - e^{-\int_0^T g(t)}\right)\right)^2\right)\left(1 - e^{-\int_0^T g(t)}\right)}{2\left(X_n + S_a\left(1 - e^{-\int_0^T g(t)}\right)\right)^2} - \frac{(T+\delta)\left(X_n + S_a\left(1 - e^{-\int_0^T g(t)}\right)\right)}{\left(X_n + S_a\left(1 - e^{-\int_0^T g(t)}\right)\right)} g(X_n) \zeta(\delta) \right\} \tag{14}$$

Then,

$$\frac{\vartheta Q(S_a)}{\vartheta S_a} = (h+z) \left[\int_{0\ S_a}^{\infty} \int_{e^{-0Tg(t)}}^{\infty} \frac{(T+\delta) \,.\, {X_n}^2}{\left(\left(X_n + S_a \quad \left(1 - e^{\int_0^T g(t)dt}\right)\right)\right)^3} \cdot g(X_n) \,.\, \zeta(X_n) d(X_n) d(\delta) \right]. \tag{15}$$

Every term in the r.h.s. of (15) is grater than or equal to 0. Therefore, $\frac{\vartheta Q(S_a)}{\vartheta S_a} \geq 0$
Finally, theorem 1 can be obtained as follows:

Theorem 1: *for fulfill the objective function of total cost, there is only one* $S_{\max}$ *and* S_a *which makes total cost minimum if:*

$$z \,.\, T + h \,.\, \int_{-\infty}^{0} \delta \,.\, \zeta(\delta) d\delta \,.\, e^{-\int_0^T g(t)dt} > 0 \tag{16}$$

Proof of theorem: for other parameters are given, the $S_{\max}$ and S_a are a simple kind of linear relation (11). Where

$$S_{\max} \geq n + n \,.\left[H\left(T + \int_0^{\infty} \delta \,.\, \zeta(\delta) d\delta\right) - H(T - \tau) \right] \tag{17}$$

Algorithm 1:

Step 1: Identify the given parameters ($n, \mu, \sigma, \tau_0, p, c, K, s, h, z, \alpha, \beta$)
Step 2: Calculating $M(T)$ and $Var(Xn)$ respectively
Step 3: Determine minimum $S_{\max}$ in each T using equation (17), where $S_{\max\ 1} = S_{\max}$, $S_{\max\ 2} = S_{\max\ 1} + 1$, then round down
Step 4: Calculating S_a using equation (6)
Step 5: Subtitusing S_a into equation (8), (11), (13) to get C_T
Step 6: If $\tau_0 \leq T \leq \mu$, then go to step 6, else, go back to step 3
Step 7: The optimal Solution is $(S_{\max}, T) = \{(S_{\max\ n}, T_n), \text{Min } C_T\}$, end

3 NUMERICAL EXAMPLE

Referring to Brezavscek and Hudoklin (2003), parameters are given as follows:

$n = 120$; $f(t) \sim N(\mu, \sigma)$ is a normal distribution with $\mu = 44$ weeks and $\sigma = 12$ weeks; $\tau_0 = 12$ weeks; $p = 58.2$ units; $c = 800.5$ units; $K = 20$ units; $s = 43$ units; $h = 0.6$ units; $z = 5196$ units.

Spare part deterioration distribution follows Weibull distribution from Jiang *et al.* (2015) with parameter α = 5,972E-05 and β = 2,575. Leadtime delay distribution follows uniform distribution from Huang *et al.* (2008).

The graphic of global minimum and local minimum, respectively, can be shown on Figures 3 and 4. The optimal solution $(S_{\max}, T) = (130, 19)$ with C_T = 716,120 ($T_{index} = 20$).

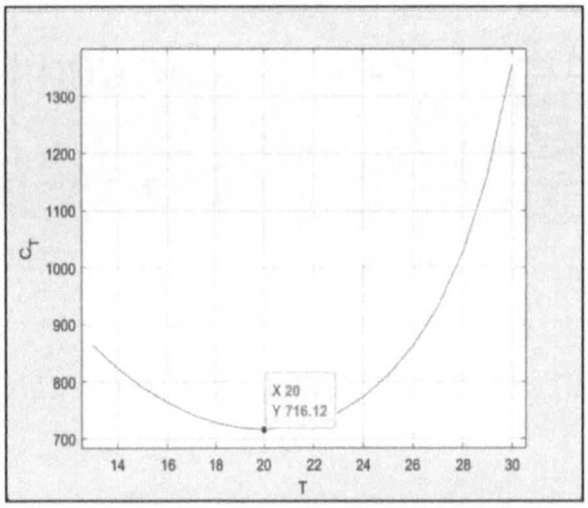

Figure 3. Variation of global minimum C_T over T.

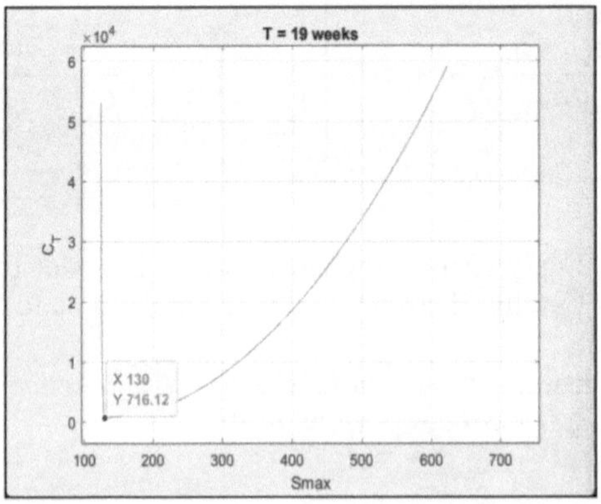

Figure 4. Variation C_T over $S_{\max}$ ($T = 19$).

When $S_{\max} = 130$, shortage will not occur due to $(X_n + Y) \leq S_a e^{-\int_0^T g(t)}$), the sum of failed spare part and failed component less than the stock availability. The comparison between the l.h.s and the r.h.s of the above conditions are shown in Table 1.

At the point $S_{\max} = 130$, it is the starting point when no shortage or shortage cost is zero. 130 unit is the minimum number of spare parts need to keep for no shortage of spare parts. The respective graphic is provided on Figure 5.

Table 1. Comparison between sparepart and stock availability for shortage status.

S_{max}	$Xn + Y$	$Sae^{-\int_0^T g(t)}$	Status
124	3,994	0,000	Shortage
125	4,119	0,875	Shortage
126	4,229	1,653	Shortage
127	4,342	2,441	Shortage
128	4,454	3,228	Shortage
129	4,566	4,015	Shortage
130	4,678	4,802	No Shortage
131	4,790	5,589	No Shortage
132	4,902	6,377	No Shortage
–	–	–	–
157	7,703	26,056	No Shortage
158	7,815	26,843	No Shortage
			No Shortage

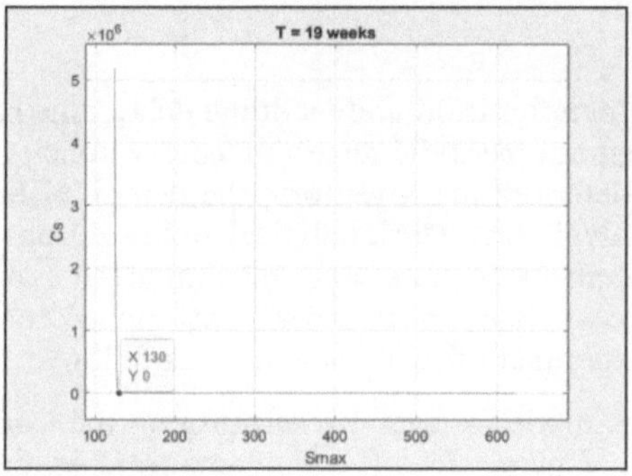

Figure 5. Variation C_S over $S_{\max}$ ($T = 19$).

4 SENSITIVITY ANALYSIS

The sensitivity analysis carried out to determine the effect or behavior of the total costs when the value of the lead time and deterioration speed are changed. The leadtime effect to the total cost provided in Table 2. The larger the delay δ, the larger the total cost due to larger holding cost. At point start with $\delta > 3$, the maintenance interval T and maximum stock $S_{\max}$ are changed.

Table 2. Effect of leadtime delay over total cost.

δ	T	S_{max}	Ch	Total Cost
1	20	133	2,610	706,807
2	20	133	10,763	707,162
3	20	133	24,940	707,778
4	19	130	26,138	708,572
5	19	130	42,042	709,295
6	19	130	62,271	710,214
7	19	130	87,112	711,343
8	19	130	116,854	712,695
9	19	130	151,785	714,283
10	19	130	192,194	716,120
11	19	130	238,368	718,219
12	19	130	290,597	720,593

Sensitivity analysis on deterioration speed is carried out by changing the μ value in the failure function. Table 3 drive the following result:

- The greater the μ value, the smaller the total costs
- The greater the μ value, the S_{max} will decrease and time interval T will increase cause higher reliability of sparepart.

Table 3. Effect of spare parts deterioration speed.

		Parameter				
Mean (μ)	distribution	α	β	Total Cost	T	S_{max}
60	weibull	8,237E-05	2,206	965,789	21	143
80	weibull	6,287E-06	2,695	557,245	30	133
120	weibull	8,830E-09	3,751	243,891	57	125

5 CONCLUSION

By considering spare part deterioration and leadtime delay, this paper proposed integrated model of block replacement and periodic review inventory policies for multi-unit system. In this paper, for modeling deteriorating inventory, the deterministic deterioration inventory was used when the data available. The analytical optimization model were set up. The optimal preventive replacement interval and maximum stocking were adjusted to get expected minimum total cost. Numerical experiment applied to implement the proposed model. From numerical experiment found some note as follows:

- The development of this model is done by substituting the leadtime delay function and spare parts deterioration function into the maintenance cost, ordering cost, holding cost and shortage cost equations. When $\delta > 0$, the total cost value will be greater due to increasing shotage costs and holding costs along with the random lead time value. Particular change in the maintenance interval and maximum stock was found when $\delta > 3$. The greater the μ value of failure function, the smaller the total cost, with the consequence that the interval T will be increase, maximum inventory S_{max} will be decrease, means sparepart have higher reliability.
- For the selected set of parameters, the minimum total cost by considering spare part deterioration and leadtime delay obtained when maintenance interval conducyed at $T = 19$ weeks and maximum stock at the point $S_{max} = 130$ units with a total cost of $C_T = 716{,}120$.

REFERENCES

Armstrong, M. J. and Atkins, D. R. (1996). Joint optimization of maintenance and inventory policies for a simple system. *IIE Trans. (Institute of Industrial Engineers)*, vol. 28, no. 5, pp. 415–424.

Armstrong, M. J. and Atkins, D. R. (1998). Note on joint optimization of maintenance and inventory. *IIE Trans. (Institute of Industrial Engineers)*, vol. 30, no. 2, pp. 143–149

Brezavscek, A. and Hudoklin, (2003). A. Joint optimization of block-replacement and periodic-review spare-provisioning policy. *IEEE Trans. Reliability*, vol. 52, pp. 112–117.

Chelbi, A., and Ait-Kadi, D. (2001). Spare provisioning strategy for preventively replaced systems subjected to random failure. *International Journal of Production Economics*, 74(1–3), 183–189.

Huang, R., Meng, L., Xi, L., and Liu, C. (2008). Modeling and analyzing a joint optimization policy of block-replacement and spare inventory with random-leadtime. *IEEE Transactions on Reliability*, 57(1), 113–124.

Hu, R., Yue, C., and Xie, J. (2008). Joint optimization of age replacement and spare ordering policy based on genetic algorithm. *Computational Intelligence and Security*.

Ilgin, M. A., and Tunali, S. (2006). Joint optimization of spare parts inventory and maintenance policies using genetic algorithms. *The International Journal of Advanced Manufacturing Technology*, *34*(5–6), 594–604.

Jiang, Y., Chen, M., and Zhou, D. (2015). Joint optimization of preventive maintenance and inventory policies for multi-unit systems subject to deteriorating spare part inventory. *Journal of Manufacturing Systems*, *35*, 191–205.

Kabir, A. B. M. Z. and Farrash, S. H. A. (1996). Simulation of an integrated age replacement and spare provisioning policy using SLAM. *Reliability Engineering & System Safety*, vol. 52, no. 2, pp. 129–138.

Kabir, A. B. M. Z. and Al-Olayan, A. S. (1996). Stocking policy for spare part provisioning under age based preventive replacement. *European Journal of Operational Research*, vol. 90, no. 1, pp. 171–181.

Kabir, A. B. M. Z. and Farrash, S. H. A. (1997). Fixed interval ordering policy for joint optimization of age replacement and spare part provisioning. *International Journal of Systems Science*, vol. 28, no. 12, pp. 1299–1309.

Ke, H. and Yao, K. (2016). Block replacement policy with uncertain lifetimes. *Reliability Engineering & System Safety*, *148*, 119–124.

Liao, W. P. and Yuan, J. (1998). Optimal replacement for a one-unit system subject to delivery and test. *Joural of Quality in Maintenance Engineering*, vol. 4, no. 1, pp. 51–65.

Nakagawa, T. (1982). A modified block replacement with two variables. *IEEE Transactions on Reliability*, *R-31*(4), 398–400.

Park, Y. T. and Park, K. S. (1986). Generalized spare ordering policies with random lead time. *European Journal of Operational Research*, vol. 23, no. 3, pp. 320–330.

Shafiee, M. and Finkelstein, M., (2015). An optimal age-based group maintenance policy for multi-unit degrading systems. *Reliability Engineering & System Safety 134*, 230–238.

Sheu, S. H. and Chien, Y. H. (2004). Optimal age-replacement policy of a system subject to shocks with random lead-time. *European Journal of Operational Research*, vol. 159, no. 1, pp. 132–144.

Sheu, S. H. and Griffith, W. S. (2001). Optimal age-replacement policy with age-dependent minimal-repair and random-leadtime. *IEEE Trans. Reliability*, vol. 50, no. 3, pp. 302–309.

Van Horenbeek, A., Buré, J., Cattrysse, D., Pintelon, L., and Vansteenwegen, P. (2013). Joint maintenance and inventory optimization systems: A review. *International Journal of Production Economics*, *143*(2), 499–508.

Wang, W., (2012). A stochastic model for joint spare parts inventory and planned maintenance optimisation. *European Journal of Operational Research 216*, 127–139.

Zahedi-Hosseini, F., Scarf, P., and Syntetos, A. A. (2017). Joint optimisation of inspection maintenance and spare parts provisioning: A comparative study of inventory policies using simulation and survey data. *Reliability Engineering & System Safety*, *168*, 306–316.

Recent Advances in Reliability and Maintenance Modeling – Okamura, Inoue & Xiao (Eds)
© 2025 the Author(s), ISBN: 978-1-032-79098-5

Advancing Layer of Protection Analysis (LOPA) and augmented analytics system to optimize SIL in train automatic door systems

R. Kusumawardani, N. Kurniati & P. Vorenza
Institut Teknologi Sepuluh Nopember, Surabaya, Indonesia

ABSTRACT: Determining safety requirements for safety systems is crucial for hazard prevention. The Indonesian train manufacturer industry, adheres to the International Electrotechnical Commission (IEC) standard to assess product safety, using Safety Integrity Level (SIL) certification. Validating product test results through SIL Assessment calculation ensures compliance with safety standards. Accelerating product failure detection is essential for enhancing safety. This study focuses on validating security levels and expediting failure detection within the Door System. The SIL Assessment yields SIL 0, below the company's standard, prompting a comprehensive evaluation of the Safety Instrumented System (SIS) for the Door System. Layer of Protection Analysis (LOPA) suggests enhancing SIL values. SIL 0 indicates adequate risk reduction, No Risk (NR) indicates no potential risk, and SIL 1 suggests increasing SIL for risk reduction. Additionally, a Machine Learning classification algorithm is developed to predict failures swiftly. This integrated approach enhances safety and operational efficiency.

1 INTRODUCTION

Safety is a state away from danger (hazard) caused by natural disasters or due to random human error (Nas 2015). A system with a high level of safety will provide a high-security guarantee as well. On the other hand, a system with a low level of safety tends not to be able to provide security guarantees at high levels of hazard. Determining the specific safety requirements of the safety system is an important part of ensuring that potential weaknesses, initial events, internal hazards, and situations that have potential hazards can be prevented (Sliwinski 2018). Therefore, the demand for a high level of safety from a system is very important to overcome hazards that harm the environment or even threaten human safety.

The door System or door function is a component of trains that is often used so that it is categorized as a high-demand component. In addition, according to data from the Ministry of Transportation, the Directorate General of Railways, until 2021 the number of trains is 3,239 units (Directorate General of Railways 2021). If it is related to the number of Door Systems in each carriage, it reaches 51,824 units. This figure is assumed to be large enough and impactful to be the focus of failure identification. Supported by many Door Systems and high frequency of use, Door Systems are also not free from various possible failures that can even lead to fatalities. On Wednesday, February 9, 2022, a door malfunction occurred which caused the train to run with the door open (Muhammad 2022). This of course will threaten the safety of passengers and even potentially cause injury to passengers and even death if the passenger is pulled or falls from the train. Therefore, the Door System is one of the most crucial parts to become an important focus in improving safety on trains.

The Safety Integrity Level (SIL) is an instrumented safety system designed according to the IEC 61508 (International Electrotechnical Commission) standard, as per the management regulations of the Petroleum Safety Authority (PSA) in Norway (Vinnem and Røed 2019).

DOI: 10.1201/9781003491309-10

The IEC standard delineates various requirements for planning and implementing safety systems, encompassing equipment under control (EUC) definition, hazard identification, EUC interactions, and risk assessment. One of the train manufacturing companies in Indonesia has adopted the IEC 61508 standard to ascertain the safety level of its products. During implementation, the company calculates the SIL Allocation derived from agreed-upon product safety standards, which are reflected in the warranty certification. Interviews with users reveal that the standard SIL Allocation for train products varies based on sub-assembly and potential risks. Presently, the SIL Allocation for the Door System averages SIL 2. However, a significant challenge in the train manufacturing industry arises when the SIL Allocation does not align with the SIL of the finished product, commonly known as SIL Assessment. This discrepancy could lead to future consumer complaints due to compromised safety and reliability, potentially resulting in higher-risk accidents. This study aims to conduct SIL Assessment calculations to verify whether the company's certification aligns with the actual product performance in the field, focusing on the Door System component calculation of SIL assessment done using test drive/road test results in the form of Output Data Logger. Data loggers are electronic equipment that can record the formation of data with a high degree of accuracy (Guragai *et al.* 2018). Data loggers work by acquiring data and then converting electrical signals into data and storing them in internal memory (Yonghua 2012). The process of recording data by measuring most sensors with electrical output. This data logger is large and recorded in real time. Output Data Logger test results will be processed using Machine Learning. Machine Learning is the application and science of algorithms that understand data. Machine Learning uses self-learning algorithms from the field of machine learning by converting data into knowledge. Machine Learning offers a high level of accuracy in representing data (Raschka 2022). Therefore, in this study predictions will be made using Machine Learning to find out the exact factors in knowing the cause of the door's malfunction.

Data loggers work by acquiring data and then converting electrical signals into data and storing them in the internal memory. They typically measure various sensors with electrical outputs and operate in real time. The collected data from the Output Data Logger will undergo analysis using Machine Learning. This discipline involves algorithms that comprehend and learn from data, transforming it into actionable insights. Machine Learning excels in accurately representing data (Raschka 2022). Thus, this study employs Machine Learning algorithms to analyze patterns in Output Data Logger data, aiming to predict potential failures in the Automatic Door System.

2 RESEARCH METHODOLOGY

The research begins with the identification of Door System problems based on the results of field studies with experts and a review of the schematic door system design owned by the train manufacturer industry. Based on these problems, a literature study was conducted on the Safety Integrity Level, IEC 61508 Standardization, 61511 Standardization, Layer of Protection Analysis (LOPA), and Automated Analytics System. The next stage is the process of collecting data from the results of the railroad test. The data is in the form of Data Logger output which is recorded from the components that make up the Door System during the road test. The data preparation process is carried out to obtain data that meets the requirements of predictive classification. Data preparation was carried out using various analytic data concepts and calculated the duration of each class label that had identified the duration of time to failure, time to diagnostic, and time to repair, calculated with the help of VBA Excel. The next step is to calculate the SIL Assessment value which is one of the SIL Safety. Instrumented System (SIS) values, which is calculated quantitatively through the process of calculating the PFD (Probability of Failure on Demand) value of the various Safety Instrument Function (SIF) components that make up the Architecture Door System. The SIL value is the level of PFD value conversion results, Table 1 shows a list of SIL level values.

Table 1. PFD and RRF each SIL level.

Safety Integrity Level (SIL)	Target Average Probability of Failure on Demand	Target Risk Reduction
4	≥ 10-5 to < 10-4	> 10.000 to ≤ 100.000
3	≥ 10-4 to < 10-3	> 1.000 to ≤ 10.000
2	≥ 10-3 to < 10-2	> 100 to ≤ 1.000
1	≥ 10-2 to < 10-1	> 10 to ≤ 100

The SIL Assessment value is used as a reference in assessing the safety of the actual situation and compared with the SIL Allocation value. If the SIL Assessment score does not reach the minimum target value from the SIL Allocation standard, it is necessary to carry out an improvement process on the system to increase the SIL value to match the standard guarantee that has been offered. In this study, if this happens, recommendations are made to increase layered safety using the Layer of Protection Analysis (LOPA) method. LOPA is formed based on hazard identification information from the Subsystem Hazard Analysis document which has been adapted to the design of the object being analyzed. However, if the SIL Assessment value is greater than the SIL Allocation, an Augmented Analytic System will only be developed to increase the agility of detecting system component failures. The Augmented Analytic System uses the logic of predicting the classification of Decision Tree Machine Learning and is displayed in real-time on the dashboard. The running model is attempted to be developed on the train cabin HMI so that the prediction results can be displayed automatically and agile in conveying information to the driver. Machine Learning prediction models are integrated into the dashboard in the form of Pickle models for recall. The purpose of establishing this Pickle model is to improve the model selection with high accuracy and in a fast time.

3 RESULT AND DISCUSSION

3.1 *Calculation of SIL assessment*

The first step in calculating the PFD value is to find the hazard rate value of each door system instrument. The hazard rate is calculated from the data set duration on the time limit between arrivals of failures, that is if the door system can still perform its functions according to the request until it fails and undergoes repair or time between failure (TBF). This time limit is determined according to the characteristics of the repairable object. Table 2 is an example of the results of the identification of the duration of TTF, TTD, and TTR from the door system.

Table 2. Maintenance data on each component of the door system.

No	Begin Time	End Time	TTF (ms)	TTD (ms)	TTR (ms)	TBF (ms)
Logic Solver						
1	13:12:43.667	14:37:28.067	4.778.012	200	30.000	4.808.212
2	14:37:28.167	16:09:25.527	4.605.213	0	900.000	5.505.213
....	....	....	....	....	....	....
33	07:25:00.940	16:43:23.920	5.900	78.509	290.401	374.810
Sensor						
1	13:12:43.667	14:37:28.067	4.777.512	700	30.000	4.808.212
2	14:37:28.167	16:09:25.527	4.605.213	0	900.000	5.505.213
....	....	....	....	....	....	....
32	07:10:35.853	07:25:00.840	648.677	128.252	86.203	863.132
Actuator						
1	20:17:29.650	20:53:35.133	1.800.677	12.300	352.306	2.165.283
2	20:53:35.233	21:08:59.233	12.000	12.100	900.000	924.100
....	....	....	....	....	....	....
33	07:25:00.940	16:43:23.920	49.400	35.009	290.401	374.810

Based on the data processing that has been done, the results show that the Door System is in the Constant Failure Rate (CFR) phase. This condition shows the suitability of field conditions with the theoretical concept used, namely the test drive is carried out after the design process is declared feasible and tested under Normal Life Testing conditions. After obtaining the value of the hazard rate, the value of the PFD is calculated. Figure 1 shows the architecture of the door system.

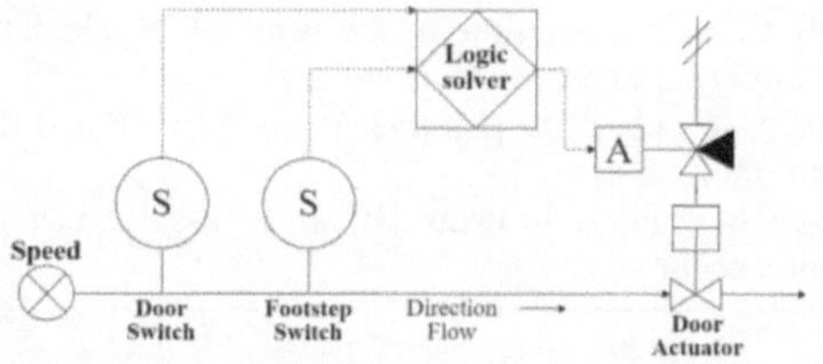

Figure 1. Architecture door systems.

In the Architecture Door System, as in Figure 1, the Sensor has a 2oo2 architecture while the Logic Solver and Actuator have a 1oo1 architecture, then the calculation of the PFD value is following Architecture 2oo2 in Equation 1 and Architecture 1oo1 in Equation 2.

$$PFD_{avg} = \lambda \times \frac{Ti}{2} \tag{1}$$

$$PFD_{avg} = \lambda \times \text{Ti} \tag{2}$$

PDF_{avg} calculations are performed at object testing time intervals. From the calculation results, the PFD_{avg} Sensor value is 1.516608, PFD_{avg} Logic Solver is 2.833008, and Actuator PFD_{avg} is 1.516608. Then the value of PFD_{avg} SIF is calculated using Equation 3.

$$PFD_{avg}\ \text{SIF} = PFD_{avgs} + PFD_{avgl} + PFD_{avgfe} \tag{3}$$

From the calculation of the total PFD value of SIF, a value of 5.866224 is obtained, if it is converted to the SIL value, then the SIL value has not reached the SIL value of 1 or is still at the SIL value of 0.

3.2 *Adjustment of subsystem hazard analysis*

In this section, an evaluation and adjustment process is carried out on the Subsystem Hazard Analysis document which will then be developed into a Layer of Protection Analysis (LOPA) recommendation analysis to increase the SIL value to match the existing standards. Tables 3–5 show the legend in providing an assessment of the Subsystem Hazard Analysis document.

Table 3. Company's severity level criteria.

Cat	Severity Level	Consequences to Persons / Environment
I	Catastrophic	Multiple fatalities and/or multiple severe injuries or extreme damage to the environment
II	Critical	Single fatality and/or severe injury or major damage to the environment
III	Marginal	Minor injuries (required medical treatment) or minor damage to the environment
IV	Insignificant	Possible minor injury (first aid)

Table 4. Company's Frequency Level Criteria

Cat	Frequency Level	Description
A	Frequent	Likely to occur frequently. The hazard will be frequently experienced.
B	Probable	Will occur several times. The hazard can be expected to occur often.
C	Occasional	Likely to occur several times. The hazard can be expected to occur several times.
D	Rare	Likely to occur sometime in the system lifecycle. The hazard can reasonably be expected to occur.
E	Improbable	Unlikely to occur but possible. It can be assumed that the hazard may exceptionally occur.
F	Highly Improbable	Extremely unlikely to occur. It can be assumed that the hazard will not occur.

Table 5. Risk ranking from manufacturer.

	IV	III	II	I
A	UND	INT	INT	INT
B	TOL	UND	INT	INT
C	TOL	UND	UND	INT
D	NEG	TOL	UND	UND
E	NEG	NEG	TOL	UND
F	NEG	NEG	NEG	TOL

Based on the results of the adjustment of the Subsystem Hazard Analysis, 20 hazard identifications may occur in the door system. The operating mode of each possible condition is divided into three, namely normal, degraded, and in an emergency. Based on the severity level, there are 7 events with a value of I is meaning catastrophic or causing disaster, 7 events with critical or critical value, and 6 events with insignificant value or less significance. Meanwhile, based on the frequency of hazard arrivals, there is 1 event with a value of B meaning probable or frequent, 3 events with a value of D meaning rare or rare, and 16 events with a value of C meaning occasional or occasional. So based on the severity and frequency assessment, an initial result risk ranking is obtained for each event, i.e., there are 6 events with a risk ranking of INT meaning intolerable or intolerable, 8 events with a risk ranking of UND meaning undesirable or undesirable, 6 events with a risk ranking of TOL means tolerable or can be tolerated.

3.3 *Formation of layer of protection analysis (LOPA)*

In the Layer of Protection Analysis (LOPA) hazard analysis is identified in each component of the main architecture components and instruments supporting the safety of other door systems. The causes and consequences for each instrument component of the door system are obtained from the information in the previous Subsystem Hazard Analysis. In the development of the Layer of Protection Analysis, the protection layer consists of many IPLs. In general, IPL consists of General Process Design, Basic Process Control System (BPCS), Alarm, Additional Mitigation (restricted access), and Additional Mitigation Dikes/ Pressure Relief. In the General Process Design section in some companies, it is assumed that for some scenarios the failure cannot occur due to the safe design of equipment and processes.

After formulating causes and consequences, the Initiation Cause Likelihood (ICL) value will be calculated for each instrument component. The ICL value is obtained from the process of calculating the probability of failure (F(t)). The following Figure 2 shows a summary of the results using a tree diagram. In the tree diagram, the components have values of SIL 0, SIL 1, and NR which are colored according to the severity of the values. The SIL 0 value is in the LS1, LS2, and LS7 scenarios, meaning that the reduction target and actual conditions are appropriate in other words each existing protective layer can reduce the risk hazard from the door system. It is necessary to develop the SIL value in scenarios that have a SIL value of 1. The types of recommendations given are adjusted to the scenario to be derived. In the S1 scenario, it is an anti-trap to anticipate door failures that cause passengers to be squeezed. Whereas in the LS3, LS4, LS5, LS6, LS8, and A3 scenarios, recommendations for security technology in the form of alarms can be given which are connected to a set of electronic devices in the form of sensors for detecting the fall of passengers on the platform/rail. The recommended sensor can be a Motion sensor with a range according to the width of the door with tolerances that are adjusted to the company's design.

Based on the calculation of the SIL value after adding the recommendation layer, it was found that the SIL value had reached SIL 0. The recommendation for investment in protection layer technology is in the form of alarm and anti-trap with a target of equipment damage once for 10 years so that it is assessed as 0.1. The result summary tree diagram is displayed again after being given recommendations, shown in Figure 3.

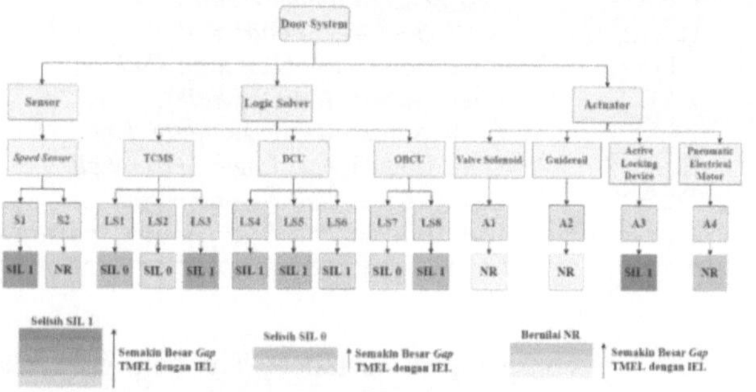

Figure 2. Tree diagram of SIL assessment results for door system components.

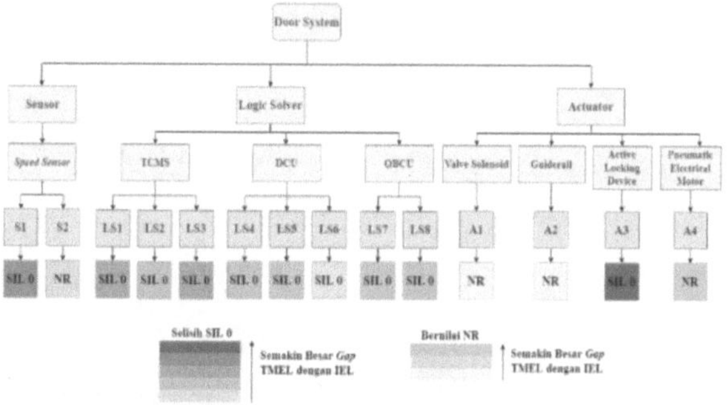

Figure 3. Tree diagram of SIL recommendation results for door system components.

3.4 *Model development*

If it is related to the level of robustness of the model, in this section, the process of developing the model will be carried out in conditions that have not been studied by Machine Learning before but have the possibility of happening in the future. Based on the analysis of the combination of failed relationships between the components of the logic solver, sensor, and actuator, the possibilities that may appear are shown in Table 6. The model has studied 26 combinations of failures, and it is necessary to study 17 combinations that have not yet emerged. In providing model learning/training in these conditions, dummy data is made with the logic of the combination of 17 failures as a form of model development. In this process, the data is expanded to 60,000 rows.

Table 6. Failure combination supplementary label.

No	Label	Condition
1	LF	*Logic Solver Fail*
2	AF	*Actuator Fail*
3	L-AF	*Left Actuator Fail*
4	R-AF	*Right Actuator Fail*
5	RL-AF	*Right-Left Actuator Fail*
6	LSF	*Logic Solver-Sensor Fail*
7	L-LSF	*Left Logic Solver-Sensor Fail*
8	R-LSF	*Right Logic Solver-Sensor Fail*
9	RL-LSF	*Right-Left Logic Solver-Sensor Fail*
10	L-SAF	*Left Sensor-Actuator Fail*
11	R-SAF	*Right Sensor-Actuator Fail*
12	RL-SAF	*Right-Left Sensor-Actuator Fail*
13	LAF	*Logic Solver-Actuator Fail*
14	LSAF	*Logic Solver-Sensor-Actuator Fail*
15	DLF-LF	Data Logger *Fail-Logic Solver Fail*
16	DLF-AF	Data Logger *Fail-Actuator Fail*
17	DLF-LAF	Data Logger *Fail-Logic Solver-Actuator Fail*

There are total 60,000 data consisting of 43 labels. The data is split again to divide the data into training data, development data, and testing data, with a ratio of 70:15:15. So for each number, training is 42,000 data, development is 9,000 data, and testing is 9,000 data. Figure 4 shows the Decision Tree Model from the classification results of the development model stage.

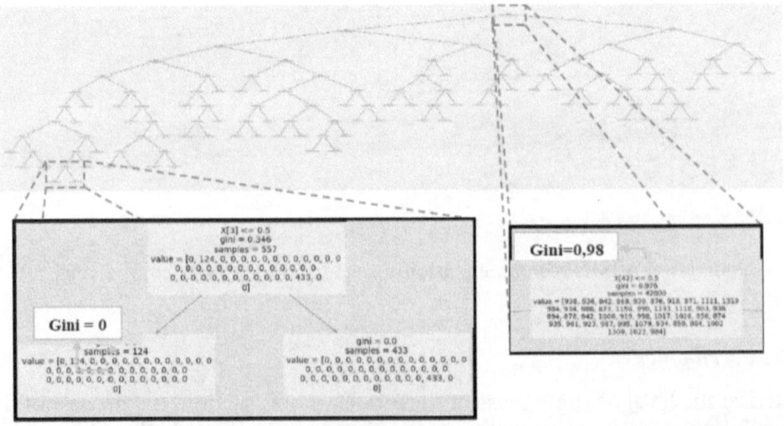

Figure 4. Model development decision tree classification.

Based on the results of the development model, a decision tree with a depth of 11 was formed with the result that the number of classes was 43 according to the predetermined label. In the Decision Tree model, the process of evaluating the depth of the model is carried out using K-Fold Cross Validation. Cross Validation is a statistical method that can be used to evaluate the performance of the model by separating the data into two subsets, namely, data on the learning process (training) and data on validation/evaluation. The model or algorithm is trained by the learning subset and validated by the validation subset. Meanwhile, K-fold is used to reduce computation time while maintaining the accuracy of the estimate. Table 7 shows the results of the K-fold Cross Validation calculation on the model.

Table 7. Failure accuracy assessment at each depth.

Decision Tree Model Depth	Average Model Accuracy
7	0,8779571429
8	0,9507238095
9	0,9798476190
10	0,9950666667
11	0,9980206349
12	0,9980142857
13	0,9980111111

As the results of the K-Fold Cross Validation calculation, it is stated that the model with a depth of 11 already has a decision with the best accuracy rate, which is 99.802%, and does not experience overfitting. This is evidenced by the accuracy value which is at its peak. Furthermore, the model accuracy calculation process is carried out using precision, recall, and f1-score metrics to determine the type of failure in the model. Based on the results of the evaluation of the precision, recall, and f1-score metrics on the developed model, it is known that the model has an accuracy rate of up to 100% in predicting the class on most labels. This is because the combinations that have not previously been studied at the pilot stage have now been studied by Machine Learning so that the labels on the class can be found correctly. However, it is possible for Machine Learning to experience prediction errors in classes that tend to be similar. There is a label with a precision value (False Positive) of 93% on the L-SAF label and a recall value (False Negative) of 91% on the L-AF label.

3.5 *Dashboard development*

The monitoring system in this research has been developed in the form of a dashboard prototype integration of Artificial Intelligence logic information. Specifically, the machine learning used is the prediction of the failure classification of the Door System component using the Decision Tree algorithm in the current condition. The model was developed on 60,000 raw data with 43 types of labels and varied in Decision Tree depths 7, 8, 9, 10, 11, 12, 13. Based on this depth variation, seven models were formed and stored in the form of a Pickle Module. Pickle is a module in the Python standard library, which can be used to store and read data into/from a document. The purpose of the formation of these models is to obtain predictive results with high accuracy but in a fast time. Figure 5 shows the prototype Dashboard for Automatic Door System Failure Detection of Trains.

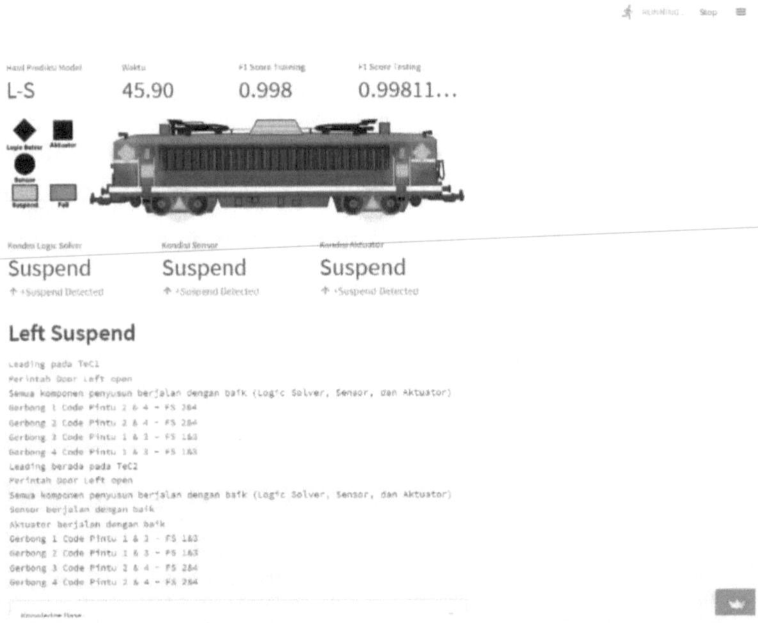

Figure 5. Dashboard for monitoring system.

4 CONCLUSION

The SIL Assessment results for the Door System indicate a significantly lower value than the train manufacturer's standard, SIL 0, prompting a comprehensive evaluation of the entire Safety Instrumented System (SIS) using Layer of Protection Analysis (LOPA). The LOPA analysis reveals three scenarios with SIL 0, indicating an appropriate risk reduction alignment, four scenarios with No Risk (NR), suggesting manageable risk levels but potential overdesign concerns, and seven scenarios with SIL 1, indicating the need to enhance safety levels. Recommendations include adding layers such as alarms and anti-trap mechanisms to match target SIL levels. Additionally, a prototype dashboard, primarily developed using Jupyter Notebook and Python, is created to integrate with the train's Human-Machine Interface (HMI) and Machine Learning classification model. This dashboard serves as an agile information hub for predicting failure classes and providing insights into the door system's component status, further enhancing the Augmented Analytics System.

REFERENCES

Ahn, J., Noh, Y., Joung, T., Lim, Y., Kim, J., Seo, Y., and Chang, D. (2019). Safety integrity level (SIL) determination for a maritime fuel cell system as electric propulsion in accordance with IEC 61511. *International Journal of Hydrogen Energy*, 44(5), 3185–3194. https://doi.org/10.1016/j.ijhydene.2018.12.065

Allen Hamilton, B. (1999). *Department of Defense Joint Software Systems Safety Engineering Handbook Developed By The Joint Software Systems Safety Engineering Workgroup Naval Ordnance Safety and Security Activity.*

Badhiye, S., Badhiye, S. S., Chatur, P. N., and Wakode, B. v. (2011). Data logger system: A survey. *In International Journal of Computer Technology and Electronics Engineering*. https://www.researchgate.net/publication/271964052

Guragai, B., Hashimoto, T., Oguma, K., and Takizawa, S. (2018). Data logger-based measurement of household water consumption and micro-component analysis of an intermittent water supply system. *Journal of Cleaner Production*, 197, 1159–1168. https://doi.org/10.1016/j.jclepro.2018.06.198.

Muhammad, T. V. A. (2022). Pintu KRL solo-yogya terbuka saat berjalan, *KAI Commuter Berikan Penjelasan*.

Nas, S. (2015). The Definitions of Safety and Security. *Journal of ETA Maritime Science*, 3(2), 53–54. https://doi.org/10.5505/jems.2015.42713.

Rachman, A., Zhang, T., and Ratnayake, R. M. C. (2021). Applications of Machine Learning in pipeline integrity management: A state-of-the-art review. *International Journal of Pressure Vessels and Piping*, *193*, 104471. https://doi.org/10.1016/j.ijpvp.2021.104471

RASCHKA, SEBASTIAN. L. (2022). *Machine Learning with pytorch and scikit-learn…: develop Machine Learning and deep learning... models with scikit-learn and pytorch*. Packt Publishing Limited.

Sliwinski, M. (2018). Safety integrity level verification for safety-related functions with security aspects. *Process Safety and Environmental Protection*, 118, 79–92. https://doi.org/10.1016/j.psep.2018.06.016.

Vinnem, J.-E., and Røed, W. (2019). *Springer Series in Reliability Engineering Offshore Risk Assessment Vol. 2 Principles, Modelling and Applications of QRA Studies Fourth Edition*. http://www.springer.com/series/6917.

Yonghua, C. (2012). Integration of SentryTM Visibility Sensor into Campbell Scientific Data Logger CR1000. *Procedia Environmental Sciences*, 12, 1137–1143 https://doi.org/10.1016/j.proenv.2012.01.399.

A physics-informed predictive scheme for predicting low-cycle fatigue life and reliability of aerospace materials under multiaxial loading conditions

Butong Li, Junjie Zhu & Xufeng Zhao
College of Economics and Management, Nanjing University of Aeronautics and Astronautics, Nanjing, Jiangsu, PR China

ABSTRACT: In our previous work, we developed a highly reliable FE model for calculating low circumferential fatigue under multiaxial loading. Subsequently, we generated high-confidence datasets using the FEA models. By leveraging the strengths of both deep learning methods and LightGBM, we proposed a fusion surrogate model called DL-LGBM-DRS. The DL-LGBM-DRS can efficiently and accurately predict low-cycle fatigue life under various multiaxial loading conditions.

Keywords: Fatigue life prediction, DL-LGBM-DRS, multiaxial loading, aerospace alloy

1 INTRODUCTION

Aiming for higher efficiency and reliability, researchers have attempted to develop alternative methods on the premise of avoiding the numerous non-linear operations of FEA. Considering that uncertain sources (such as material property, geometry, and load) will lead to a scatter of fatigue life, statistical methods have been proposed to achieve a more plausible fatigue life distribution [1]. The Monte Carlo simulation (MCS) is generally used as a universal prediction framework in probabilistic fatigue life prediction cases. For multivariate uncertainty assessment, the large number of input influential parameters and output results are challenging to deal with, which can be solved by MCS in combination with weighting laws in structural reliability, e.g., most probable point (MMP) [2]. Bai *et al.* proposed a probabilistic high and low cycle fatigue (H-LCF) life prediction framework under MCS-generated random load history and confirmed the feasibility of which on a practical turbine shaft case [3]. He *et al.* considered a novel asymptotic stochastic response surface approach under multi-source heterogeneous uncertainty, demonstrating that statistic methods are highly accurate in real-world problems [4].

Though statistical knowledge-based methods can predict fatigue life with high accuracy and efficiency, there are still some limitations in the implementation process and prediction results: (i) difficulty in obtaining sufficient quantities of statistical data, (ii) low interpretability and fidelity compared to physics-based methods (iii) expensive computational costs in the presence of considerable uncertainty. For this reason, there is a growing demand for fatigue prediction, which requires novel hybrid methodologies [5]. Machine learning approaches could be desirable in problems with nonlinearities and non-convergence properties, which are appropriate for predicting problems involving physically intangible information [6–8]. Wang *et al.* proposed a surrogate-model-assisted fatigue assessment framework, which combined the practical FEA method and XGboost technique to estimate the creep-fatigue reliability in low-pressure turbine disks [9]. Roy and Chakraborty developed a hybrid metamodel for structure reliability using support vector regression and

 DOI: 10.1201/9781003491309-11

sequential adaptive sampling [10]. Physics-informed machine learning (PIML) is an emerging paradigm that uses physical principles as the 'prior' knowledge, thus significantly improving the machine learning model's performance [11]. One approach to PIML is by embedding physical or empirical laws (e.g., ordinary differential equations (ODEs)) into machine learning networks [12]. Thus, in practical engineering applications, the insufficient dimension and amount of data will severely affect the training performance of PIML [13]. As mentioned above, FEA can be employed as a proper alternative for physics law-based data acquisition, which is prevalent in plenty of engineering projects [14]. PIML based on FEA proved to be a highly reliable method for situations with a lack of data [15].

The primary works are as follows:

A fusion machine learning method is developed to predict the fatigue life based on the inputs from stress/strain and load spectra. The proposed method can obtain higher prediction accuracy than the individual model under small sample prediction without losing any input features.

2 SURROGATED MODEL DL-LGBM-DRS

In this work, we aimed at the fatigue behavior of alloy components under complex multiaxial loading. Prediction of fatigue behavior under multiaxial variable amplitude is challenging owing to the involvement of multiple covariates, including continuous correlation (e.g., loading spectrum) coupled with discrete variables (e.g., physical states of the material). This imposes high demands on the generalizability of the surrogate model and the performance of the model on the temporal continuity task. In this regard, we propose a novel data-driven model, DL-LGBM-DRS, based on a dynamic optimization approach, DRS, for the prediction of fatigue life under multiaxial loading. The scheme of proposed model is shown in Figure 1.

2.1 *LightGBM model*

The gradient boosting decision tree (GBDT) algorithm is the integration of multiple decision trees and has been widely used for its favorable generalization. Models based on GBDT, including random forest, XGBoost, and LightGBM, have respective features and capabilities, with differentiated performance in terms of time efficiency or regression performance [16]. Out of concern for computational cost and efficiency, we introduced LightGBM as the implementation of discrete data processing. The GBDT algorithm achieves outstanding prediction of dense discrete data with strong generalization and expression capabilities.

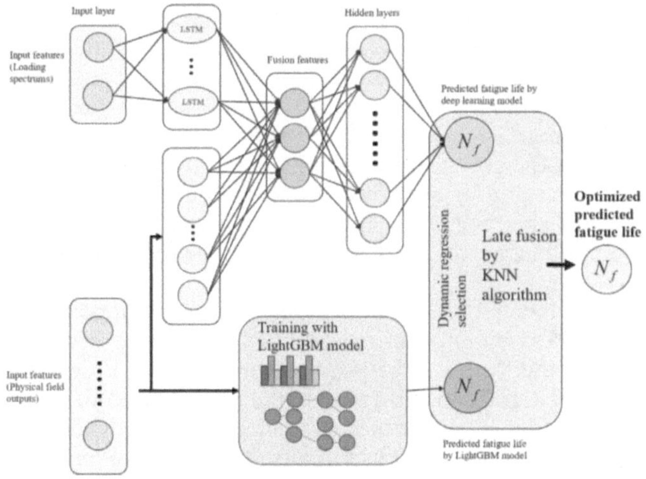

Figure 1. Scheme of the DL-LGBM-DRS model.

LightGBM is a widely adopted model based on the model XGBoost, which was proposed by Ke [17]. Facilitated by the introduction of gradient-based one-side sampling (GOSS) and exclusive feature bunding (EFB), LightGBM can handle cases with greater complexity at a faster rate. With the proposed GOSS algorithm, the model can disaggregate large data into smaller gradients. The EFB technique can contribute to the model's ability to improve model efficiency without losing information. In addition, LightGBM utilizes a Histogram-based Algorithm and a leaf-wise algorithm with depth limitations. Processing of dense discrete data that requires high computational cost allows for faster training with LightGBM, which is favorable for the cases of the FEA data accumulated in the previous work.

2.2 *Deep learning model*

LightGBM has been widely used for industrial cases. Nevertheless, LightGBM lacks the ability to handle sequential variables, resulting in its incapability to process critical time-series data. With the purpose of being able to analyze information with strong temporal correlations in multiaxial fatigue, Long short-term memory (LSTM) is introduced in this work.

LSTM is a recurrent neural network (RNN) with a long-term memory function. In the hidden layer neuron nodes of an ordinary RNN, we added a memory gate to help the model capture and forecast time-coherent information.

2.3 *DL-LGB-DRS fusion model*

In the previous section, we have delineated the separate LGBM model and Deep learning model. As machine learning models based on different algorithms, LGBM has better performance in handling discrete inputs, and LSTM is able to handle the time-sequential data. With a view of combining and utilizing the unique advantages of both, we will focus on the fusion of the two models. Research on model fusion has been longstanding. In fact, ensemble learning methods, including GBDT, leverage Bagging or Boosting to fuse simple and large sub-models into a final model with high predictive power. However, a limitation of these methods is that they only consider global weights and are not capable of analyzing the local performance of the sub-models. Schemes have been investigated for the fusion of more complex sub-models, such as the introduction of the Teacher Forcing mechanism to guide the fusion of GBDT with Bi-LSTM [18]. Researchers have also proposed a fusion method based on local relation mining, Dynamic Classifier Selection (DCS) [19]. Further, Dynamic Ensemble Selection (DES) has been implemented for the better-localized performance of the sub-models in prediction [19]. Some literature suggests that DCS using only accumulated errors is not accurate enough to differentiate the performance of classifiers [20,21]. Dynamic regression selection (DRS) is an algorithm capable of selecting the best regressors to predict the given query pattern. DRS approaches can calculate the error of the regressors in the region of competence and define the optimal regressor with the minimum error [22]. In this work, the integration of the proposed strong models can benefit from the employment of DRS to improve the performance of the resulting prediction.

The DRS can be implemented in several methods, including Support Vector Machine and Neural Networks [23]. Rooney *et al.* proposed DRS algorithms with K-nearest neighbor (KNN) and proved to be effective in generating homogeneous ensembles using Random Subspace [24]. The KNN algorithm can evaluate the contribution of the k data to the prediction in both models, guided by the defined error index. The predicted values of K_{th} models with better performance will be selected as outputs. The fusion model allows for a sensible proportion of weights, simultaneously exploiting the advantages of LSTM and LightGBM.

3 PREDICTION OF LOW-CYCLE FATIGUE LIFE UNDER MULTIAXIAL LOADING BASED ON THE PROPOSED MODEL

3.1 *Generation of datasets*

Fatigue behavior is primarily influenced by the loading conditions. The external environment factor, as determined by the loading conditions, was incidentally the only variable in this study. In the previous research on FEA validation, we noted that physical states such as stress and strain of the specimens have a strong correlation with fatigue life. For the consideration of improving the predictive performance of the model, we augmented the physical states to the input data to increase the dimension of the data features.

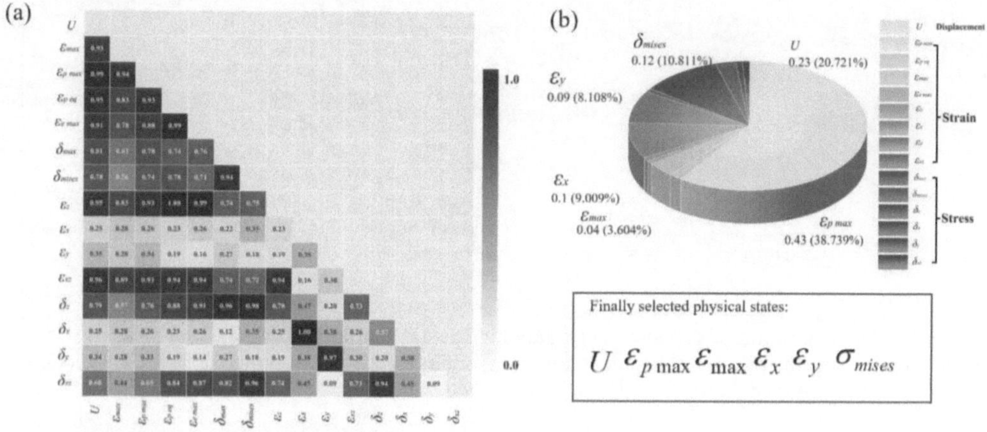

Figure 2. Feature analysis of candidate inputs: (a) correlation analysis using Pearson matrix; (b) feature importance of physical states.

Prior to generating the datasets, we preprocessed the input features to minimize redundant features and thus improve training efficiency. For the physical states involved, a Pearson matrix was employed to analyze their correlation, while the contribution of each feature to the fatigue life will be ranked in terms of feature importance available for GBDT-based algorithms. The results are depicted in Figure 2. The candidate physical states are classified into three categories: displacement, stresses, and strains. It can be concluded from the Pearson correlation matrix that there is a high degree of correlation between the sub-items within each category. Considering the conversion relevance of the subitems to each other (e.g., equivalent plastic strain versus maximum principal plastic strain), it is reasonable to exclude the highly correlated inputs in the generated datasets. Figure 2b shows the contribution of each feature to the output value, i.e., fatigue life, in the form of a pie chart. It is noteworthy that the feature importance method based on the GBDT algorithm performs a direct evaluation of highly correlated input features. Assuming that several highly correlated feature values have the same contribution to the output, the method will only select the highest contributing feature among them for weighted ranking. The pie chart shows the most important physical states, including displacement, max principal plastic strain, max principal strain, and von Mises stress. We similarly analyzed the importance of stress and strain in each direction component and found that the stress-strain in the unidirectional component is highly correlated. Strain in the non-load applying direction, i.e., x-axis and y-axis, is of significant importance. Eventually, by combining the correlation matrix with feature importance analysis, we are capable of selecting the final input features. The physical states that can be utilized as input features are selected, as shown in Figure 2b.

Consequently, all the nodes in the FEA simulated fatigue occurring area will be considered in the inputs. Based on this, 64 datasets were generated, and each dataset contained 29160 pieces of data. The inputs contain multiple categories of loading conditions. The construction of training sets with the same input features as in the practical case leads to a potential degradation of the model's extrapolation performance. Thus, we will exclude the loading conditions that are the same as the validation scenarios in the generation of training sets.

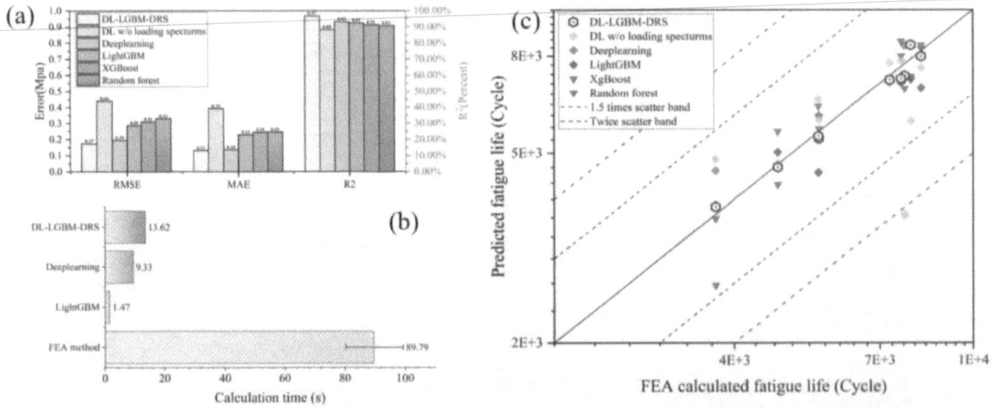

Figure 3. The evaluation of proposed surrogate models: (a) the RMSE, MAE, and R-square of utilized models; (b) the average calculation time of models; (c) the performance of proposed models in the fitness with FEA calculation.

3.2 *Model performance*

The K-Fold cross-validation method is used to evaluate the performance of the trained models. The datasets are divided into five folds. In each evaluation process, four folds are used as training sets, and the rest is used as the validation set. After times of repetitions, the cross validation mean absolute error, root mean square error, and R-square are used to assess the prediction accuracy of employed surrogate models, as shown in Figure 3a. Similarly, the time cost for the prediction in each method is displayed in Figure 3b. It can be summarized intuitively that the fusion model DRS proposed in this work has better performance in predicting simulated fatigue life. The LGBM has a sub-some predictive performance, showing its high learning efficacy on discretized feature data, and is a potential model for fatigue behavior prediction. The results from deep learning are noteworthy. The accuracy of the deep learning model, which utilizes a fusion of loading spectrums and physical states as inputs, is slightly lower compared to the proposed DRS model. However, upon removal of the loading spectrums, there is a significant decrease in the accuracy of deep learning, accompanied by a substantial increase in errors. This disparity underscores the importance of load spectrum information as an input. In addition, we compare other GBDT-based algorithms, including XGBoost and random forest, and the stepwise decrease in their predicted performance is consistent with the pattern we found in our previous study [16]. Generally, the errors of the machine learning models are within acceptable limits, and the fusion model has a higher prediction accuracy compared to the individual model. In terms of time cost, the fusion model is slower compared to the standalone model due to the addition of the decision layer that plays the role of late fusion. However, machine learning models for the environmental applications in this work have an unparalleled speed advantage over FEA methods. Figure 3c illustrates a comparison between the fatigue life predicted by the machine learning models and the fatigue life calculated by the FEA method. It is evident that the proposed DL-LGBM-DRS model exhibits highly favorable results compared to other models when predicting fatigue life from the generated datasets. This supports the feasibility of being able to use DL-LGBM-DRS as the surrogate model for fatigue life prediction with greater efficiency while maintaining high accuracy.

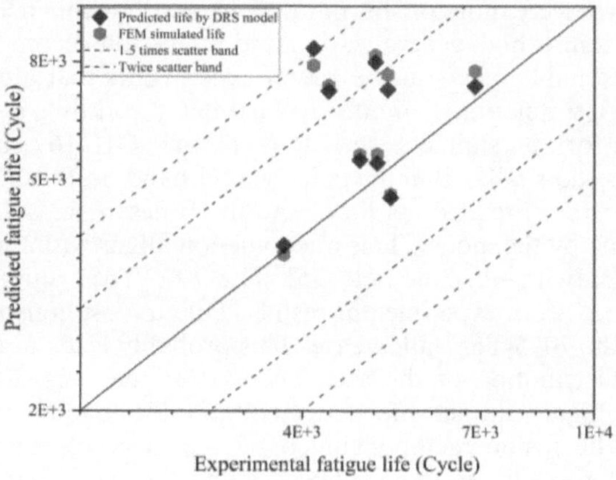

Figure 4. The performance of the DL-LGBM-DRS model in the prediction of fatigue life.

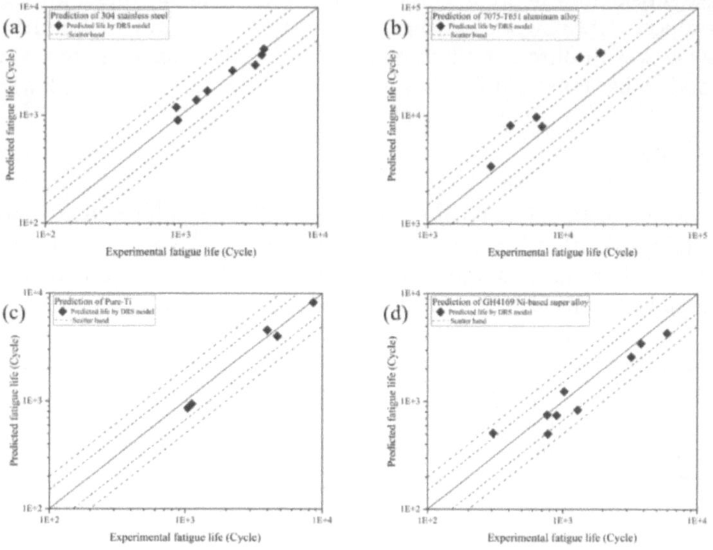

Figure 5. The performance of DL-LGBM-DRS model in fatigue life prediction of other metals.

3.3 *The prediction of fatigue life*

In this section, the fatigue life predicted by the proposed framework is analyzed and discussed. The fitness of results is demonstrated, and the performance is evaluated using 1.5 times and twice scatter bands. Figure 4 shows the comparison of the predicted and calculated fatigue lives and experimental lives in TC4 samples. Of particular note is that the prediction in this chapter is a prolongation of the FEA simulation, and a combination of the accuracy of the FEA simulation and the effectiveness of the surrogate model determines the accuracy of the predicted values. It can be demonstrated that the prediction capability for overall cases is excellent, as shown in all the results that fall within the twice-scatter band. In the case of several constituted loading conditions, the fitness of the results is remarkably high, and the prediction lives are practically identical to the experimental lives. This is attributed to the relative concentration of test results and the highly linearized relationship between the physical tests involved, as described previously. In scenarios with other multiaxial loading paths, the predicted lives were more off-fit, but they were still within acceptable limits. Since the surrogate model exhibits high generalization and prediction accuracy, it can be assumed that this is due to reasonable biases in the actual experiments that are not considered and inevitable by the FEA simulations. Figure 5 illustrates the fitting results for other alloy materials. The results for 304 stainless steel, Pure-Ti, and GH4169 super alloy fitted reasonably well; the fitness dots fall within the twice scatter band. In the GH4169 case, it can be observed that most of the fitted points lie below the fitness line, indicating a pessimistic prediction of fatigue life by the model. This phenomenon aligns with the trend in fatigue life calculated using the Brown-Miller method [25]. The 7075-T651 aluminum alloy case has deviations of predictions from experimental results. Due to cost constraints, we conducted fewer experiments in the 7075-T651 alloy case. This probably leads to the consequence of a certain probabilistic distribution of the experimental fatigue lives. Moreover, 7075-T651 alloy demonstrates a higher fatigue life under comparable loading conditions due to its material properties. The fusion method guided by the physical model is inadequate for predicting high-cycle fatigue life in this context. Due to the exotic introduction of material mechanical properties in this experiment to save costs, there will be minor differences in the material properties, leading to discrepancies in the final predictions. In summary, the proposed method is satisfactory in predicting the low cycle fatigue life under multiaxial loading. Further, the proposed framework delivers results that are a tight fit for the experimental fatigue lives in cases with different alloys employed. This reveals the potential ability of the proposed method to be replicated for fatigue life prediction of other metallic materials.

4 CONCLUSION

In this paper, a surrogate model based on deep learning, LightGBM, and a dynamic regression selection algorithm is proposed. This fusion model, DL-LGBM-DRS, has better prediction accuracy than the standalone model. With the highly reliable datasets generated by the FEA model, the model can provide an accurate prediction of fatigue life under multiaxial loading conditions. The fit points for predicted and experimental fatigue life are within the twice-scatter bands. Predictions of low-cycle fatigue life under multiaxial loading for other alloys have additionally been implemented. The results suggest that the DL-LGBM-DRS model is capable of performing fatigue life prediction at a much faster speed while maintaining high accuracy. It is simultaneously a potentially generalized model that can be used for the prediction of other metals and alloys.

ACKNOWLEDGEMENT

This work was supported by the Aeronautical Science Foundation of China under Grant 2022Z057052002, the Key Laboratory of Intelligent Decision and Digital Operations, Ministry of Industrial and Information Technology under Grant NJ202307.

REFERENCES

[1] Tabandeh A., Sharma N. and Gardoni P. (2022). Uncertainty propagation in risk and resilience analysis of hierarchical systems. *Reliab Eng Syst Saf*;219:108208.

[2] Bichon B. J., McFarland J. M. and Mahadevan S. (2011). Efficient surrogate models for reliability analysis of systems with multiple failure modes. *Reliab Eng Syst Saf*;96:1386–95.

[3] Bai S., Li Y.-F., Huang H.-Z., Ma Q. and Lu N. (2022). A probabilistic combined high and low cycle fatigue life prediction framework for the turbine shaft with random geometric parameters. *Int J Fatigue*;165:107218.

[4] He J., Huang M., Wang W., Wang S. and Guan X. (2021). An asymptotic stochastic response surface approach to reliability assessment under multi-source heterogeneous uncertainties. *Reliab Eng Syst Saf*;215:107804.

[5] Xu Y., Kohtz S., Boakye J., Gardoni P. and Wang P. (2023). Physics-informed machine learning for reliability and systems safety applications: State of the art and challenges. *Reliab Eng Syst Saf*;230:108900.

[6] LeCun Y., Bengio Y. and Hinton G. (2015). Deep learning. *Nature*;521:436–44.

[7] Xu Z. and Saleh J. H. (2021). Machine learning for reliability engineering and safety applications: Review of current status and future opportunities. *Reliab Eng Syst Saf*;211:107530.

[8] Saraygord Afshari S., Enayatollahi F., Xu X. and Liang X. (2022). Machine learning-based methods in structural reliability analysis: A review. *Reliab Eng Syst Saf*;219:108223.

[9] Wang R.-Z., Gu H.-H., Zhu S.-P., Li K.-S., Wang J., Wang X.-W., *et al.* (2022). A data-driven roadmap for creep-fatigue reliability assessment and its implementation in low-pressure turbine disk at elevated temperatures. *Reliab Eng Syst Saf*;225:108523.

[10] Roy A. and Chakraborty S. (2020). Support vector regression based metamodel by sequential adaptive sampling for reliability analysis of structures. *Reliab Eng Syst Saf*;200:106948.

[11] Karniadakis G. E., Kevrekidis I. G., Lu L., Perdikaris P. Wang S. and Yang L. (2021). Physics-informed machine learning. *Nature Reviews Physics*;3:422–40.

[12] Lu L., Jin P., Pang G., Zhang Z. and Karniadakis G. E. (2021). Learning nonlinear operators via DeepONet based on the universal approximation theorem of operators. *Nat Mach Intell*;3:218–29.

[13] Hu M., Li W., Yan K., Ji Z. and Hu H. (2019). Modern machine learning techniques for univariate tunnel settlement forecasting: A comparative study. *Math Probl Eng*;2019:7057612.

[14] Lin P., Zhang L. and Tiong R. L. K. (2023). Multi-objective robust optimization for enhanced safety in large-diameter tunnel construction with interactive and explainable AI. *Reliab Eng Syst Saf*;234:109172.

[15] Olleak A. and Xi Z. (2020). Calibration and validation framework for selective laser melting process based on multi-fidelity models and limited experiment data. *Journal of Mechanical Design*;142.

[16] Li B., Zhu J., Zhao X. and Yao H. (2020). Residual stress prediction in laser shock peening induced LD-TC4 alloy by data-driven ensemble learning methods. *Opt Laser Technol*x;176:110946.

[17] Ke G., Meng Q., Finley T., Wang T., Chen W., Ma W., *et al.* (2017). *LightGBM: A Highly Efficient Gradient Boosting Decision Tree.*

[18] Wang S. and Ma J. (2023). A novel GBDT-BiLSTM hybrid model on improving day-ahead photovoltaic prediction. *Sci Rep*;13:15113.

[19] Jing Y., Hu H., Guo S., Wang X. and Chen F. Short-term prediction of urban rail transit passenger flow in external passenger transport hub based on LSTM-LGB-DRS. *IEEE Transactions on Intelligent Transportation Systems* 2020;PP:1–11.

[20] Cruz R. M. O., Sabourin R. and Cavalcanti G. D. C. (2023). META-DES. Oracle: Meta-learning and feature selection for dynamic ensemble selection. *Information Fusion*x;38:84–103.

[21] Cruz R. M. O., Sabourin R., Cavalcanti G. D. C. and Ing Ren T. (2015). META-DES: A dynamic ensemble selection framework using meta-learning. *Pattern Recognit*;48:1925–35.

[22] Mendes-Moreira João and Jorge AM and SC and de SJF. (2009). Ensemble learning: A study on different variants of the dynamic selection approach. In: Perner P, editor. *Machine Learning and Data Mining in Pattern Recognition*, Berlin, Heidelberg: Springer Berlin Heidelberg; p. 191–205.

[23] Partalas I., Tsoumakas G., Hatzikos E. V. and Vlahavas I. (2008). Greedy regression ensemble selection: Theory and an application to water quality prediction. *Inf Sci (N Y)*;178:3867–79.

[24] Ho T. K. (1998). The random subspace method for constructing decision forests. *IEEE Trans Pattern Anal Mach Intell*;20:832–44.

[25] Liu J., Ran Y., Wei Y. and Zhang Z. (2021). A critical plane-based multiaxial fatigue life prediction method considering the material sensitivity and the shear stress. *International Journal of Pressure Vessels and Piping*;194:104532.

Recent Advances in Reliability and Maintenance Modeling – Okamura, Inoue & Xiao (Eds)
© 2025 the Author(s), ISBN: 978-1-032-79098-5

Estimating metrics-based software reliability models with elastic net

Y. Li, H. Okamura, J. Zheng & T. Dohi
Graduate School of Advanced Science and Engineering, Hiroshima University, Japan

ABSTRACT: This paper presents a novel framework for Software Reliability Growth Models (SRGMs) that incorporates software metrics. We integrate classical Poisson regression-based fault prediction with Non-Homogeneous Poisson Process (NHPP)-based SRGMs. A key feature of our approach is the simultaneous handling of time-series data for fault detection and software metrics across multiple modules. We outline a modeling framework that combines Poisson regression-based fault prediction with SRGMs and develop an efficient parameter estimation algorithm using the Expectation-Maximization (EM) method.

1 INTRODUCTION

Software reliability is a critical factor for high-reliability systems. Over the past four decades, numerous studies have proposed software reliability growth models (SRGMs) to quantitatively assess software reliability based on fault data observed during the testing phase. These traditional SRGMs model the dynamic behavior of fault detection, estimating model parameters from the fault detection history of a given software project and providing predictions for the number of faults likely to be detected in the future. Because SRGMs primarily require data on the number of faults during the testing phase, they are highly tractable and have been widely used to quantitatively assess software reliability.

However, a notable limitation of SRGMs is that they rely solely on fault detection data. In reality, software faults are influenced by a variety of environmental factors. Factors related to the software code, such as the size of the codebase and the complexity of the program logic, as well as factors related to the software process, such as test coverage and review results, collectively contribute to the occurrence of software faults. These complex factors interact to introduce faults, highlighting the need for SRGMs to integrate broader environmental considerations to improve their predictive accuracy and reliability.

There are several challenges associated with using software features extracted directly from the software itself. Okamura & Dohi (2014) proposed a novel software reliability framework that integrates traditional SRGMs with Poisson regression, termed PR-NHPP, to elucidate the relationship between the number of inherent faults and software features. A significant aspect of PR-NHPP is its ability to simultaneously handle time-series data for fault detection and software metrics across multiple modules. However, a common issue arises because the number of software features typically exceeds the number of software modules, posing a challenge in effectively determining the coefficients of Poisson regression. In this paper, we explore regularization as a solution for PR-NHPP. Specifically, we discuss the applicability of the elastic net, which is a generalization of the well-known L1 and L2 regularization techniques. This method combines the benefits of both to provide a more robust framework for dealing with the dimensionality and collinearity of software metrics in the context of software reliability modeling.

 DOI: 10.1201/9781003491309-12

Table 1. Typical NHPP-based SRGMs.

NHPP-based SRGMs	Distribution
Goel & Okumoto model 14	Exponential
Gamma model 12	Gamma
Modified Duane model 6	Pareto
Truncated-normal model 10	Truncated-normal
Log-normal model 1, 10	Log-normal
Infection S-shaped model 7	Truncated-logistic
Log-logistic model 5	Log-logistic
Gompertz model 8	Truncated-extreme-value-max
Frechet model 8	Log-extreme-value-max
Gompertz model 8	Truncated-extreme-value-min
Generalized exponential 3	Weibull
Phase-Type model	Phase-Type

2 SOFTWARE RELIABILITY GROWTH MODEL

2.1 *NHPP-based software reliability growth model*

A software reliability growth model (SRGM) is a stochastic model used to represent the number of faults during software testing, originally developed to quantify software reliability that is defined as the probability that software will not fail within a specified time period. Nowadays, SRGMs also assess the maturity of software testing by estimating the remaining faults, which can guide decisions about software release timing, etc.

SRGMs are classified into different types based on their mathematical foundations. This paper focuses on SRGMs characterized by non-homogeneous Poisson processes (NHPPs) with finite number of faults, which are favored for their mathematical tractability. NHPP-based SRGMs effectively capture the dynamic behavior of fault detection during the software testing phase.

Langberg and Singpurwalla introduced a modeling framework that encompasses nearly all NHPP-based software reliability growth models (SRGMs) under two key assumptions:

- (A-1) The total number of software faults is modeled as a Poisson random variable N with mean ω.
- (A-2) The times at which faults are detected are independent and identically distributed (IID) random events, each described by a cumulative distribution function (CDF) $F(t; \mathbf{q})$, where $\mathbf{q}$ is a parameter vector.

Let $\{X(t); t \geq 0\}$ be the cumulative number of software faults detected up to time t. Provided that the total number of software faults is $N = n$, the conditional probability mass function (p.m.f.) of $X(t)$ is described by the binomial distribution:

$$P(X(t) = x|N = n) = \binom{n}{x} F(t; \theta)^x \overline{F}(t; \theta)^{n-x}, \quad x = 0, 1,; n, \tag{1}$$

where $\overline{F}(t; \mathbf{q}) = 1 - F(t; \mathbf{q})$. Since N follows a Poisson distribution with mean ω, the unconditional p.m.f. of $X(t)$ is

$$P(X(t) = x) = \frac{(\omega F(t; \mathbf{q}))^x}{x!} e^{-\omega F(t;\mathbf{q})}, \quad x = 0, 1, \ldots \tag{2}$$

This formula states that the cumulative number of faults follows a NHPP with mean $\omega F(t;\mathbf{q})$, whose mean eventually converges to the expected total number of faults ω. Furthermore, NHPP-based SRGMs with finite number of faults are primarily characterized by the fault detection time distribution $F(t)$, which determines the pattern of fault detection. Table 1 lists typical NHPP-based SRGMs and their associated fault detection time distributions.

In NHPP-based SRGMs, the model parameters are estimated using fault detection data, typically a time series of the number of faults detected during the test. The data are represented as $\mathcal{D} = \{(t_1, x_1), \ldots, (t_K, x_K)\}$, where each pair (t_k, n_k) corresponds to the test interval and the number of defects detected in the k-th test interval. Maximum likelihood estimation (MLE) is often used in this context. The log-likelihood function for NHPP-based SRGMs can be expressed as

$$\mathcal{L}(\omega, \mathbf{q}) = \sum_{k=1}^{K} n_k \log \omega + \sum_{k=1}^{K} n_k \log(F(u_k;\mathbf{q}) - F(u_{k-1};\mathbf{q})) - \sum_{k=1}^{K} \log n_k! - \omega F(u_K;\mathbf{q}) \quad (3)$$

where $u_k = \sum_{l=1}^{k} t_l$. The MLE for NHPP-based SRGMs is obtained by maximizing this log-likelihood function. Since the function is non-linear with respect to the parameters ω and $\mathbf{q}$, numerical techniques are required to find the MLE.

Okamura *et al.* (2003) presented the use of the expectation-maximization (EM) algorithm for this purpose in NHPP-based SRGMs. The EM algorithm provides stable MLEs over different types of fault detection data compared to general purpose maximization techniques such as Newton's method or the Nelder-Mead simplex method.

2.2 *NHPP-based SRGM with Poisson regression*

The primary weakness of the conventional SRGM is its inability to incorporate information beyond fault detection data. Even when information about the software structure and other relevant aspects is available, the estimation of the fault detection process relies solely on the project's own historical fault detection data. While this limitation can be advantageous in some cases, it also presents a significant drawback. When fault detection data is available for multiple software modules, the SRGM requires that the fault detection process be estimated independently for each module, ignoring any potential correlations or insights that might be gained from analyzing the data. This approach fails to take full advantage of the available information, potentially resulting in less accurate predictions and a more fragmented understanding of the overall software reliability growth process.

To address this issue, Okamura & Dohi (2014)introduced the NHPP-based SRGM with Poisson regression (PR-NHPP). The main idea behind this model is to combine Poisson regression on the total number of faults with the NHPP-based SRGM, which assumes a finite number of faults. The model is constructed based on the following assumptions:

- (B-1) The total number of software faults in the i-th software module is a Poisson random variable N_i with mean $\mu(\mathbf{x}_i)$, where:

$$\log \mu(\mathbf{x}_i) = \mathbf{b}^{\mathrm{T}}\mathbf{x}_i, \quad (4)$$

  and $\mathbf{x}_i$ is a vector of software metrics for the i-th module.
- (B-2) All fault detections in the i-th module occur at independent and identically distributed (IID) random times, which follow a cumulative distribution function (CDF) $F_i(t;\mathbf{q}_i)$, where $\mathbf{q}_i$ is a parameter vector of the fault-detection time distribution.

The essential difference from Langberg and Singpurwalla's assumption is that the expected total number of faults is modeled by the regression with software metrics.

Under these assumptions, the p.m.f. for the cumulative number of faults detected in the i-th module by testing time t is modeled as follows:

$$P(X_i(t) = x) = \frac{(\mu(\mathbf{x}_i)F_i(t; \mathbf{q}_i))^x}{x!} e^{-\mu(\mathbf{x}_i)F_i(t;\mathbf{q}_i)}, \quad x = 0, 1, \ldots \tag{5}$$

This expression follows the form of an NHPP, where $\mu(\mathbf{x})$ represents the total number of faults modified by software metrics specific to module i, and $F_i(t; \mathbf{q}_i)$ is the fault detection function parameterized by $\mathbf{q}_i$. While the form of the p.m.f. maintains the characteristic structure of an NHPP, the total number of faults across different modules are interconnected due to the influence of software metrics, indicating a correlation in fault occurrences among modules. In (Okamura & Dohi 2014), Okamura and Dohi developed the EM algorithm for PR-NHPP to obtain MLE from observed fault detection data and software metrics, by combining the EM procedure for the ordinary SRGMs and the parameter estimation for generalized linear model (GLM).

3 PR-NHPP WITH ELASTIC NETS

3.1 *Parameter estimation*

This paper proposes a variant of the PR-NHPP model by incorporating elastic net regularization. PR-NHPP combines SRGMs with Poisson regression. However, Poisson regression may become inefficient when the number of software metrics exceeds the number of software modules. Typically, software projects have a large number of metrics, such as lines of code and complexity measures, while the number of modules in practical scenarios rarely exceeds ten. As a result, the standard PR-NHPP model often requires variable selection using criteria such as the Akaike Information Criterion (AIC) to avoid overfitting.

An alternative approach to dealing with overfitting is regularization. Well-known regularization techniques include Ridge and Lasso regression, which impose penalties on the L2 and L1 norms of the regression coefficients, respectively, to control the magnitude of these coefficients. The elastic net is a hybrid of these methods, effectively blending Ridge and Lasso penalties, and serves as a generalization of both, providing a robust framework for handling cases with numerous metrics relative to the number of modules.

Suppose there are m software modules and their software metric vectors $\mathbf{x}_1, \ldots, \mathbf{x}_m$ are measured. Let $\mathcal{D}_i = \{(t_{i,1}, n_{i,1}), \ldots, (t_{i,K_i}, n_{i,K_i})\}$ be the fault detection data for i-th module, i.e., a time series of the number of faults detected in the i-th module, where $(t_{i,k}, n_{i,k})$ denotes a pair of the interval time of the k-th test and the number of faults detected by the k-th test in the i-th module. For given observed data $\mathbf{x}_1, \ldots, \mathbf{x}_m$ and $\mathcal{D}_1, \ldots, \mathcal{D}_m$, the log-likelihood function is formulated as follows.

$$\begin{aligned}\mathcal{L}(\beta, \theta_1, \ldots, \theta_m) = &\sum_{i=1}^{m}\sum_{k=1}^{K_i} n_{i,k} \log \mu(\mathbf{x}_i) + \sum_{i=1}^{m}\sum_{k=1}^{K_i} n_{i,k} \log\big(F_i(u_{i,k}; \theta_i) - F_i(u_{i,k-1}; \theta_i)\big) \\ &- \sum_{i=1}^{m}\sum_{k=1}^{K_i} \log n_{i,k}! - \sum_{i=1}^{m} \mu(\mathbf{x}_i) F_i(u_{i,K}; \theta_i),\end{aligned} \tag{6}$$

where $u_{i,k} = \sum_{l=1}^{k} t_{i,l}$. In the context of MLE, we find the model parameter vector that maximizing the above log-likelihood function, i.e.,

$$(\widehat{\beta}, \widehat{\theta}_1, \ldots, \widehat{\theta}_m) = \underset{\beta, \theta_1, \ldots, \theta_m}{\operatorname{argmax}} \mathcal{L}(\beta, \theta_1, \ldots, \theta_m). \tag{7}$$

In the case of elastic net, we find the solution of maximization problem

$$(\widehat{\beta}, \widehat{\theta}_1, \ldots, \widehat{\theta}_m) = \underset{\beta,\theta_1,\ldots,\theta_m}{\operatorname{argmax}} \mathcal{L}(\beta, \theta_1, \ldots, \theta_m) + \lambda\left(\frac{1-\alpha}{2}\|\beta\|_2^2 + \alpha\|\beta\|_1\right), \tag{8}$$

where $\lambda \geq 0$ and $0 \leq \alpha \leq 1$ are hyperparameters to determine the magnitude of regularization.

Next we develop the EM algorithm for PR-NHPP with elastic net to obtain the solution of Equation (8). The EM algorithm is an iterative method for computing ML estimates with incomplete data (Dempster *et al.* 1977). In general, let $\mathcal{D}$ and $\mathcal{U}$ be observable and unobservable data, respectively, and we wish to estimate a model parameter vector λ from the observable data $\mathcal{D}$.

The EM algorithm consists of E-step and M-step. E-step computes the expectation of log-likelihood function of the complete data $(\mathcal{D},\mathcal{U})$ with the distribution for unobservable data when a provisional parameter vector λ' is given. Formally, the expected log-likelihood function to be computed in the E-step can be written as

$$Q(\lambda|\lambda') = \int p(\mathcal{U}|\mathcal{D};\lambda')\log p(\mathcal{D},\mathcal{U};\lambda)d\mathcal{U}, \tag{9}$$

where $p(\cdot)$ is any probability density or mass function. In the above equation, $p(\mathcal{U}|\mathcal{D};\lambda')$ and $p(\mathcal{D},\mathcal{U};\lambda)$ correspond to the p.d.f. of unobserved data and the complete likelihood function, respectively. Furthermore, the p.d.f. of unobserved data can be derived from Bayes theorem:

$$p(\mathcal{U}|\mathcal{D};\lambda') = \frac{p(\mathcal{D},\mathcal{U};\lambda')}{\int p(\mathcal{D},\mathcal{U};\lambda')d\mathcal{U}}. \tag{10}$$

On the other hand, in the M-step, we find a new parameter vector λ'' that maximizes the expected log-likelihood function, i.e.,

$$\lambda'' := \underset{\lambda}{argmax}\ Q(\lambda|\lambda'). \tag{11}$$

The parameter vector λ'' becomes provisional parameters in the next E- and M-steps. The one EM step updates model parameters and ensures that the log-likelihood function increases. The EM steps are repeatedly executed until model parameters converges.

In the case of PR-NHPP, the unobserved data $\mathcal{U}$ is defined as a set of the total number of faults $N_1, \ldots, N_m$ and the exact fault detection time sequence for all faults $S_{i,1} < S_{i,2} < \cdots < S_{i,N_i}$ for all $i = 1,\ldots,m$. If we know the total number of faults, the coefficients of the Poisson regression can be obtained by the ordinary calculation technique to calculate the coefficients of the Poisson regression. Then the problems of calculating the SRGM parameters and the Poisson regression coefficients can be separated.

For SRGM parameters, we can apply the EM algorithm for ordinary SRGM, provided that the coefficients of the Poisson regression are given (Okamura *et al.* 2003). Also, since the expected total number of faults is given by

$$E[N_i] = \sum_{k=1}^{K_i} n_k + e^{\beta^T \mathbf{x}_i}\overline{F}_i(s_{i,K_i};\theta_i). \tag{12}$$

The coefficients of the Poisson regression can be obtained by solving the ordinary Poisson regression for the expected total number of faults. This procedure continues until the parameters of the SRGM and the coefficients converge. In the case of elastic net, we only consider the regularization for the coefficients of Poisson regression, and thus the procedure for calculating the coefficients is replaced by the elastic net algorithm.

Finally, for given hypter parameters λ and α, the EM steps for PR-NHPP is presented in Algorithm 1, where the functions last and sumn are designed to process fault detection data. Specifically, last retrieves the last time of the observation period, i.e., s_{i,K_i}, and sumn calculates the total number of observed faults during that period, i.e., $\sum_{k=1}^{K_i} n_k$. Also the functions EM and PR are utilized for parameter estimation. The EM function performs one iteration of the EM step to update the SRGM parameters, while PR calculates the coefficients of Poisson regression based on given hyperparameters.

Algorithm 1: EM-step for PR-NHPP model with elastic net.

Regression coefficients	`beta`
Distribution parameters	`theta[1], ..., theta[m]`
Module metrics	`x[1], ..., x[m]`
Fault detection data	`fd[1], ..., fd[m]`
CDF for fault detection time distribution	`F[1], ..., F[m]`

```
for i = 1:m
  omega[i] <- exp(beta * x[i])
  Fbar <- 1 - F[i](last(fd[i]),     theta[i])
  N[i] <- sumn(fd[i]) + omega[i] * Fbar theta[i] <- EM(omega[i],
  theta[i], fd[i], F[i])
end
beta <- PR(N[1], ..., N[m], x[1], ... x[m], gamma, alpha)
```

Table 2. Software modules.

Module	Faults 7/2009	7/2012	Description
Catalina	237	272	The servlet container core
Connector	75	89	Coyote connectors
Jasper	70	74	The JSP (JavaServer Pages) compiler and runtime engine
Servlets	51	57	Servlet API and support programs for CGI, SSI and WebDAV
Tester	1	1	Unit testing framework
Webapps	53	65	Web application for administration, documentation and examples.

3.2 *Determining hyperparameters*

Next, we consider the process of determining the hyperparameters λ and α. In general, the choice of hyperparameters is critical because they directly affect the strength of the regularization and the complexity of the model in predictive modeling.

Hyperparameter tuning can be approached in several ways, depending on the sensitivity of the model to these parameters and the desired balance between bias and variance. Common methods include Grid search, Random search and Bayes optimization. The grid search involves systematically varying the hyperparameters over a predefined range and evaluating the model's performance for each combination. The random search randomly selects combinations of hyperparameters to test, which can be more efficient and often produces better results when dealing with a high-dimensional space. Bayes optimization is also called the black-box optimization. This technique uses a probabilistic model to predict the performance of different hyperparameter settings and intelligently explores more promising settings. It is particularly useful for optimizing complex models because it can significantly reduce the number of evaluations required. In this experiment, we use a package mlrMBO for Bayes optimization in R.

In all approaches, it is necessary to have a measure to evaluate the effectiveness of given hyperparameters. In this paper, the prediction error is used as a criterion to evaluate the performance of hyperparameters. The methodology involves dividing the original data set into training and test subsets at a given time. The model parameters, except the hyperparameters, are then estimated from the training data. The selection of hyperparameters is guided by the prediction error, which is calculated as the difference between the predicted number of future faults derived from the estimated parameters and the actual number of faults observed in the test data. This method ensures that the chosen hyperparameters optimize the model's ability to predict new, unseen data, thereby increasing its practical applicability and reliability.

4 EXPERIMENT

In this paper, we conduct the numerical experiment to validate the performance of PR-NHPP with elastic net. The fault detection data are collected from the Apache Tomcat project in ASF Bugzilla[1]. Apache Tomcat provides Java servlets and its components are written in Java and C languages. The experiments deal with the versions of Apache Tomcat 5.5.12 to 5.5.36, since we choose fully mature application. In the experiments, we collected reports for minor, normal, major and critical bugs detected in 6 modules; Catalina, Connectors, Jasper, Servlet, Tester and Webapps. Each module consists of many Java and C programming files. These bugs were reported between May 2004 and July 2012. Table 2 summarizes the software modules and their total number of bugs reported up to July 2009 and July 2012, respectively. From this table, almost all bugs have been fixed by July 2009, which is an intermediate point between May 2004 and May 2013, and thus all current modules are considered stable in terms of bug fixes.

Next we measure software metrics for modules. Table 3 presents these metrics for all the modules in Tomcat version 5.5.12. The software metrics are lines of code (LOC), the number of statements (St), the percent branch statements (Br), the percent lines with comments (Co), the number of functions (Fn), the maximum McCabe complexity (Mc) and the average McCabe complexity (Ac).

Table 3. Code metrics.

Module	LOC	St	Br	Co	Fn	Mc	Ac
Catalina	125462	39386	19.2	34.4	4543	90	2.79
Connector	168482	65829	19.2	26.9	5912	175	3.37
Jasper	41861	15557	19.2	26.8	1691	90	2.88
Servlets	15480	1944	8.0	66.5	400	26	1.45
Tester	12677	4750	19.7	30.2	409	54	3.36
Webapps	37782	12851	9.5	31.2	1035	39	2.32

Table 4. Estimated coefficients.

	λ	α	SE	LOC	St	Br	Co	Fn	Mc	Ac
L1 + Gamma	5.9	1.0	57.5	3.9	0.0	5.7e-2	0.0	2.8e-4	-1.1e-2	0.0
L2 + Gamma	15.0	0.0	64.1	3.5	1.4e-5	1.7e-2	4.4e-3	2.0e-4	-5.4e-5	2.7e-2
EN + Gamma	8.5	0.6	62.6	4.0	7.1e-7	0.0	0.0	3.2e-4	0.0	0.0
L1 + PH5	3.3	1.0	32.9	3.8	0.0	0.0	0.0	3.7e-4	0.0	0.0
L2 + PH5	7.2	0.0	34.3	3.5	1.6e-5	1.1e-2	4.8e-3	2.3e-4	-6.9e-4	5.1e-3
EN + PH5	4.4	0.5	33.7	3.8	8.6e-6	0.0	2.9e-4	2.9e-4	0.0	0.0

[1]https://bz.apache.org/bugzilla/

The original dataset consists of the number of faults detected over a period of 108 months. This data set is segmented into an initial period of 80 months and a subsequent period of 28 months. Model parameters are estimated using data from the first 80 months. Predictions are then computed for the number of faults in the last 28 months for each module. To fine-tune the hyperparameters, the initial 80-month period is further divided into the first 74 months and the last 6 months. The hyperparameters are then adjusted to minimize the squared prediction errors for the last 6 months. In addition, the module Tester is excluded from the analysis because it has only one fault recorded.

Table 4 presents the estimated coefficients of Poisson regression with regularization. This experiment considers two scenarios based on the underlying SRGM:

- Gamma-type model (Gamma): All modules are modeled using a gamma distribution. The Gamma model is well-regarded for its flexibility and ability to fit a wide range of practical software fault patterns.
- Phase-type model with 5 phases (PH5): All modules are modeled using a Phase-type distribution with 5 phases. PH5 is a semi-parametric model known for its high data fitting capability, covering almost all software fault patterns beyond Gamma.

In the table, L1, L2, and EN stand for L1 regularization, L2 regularization, and elastic net, respectively. Furthermore, λ and α are the hyperparameters determined by Bayes optimization. SE indicates the minimum squared prediction error for the 6 months, which is the criterion used to determine the hyperparameters. From the table, we can see that L1 and EN result in lower squared errors compared to L2. This indicates that both L1 and EN are more effective in improving the predictive accuracy of the model for this particular data set. In addition, L1 and EN have the unique ability to set some regression coefficients to exactly zero. This feature is particularly useful for feature selection and model simplification, as it helps to identify and eliminate non-influential predictors from the model. In addition, the analysis shows that the squared errors are significantly affected by the choice of the underlying SRGM. Although the Gamma model is celebrated for its high flexibility, it appears that its data fitting capabilities may not be sufficient to accurately predict failure detection patterns across all modules. This finding highlights the importance of selecting an appropriate SRGM that fits well with the specific characteristics of the defect data in each module to achieve optimal predictive performance.

Next, we present the predictions for the number of faults over the last 28 months, using the models we've estimated. Table 5 illustrates the absolute errors between the predicted values and the actual number of faults detected. Except for the predictions pertaining to the Connector module, the forecasts closely align with the actual fault counts. This accuracy is particularly significant given the relatively long duration of 28 months for making predictions. Notably, the prediction errors for PH5 are slightly smaller than those for Gamma. This difference underscores the importance of selecting an appropriate underlying SRGM, as it can significantly impact prediction accuracy.

Table 5. Prediction results.

	Catalina	Connector	Jasper	Servlets	Webapps
L1 + Gamma	9.37	2.96	0.31	1.78	2.99
L2 + Gamma	8.63	37.75	0.25	1.81	1.90
EN + Gamma	8.72	36.77	0.29	1.80	3.00
L1 + PH5	5.73	30.62	0.25	1.56	2.08
L2 + PH5	5.68	31.39	0.27	1.53	1.76
EN + PH5	5.79	31.29	0.25	1.56	2.43

5 CONCLUSION

This paper considered the PR-NHPP, which is a combined model of SRGM with Poisson regression. In particular, we have focused on the regularization using elastic net. To obtain the model parameters, we have developed the EM algorithm, which takes into account the regularization. And we have discussed how to determine the hyperparameters. In the experiment, we have demonstrated the prediction of the number of faults using PR-NHPP with open source data. In future, we will consider the approach to enhance the prediction ability in the framework of PR-NHPP.

REFERENCES

Achcar, J. A., D. K. Dey and M. Niverthi. (1998). A Bayesian approach using nonhomogeneous Poisson processes for software reliability models. In A. P. Basu, K. S. Basu, and S. Mukhopadhyay (Eds.), *Frontiers in Reliability*, pp. 1–18. Singapore: World Scientific.

Dempster, A. P., N. M. Laird and D. B. Rubin. (1977). Maximum likelihood from incomplete data via the EM algorithm. *J. Roy. Stat. Soc. B B-39*: 1–38.

Goel, A. L. (1985). Software reliability models: Assumptions, limitations and applicability. *IEEE Transactions on Software Engineering SE-11*: 1411–1423.

Goel, A. L. and K. Okumoto. (1979). Time-dependent error-detection rate model for software reliability and other performance measures. *IEEE Transactions on Reliability R-28*: 206–211.

Gokhale, S. S. and K. S. Trivedi. (1998). Log-logistic software reliability growth model. In *Proc. 3rd IEEE Int'l. High-Assurance Systems Eng. Symp. (HASE-1998)*, pp. 34–41. IEEE CS Press.

Langberg, N. and Singpurwalla, N.D. (1985). Unification of some software reliability models. *SIAM Journal on Scientific Computing*, vol.6, no.3, pp. 781–790.

Littlewood, B. (1984). Rationale for a modified duane model. *IEEE Transactions on Reliability R-33*(2): 157–159.

Ohba, M. (1984). Inflection S-shaped software reliability growth model. In S. Osaki and Y. Hatoyama (Eds.), *Stochastic Models in Reliability Theory*, pp. 144–165. Berlin: Springer-Verlag.

Ohishi, K., H. Okamura and T. Dohi. (2009). Gompertz software reliability model: estimation algorithm and empirical validation. *Journal of Systems and Software 82*(3): 535–543.

Okamura, H. and T. Dohi. (2014). A novel framework of software reliability evaluation with software reliability growth models and software metrics. In *Proceedings of The 15th IEEE International Symposium on High Assurance Systems Engineering (HASE 2014)*, pp. 97–104. IEEE CPS.

Okamura, H., T. Dohi and S. Osaki. (2013). Software reliability growth models with normal failure time distributions. *Reliability Engineering and System Safety 116*: 135–141.

Okamura, H., Y. Watanabe and T. Dohi. (2003). An iterative scheme for maximum likelihood estimation in software reliability modeling. In *Proc. 14th Int. Symp. Software Rel. Eng.*, pp. : 246–256.

Okamura, H. and T. Dohi. (2006). Building Phase-Type Software Reliability Models *2006 17th International Symposium on Software Reliability Engineering*:289–298.

Pham, Hoang. (1999). *Software Reliability*. Heidelberg:Springer-Verlag.

Yamada, S., M. Ohba and S. Osaki. (1983). S-shaped reliability growth modeling for software error detection. *IEEE Transactions on Reliability R-32*: 475–478.

Recent Advances in Reliability and Maintenance Modeling – Okamura, Inoue & Xiao (Eds)
 ISBN: 978-1-032-79098-5

A heuristic improvement of anomalous sound detection by compound scaling in deep learning

Yu-Hsun Lin & Chang-Wen Wang
Department IEEM, National Tsing Hua University, Hsinchu, Taiwan (R.O.C.)

ABSTRACT: Anomalous sound detection task identifies the anomaly sound from a target event (e.g., factory machines). By detecting the abnormal signal in advance, we can conduct predictive maintenance to prevent the possible machine failures. Therefore, the factory activities can be guarded by incorporating intelligent acoustic sensor data processing. Therefore, the improvement of anomalous sound detection plays an important role for smart manufacturing (e.g., Industry 4.0). In order to improve the performance of anomalous sound detection, we utilize an influential concept called compound scaling from computer vision research domain. With the aid of integrating compound scaling, our proposed method outperforms other AI models in the experiments. In addition to the outstanding performance, the findings of applying the concept (i.e., compound scaling) from computer vision to acoustic data are also important assets for the future studies in cross domain applications.

1 INTRODUCTION

Anomalous sound detection (ASD) is the process of identifying abnormal sounds within a given environment. This process is often utilized in various applications such as security systems, monitoring equipment, and predictive maintenance in industrial machinery. There are research works (Foggia *et al.* 2015; Li *et al.* 2018) applied ASD to surveillance applications. For example, the sound of gunshot or breaking glass is a strong sign for security concerns in surveillance applications. On the other hand, ASD plays an important role in predictive maintenance. We can address the machinery failure issues in advance by detecting the abnormal sound from machines (Kawaguchi and Endo 2017). Once we catch the anomalous sound, we can apply predictive maintenance and prevent the equipment failure.

In order to evaluate the performance among various ASD AI models, there is a benchmark dataset called DCASE (Detection and Classification of Acoustic Scenes and Events) (Koizumi *et al.* 2020, Mai *et al.* 2022), that is designed for grading the detection results. Among the challenges in DCASE, the ASD task is evaluated in the task 2 of DCASE hold in 2020. Therefore, we denote the ASD task challenge as DCASE 2020 Task 2 in short.

Among the AI models, STgram (Liu *et al.* 2022) addressed ASD task by fusing the spectral and the temporal information of audio data and achieved outstanding results as compared with other AI models. The input feature tensor in STgram is organized by two matrix shaped data, Sgram and Tgram, that represent the spectral information and temporal information, respectively. Therefore, STgram addresses ASD task from the perspective in a computer vision alike way by formulating the 1D audio data into 2D matrix data. Hence, the novel techniques in computer vision domain would also have great research potential for addressing the ASD task. And this research opportunity inspires our proposed method in this work.

Among the novel techniques developed in computer vision domain, compound scaling (Tan and Le 2019) is an elegant but powerful approach for convolutional neural network

DOI: 10.1201/9781003491309-13

Table 1. The difference of AI tasks between computer vision and this work.

Task	Computer Vision Image Recognition	This Work ASD
# of Class	≥ 2	2 (normal, abnormal)
Data Diversity	High	Mostly normal data
Resize & Crop	Yes (down-sizing usually)	No

(CNN) models. This concept is firstly proposed in EfficientNet (Tan and Le 2019) that achieved state-of-the-art results in image recognition. However, the research of applying this concept to ASD task is near to none. Since the detection process of STgram in ASD task is similar to the process of CNN models in image recognition, STgram became a natural candidate as our target baseline model for integrating compound scaling. Furthermore, STgram is one of the few top-performing AI models that the source code is publicly available. Therefore, we investigate the potential of compound scaling in ASD task by incorporating the concept into STgram heuristically. As a result, our proposed framework is called Extended STgram by integrating compound scaling into the baseline STgram.

Table 1 demonstrates that the differences between the computer vision task and the ASD task in this work. Therefore, we cannot apply compound scaling blindly without taking these differences into considerations. The computer vision task has many classes and heterogeneous data distribution. The resizing and cropping is a common pre-processing in CNN models since the input image size is larger than that of the input size of CNN models. Hence, we have inherent information loss in the pre-processing stage for CNN models. Therefore, applying compound scaling for the CNN's input size is beneficial since we can preserve more detail from an input image.

On the other hand, the ASD task has mostly normal data and the pre-processing operations (i.e., resizing and cropping) are not required for ASD AI models. Therefore, the information loss is not introduced in the pre-processing stage for ASD task. As a result, we have to apply compound scaling carefully to the ASD AI models since the application scenario is highly different to the computer vision task. Hence, we investigate several heuristic design aspects of applying compound scaling to STgram, these results will become a stepping stone for the future studies that want to apply compound scaling in audio applications. Therefore, our contributions in this paper are summarized as below.

- We improve the ASD performance by applying compound scaling to input data. The upsampling of input audio data is beneficial, whereas overly data scaling will introduce negative impact for ASD task. This phenomenon echoed the observations in Table 1 since there is no additional loss introduced in pre-processing stage for ASD AI models.
- In addition to the common aspects for model scaling (e.g., depth, width), we found that dilated convolution is helpful for ASD task. Dilated convolution is a special type of convolution that we can perceive more data samples with the same number of parameters as compared with the normal convolution.
- We improve the performance of STgram by integrating compound scaling heuristically. We can discover more potential from single model performance by using the concept of compound scaling. The performance of our proposed Extended STgram is comparable to the top-ranking models that are using ensemble learning technique in DCASE 2020. Furthermore, our method with ensemble learning outperforms the top-ranking models in the experiments.

This work is organized as follows. Related work is discussed in Section 2. Our method is described in Section 3. The experimental results and ablation studies are illustrated in Section 4 and 5, respectively. We conclude this work in Section 6.

2 RELATED WORK

We reviewed several important research topics of ASD in this section. The DCASE 2020 Task 2 evaluates ASD task by providing a benchmark dataset. We also reviewed the top performing models in DCASE 2020 Task 2.

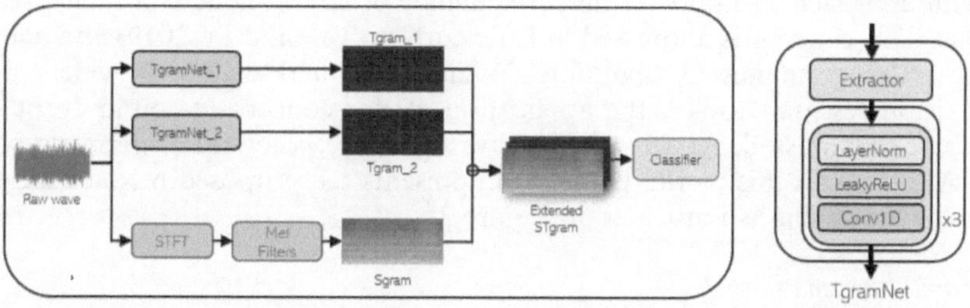

(a) (b)

Figure 1. (a) Sgram and Tgrams are extracted from the raw audio wave and ⊕ symbol denotes the concatenationof the Sgram and Tgrams. STFT denotes Short-time Fourier Transform. (b) Structure of TgramNet.

2.1 *Unsupervised learning for ASD*

Due to the difficulties of collecting abnormal data, conventional ASD usually tackles the problem with unsupervised approaches. The baseline model of DCASE 2020 Task 2 is an autoencoder based model (Koizumi *et al.* 2020). The research work Daniluk_SRPOL (Daniluk *et al.* 2020) wins second place in DCASE 2020 Task 2 is also an autoencoder based approach.

2.2 *Self-supervised learning for ASD*

Many top ranking algorithms adopt the self-supervised learning paradigm. For example, the winning algorithms Giri_Amazon (Giri *et al.* 2020) and Primus_CP-JKU (Primus 2020) win the first place and the third place in DCASE 2020 Task 2, respectively. The contrastive learning for ASD is leveraged in the study (Guan *et al.* 2023). The concept of wavegram for ASD is addressed in the research work (Chen *et al.* 2023) for fully exploiting the temporal information. A recent work addressed ASD task from the perspective of designing new margin loss for AdaCos called Sub-Cluster AdaCos (Wilkinghoff 2021). Since there are many configurations in Sub-Cluster AdaCos, we compared the version with machine learning for performance comparison. Because the top-performing results and the CNN-like operations of STgram research work (Liu *et al.* 2022), we adopt STgram as our baseline model for investigating the concept of compound scaling in ASD task.

2.3 *Compound scaling*

Compound scaling (Tan and Le 2019) is an important technique to improve the performance of AI models in computer vision. The concept of compound scaling is proposed in EfficientNet (Tan and Le 2019), which is a state-of-the-art CNN backbone for numerous computer vision applications. The design space of compound scaling involves the scaling of the input data size, the depth and the width of a neural network. The width of a neural network refers to the number of convolution filters in each layer, while the depth denotes the number of convolution layers in a neural network.

In this work, we leverage the concept of compound scaling for improving the performance of STgram. Due to the differences as listed in Table 1, we cannot apply compound scaling to STgram naively. The design aspects of our proposed Extended STgram are illustrated in the following paragraphs.

3 METHOD

The proposed method, Extended STgram, is described in this section. We integrate the concept of compound scaling into STgram, which improves the performance in ASD task.

We utilize STgram (Liu *et al.* 2022) as our base model since STgram addresses the audio clips from the perspective of CNN models. On the other hand, compound scaling is a simple but powerful approach to improve the performance of CNN models in image recognition applications. This concept is addressed in EfficientNet (Tan and Le 2019) and achieved the state-of-the-art performance at that time. While compound scaling is widely utilized in computer vision research topics, the application of this concept in sound event detection domain is rarely addressed. At a result, we investigate the potential of applying compound scaling to ASD task in this work. This section presents the proposed method for ASD, the framework of the system is illustrated in Figure 2.

3.1 *STgram: Baseline model*

STgram plays a role of our baseline model in this work. STgram consists of Sgram and Tgram that represents spectral information and temporal information, respectively. In order to address the issues of compound scaling of STgram, we denote the scaled version of STgram as Extended STgram in this work.

In STgram, we took the raw audio $x \in \mathbb{R}^{1 \times L}$ as input, then convert the audio data into two different feature representations, Sgram and Tgram. STFT is short for Short-time Fourier Transform. Sgram is the log-Mel spectrogram which is widely utilized in audio deep learning algorithms as feature representation to express time-domain and frequency-domain information. The log-Mel spectrogram $F_s \in \mathbb{R}^{M \times N}$ can be obtained as follows:

$$\mathbf{F}_S = \log\left(\mathcal{W}_M \cdot \|STFT(\mathbf{x})\|^2\right), \tag{1}$$

where M, N denotes the number of Mel-bins and time frames of the audio signal, $\mathcal{W}_M \in \mathbb{R}^{M \times B}$ denotes the Mel-filter banks and B is the number of frequency bins of the spectrogram.

Tgram, on the other hand, is a feature extracted by a CNN-based network that is complement to the information of Sgram. The dimension of Tgram is $F_T \in \mathbb{R}^{M \times N}$, which the size is the same as Sgram. Sgram and Tgram is concatenated to create a two-channel feature tensor, STgram. A classifier (e.g., MobileFaceNet in STgram (Liu *et al.* 2022)) will determine the status of audio clips, normal or anomalous, based on the feature tensor.

3.2 *Extended STgram: Compound scaling of STgram*

In order to address the compound scaling of STgram, we investigate the following aspects to improve the performance for ASD.

- Scaling of input data
  This aspect is a common element of compound scaling in image recognition. Therefore, we will address upsampling issue for the input audio clip in Extended STgram.
- Scaling of model
  Compound scaling of model will increase the depth and the width of a deep learning model. We will deal with the model scaling issue for STgram since we have an extra constraint that the feature size should be the same between Sgram and Tgram.
- Scaling of feature tensor
  We found that the return is diminishing rapidly if we just simply scaling the data sampling rate and the depth/width of deep learning model. We observed that scaling the channel number of the feature tensor is quite important to improve the performance of our proposed Extended STgram.

The overall framework of the proposed Extended STgram can be found in Figure 2.

3.2.1 *Scaling of input data*

The compound scaling in EfficientNet (Tan and Le 2019) will resize an image to a larger target size for achieving better recognition result. On the other hand, the sampling rate of an audio clip is similar to the role of resolution in an image. Inspired by the concept of compound scaling, higher sampling rate would bring more fine-grained patterns for our model to learn from. Therefore, we will upsample the input audio clip to improve the performance in sound event detection task. Empirically, we found that doubling the sampling rate (e.g., upsampling from 16*k* Hz to 32*k* Hz) of audio data is beneficial in this work. Although compounding scaling of input data is widely adopted in computer vision research works, we cannot apply this concept to audio signal blindly. Furthermore, the AI models have vastly different computation process between image data and audio data.

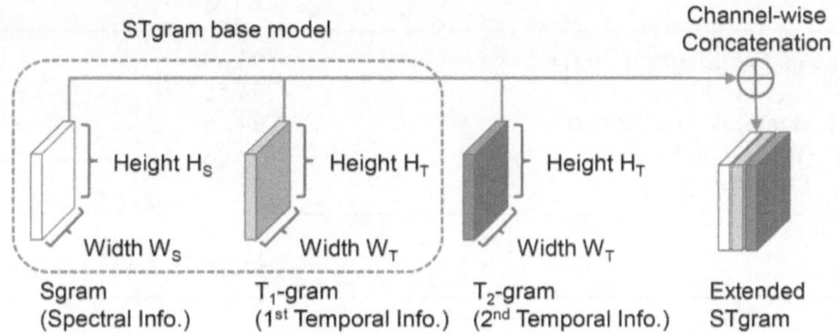

Figure 2. Illustration of the proposed Extended STgram and the relation among Sgram and Tgram.

3.2.2 *Scaling of model*

Figure 2 shows the relation among Sgram and Tgram in STgram. The size of Sgram are equal to the size of Tgram (i.e., $W_S = W_T = N$ and $H_S = H_T = M$) since we will concatenate Sgram and Tgram for our AI model. Hence, we have to change the size of Tgram if the size of Sgram is changed by audio clip upsampling.

In contrast to Sgram, the size of Tgram is determined by TgramNet as shown in Figure 1. The original TgramNet is designed to extract feature representations from audio data with 16*k* Hz. In order to improve the performance of sound event detection, we will upsample the audio data from 16*k* Hz to 32*k* Hz. Therefore, we have to modify the structure and proposed a new design of the TgramNet by adding extra depth and width to enhance the model ability to capture feature representations in audio clips with higher sampling rate.

Since the audio data is upsampled, we need more convolution layers or wider convolution kernel to cover the same original data samples. That is, we need a model with larger receptive field for the upsampled audio data. In order to address this issue of the upsampled audio data, we applied dilated convolution to build the TgramNet model. The dilated convolution is integrated into the first two layers of the TgramNet.

Dilated convolution is a special type of convolution that we can acquire a larger receptive field as compared with the traditional convolution with the same model parameters. Comparing with normal 1D-convolution layer, dilated convolution involves pixels skipping in the kernel and hence be able to cover larger receptive field that sees more information of the input with relatively small kernel size. Ablation study in experiments will show that dilated convolution is one of the keys to improve overall performance of the model.

3.2.3 *Scaling of feature tensor*

In addition to the scaling of input data and depth/width of deep learning model. We found that increasing the channel number of feature tensor by adding extra TgramNet is quite useful.

This is an interesting observation since the original design of Tgram already aims to complement the information in Sgram. However, scaling the feature tensor by adding new Tgram feature is still beneficial for sound event detection. Therefore, our Extended STgram consists of scaling of input data, model and feature tensor to improve the performance of ASD.

4 EXPERIMENTS

We compared our proposed Extended STgram to the research works in the field of anomalous sound detection. We applied our method to a widely adopted benchmark, DCASE 2020 Challenge, to demonstrate the potential of the proposed compound scaling technique.

Table 2. The performance comparison among ASD AI models in DCASE 2020 task 2 evaluation set.

Model	Mean AUC	Mean pAUC
Sub-Cluster AdaCos (Wilkinghoff 2021)	94.69	87.90
Giri_Amazon (Giri *et al.* 2020)	94.58	84.95
Daniluk_SRPOL (Daniluk *et al.* 2020)	94.56	86.24
Primus_CP-JKU (Primus 2020)	92.78	87.64
ST_gram_MFN (Liu *et al.* 2022)	92.78	86.69
Ours	93.73	88.29
Ours (ensemble)	**95.16**	**89.29**

Table 3. The ablation studies of our Extended STgram under various scenarios.

	fan		pump		slider		valve		ToyCar		ToyConveyor		Mean	
Sampling Rate	AUC	pAUC	AUC	pAUC	AUC	pAUC	AUC	pAUC	AUC	pAUC	AUC	pAUC	AUC	pAUC
2-15 16 kHz	95.63	92.31	92.84	87.87	99.19	95.92	98.46	94.41	95.32	92.10	78.41	64.28	93.31	87.82
32 kHz	96.94	93.84	94.88	88.44	98.21	94.69	97.31	94.75	95.06	90.96	79.98	67.07	**93.73**	**88.29**
64 kHz	93.01	86.94	96.51	87.16	97.72	93.90	93.00	90.94	94.17	89.69	75.64	64.39	91.67	85.50
# of Mel bins	AUC	pAUC	AUC	pAUC	AUC	pAUC	AUC	pAUC	AUC	pAUC	AUC	pAUC	AUC	pAUC
2-15 128	96.94	93.84	94.88	88.44	98.21	94.69	97.31	94.75	95.06	90.96	79.98	67.07	**93.73**	**88.29**
256	98.93	94.88	93.23	87.92	99.63	98.15	97.24	93.15	92.59	86.82	71.24	62.21	92.14	87.19
Conv. Type	AUC	pAUC	AUC	pAUC	AUC	pAUC	AUC	pAUC	AUC	pAUC	AUC	pAUC	AUC	pAUC
2-15 Dilated Conv.	96.94	93.84	94.88	88.44	98.21	94.69	97.31	94.75	95.06	90.96	79.98	67.07	**93.73**	**88.29**
Normal Conv.	93.27	91.00	95.86	89.38	98.68	95.30	96.37	93.15	94.49	91.56	70.53	61.86	91.53	87.04
# of Tgram	AUC	pAUC	AUC	pAUC	AUC	pAUC	AUC	pAUC	AUC	pAUC	AUC	pAUC	AUC	pAUC
2-15 1	92.55	88.54	93.69	85.05	98.85	96.16	97.86	94.21	94.66	90.65	82.28	68.38	93.31	87.16
2	96.94	93.84	94.88	88.44	98.21	94.69	97.31	94.75	95.06	90.96	79.98	67.07	**93.73**	**88.29**
4	92.72	84.13	91.82	82.18	99.04	95.66	97.63	93.18	94.32	90.01	80.84	66.50	92.73	85.28

4.1 *Dataset*

We describe the adopted dataset for ASD task in this section. The benchmark dataset that we utilized in this work is DCASE 2020 Task 2 dataset (Koizumi *et al.* 2019). The benchmark dataset is a combination of ToyADMOS (Koizumi *et al.* 2019) and MIMII (Purohit *et al.* 2019).

There are about 1000 normal clips for each machine type and machine ID. Evaluation set also contains three machine IDs for each machine type, the IDs are the same as additional set, each ID contains around 400 clips including both normal and anomalous samples. DCASE 2020 Task 2 evaluates the performance of each model based on the results from evaluation set.

4.2 *Training configurations*

The baseline STgram model chooses the frame size of 1024 samples with 50% overlap (hop-length equals to 512), and the number of Mel bins are 128 for the log-Mels spectrogram. Therefore, the height M of the Sgram will be 128. The width N of Sgram is linearly proportional to the sampling rate r according to the training setting (i.e., frame size = 1024 samples with hop-length = 512) of STFT in STgram (Liu *et al.* 2022). Therefore, the number of time frames (i.e., N) is calculated by $\lceil 10 \times \frac{r}{512} \rceil$ with sampling rate r since each audio clip is 10 seconds and the hop-length is 512. Since the audio clip in DCASE 2020 Task 2 has $r = 16$ kHz, the size $M \times N$ in STgram will be 128×313 for both its Sgram and Tgram. The baseline STgram adopts a CNN model as the classifier for anomaly detection. We denote the baseline STgram with Mobile FaceNet (MFN) as STgram-MFN in the following paragraph.

For applying compound scaling to STgram, We increased the sampling rate from 16 kHz to 32 kHz in STgram. We trained each model for 200 epochs, the batch size is set to 128. We applied Adam optimizer with learning rate equal to 0.0001, and utilizing cosine annealing scheduler to control the learning rate. Other hyper-parameters follow the settings in baseline STgram (Liu *et al.* 2022). Each training procedure takes about 12 hours based on two NVIDIA Tesla V100 GPU.

4.3 *Evaluation metrics*

The metrics to evaluate the task are the same as DCASE 2020 Task 2, which are area under the receiver operating characteristic (ROC) curve (AUC) and partial-AUC (pAUC). Partial-AUC is calculated as the AUC under a low false-positive-rate (FPR) range $[0,p]$ to prevent model from frequently giving false alarms. DCASE 2020 Task 2 set p to 0.1 to evaluate the performance of models.

4.4 *Performance comparison*

We compare the performance between our proposed Extended STgram and several state-of-the-art models on the leader board of DCASE 2020 Task 2. The results are illustrated in Table 1. We can see that our Extended STgram outperforms original STgram-MFN by up to 1% on mean AUC. Our Extended STgram outperforms other models on mean pAUC. Since a low FPR algorithm plays an important role in practical applications, this outstanding result demonstrates the significance of our proposed Extended STgram. Moreover, if two systems have the same average rank, DCASE 2020 Task 2 prefers the system with the higher pAUC rank.

5 ABLATION STUDY

In order to evaluate the design choices of compound scaling, we conducted ablation studies to investigate the performance of different model configurations (e.g., sampling rate, the number of Mel bins, type of convolution and the number of Tgram). The original configuration [sampling rate, # of Mel bins, Conv. Type, # of Tgram] of audio clip in DCASE 2020 Task dataset is [16 kHz, 128, normal, 1]. We investigated various settings of each design factors for our Extended STgram as shown in Table 3.

The performance is compromised if the sampling rate is higher than 32 kHz. This experimental result demonstrates that the excessively scaling of sampling rate in post-processing will have negative impact for AI model. The experimental results show that increasing the number of Mel-bins would have negative impact for ASD task. Therefore, we only conduct scaling of the width (i.e., related to sampling rate) in Sgram. The dilated convolution provides better performance as compared with the normal convolution configuration. The performance in 'ToyConveyor' is improved signification by using dilated convolution that we increase the result up to 9.45 in AUC and up to 5.21 in pAUC. Meanwhile, we can see that the best performance occurred in adding two Tgrams. Although adding three Tgrams has similar performance, the newly added Tgram will bring extra computational complexity. By taking the performance and the computational complexity into considerations, we choose adding two Tgrams in our method.

6 CONCLUSION

In this work, we applied the concept of compound scaling to the baseline STgram and achieved the best performance in ASD task. The value of [AUC, pAUC] of our proposed method with ensemble learning is up to [95.16, 89.29] for DCASE 2020 Task 2 benchmark dataset. Although the concept of compound scaling is originated from the applications in computer vision, we found that the concept is still helpful for acoustic applications. In contrast to the usage of compound scaling in computer vision, the properties of ASD task is vastly different. As a result, we have to consider the design aspects carefully by leveraging the compound scaling heuristically. Investigating the general principles of applying compound scaling to acoustic data is, of course, one of our future research works. Our research work plays a role of pilot study of applying the crucial concepts from computer vision to the audio data applications. We expected our findings can encourage more studies leveraging essential concepts from other areas for the acoustic applications.

ACKNOWLEDGMENT

This work was supported in part by the National Science and Technology Council, Taiwan under Grant NSTC 112-2634-F-006-002 and 112-2221-E-007-058. We thank to National Center for High-performance Computing (NCHC) of National Applied Research Laboratories (NARLabs) in Taiwan for providing computational and storage resources.

REFERENCES

Chen, H., L. Ran, X. Sun, and C. Cai (2023). Sw-wavenet: Learning representation from spectrogram and wavegram using wavenet for anomalous sound detection. *In IEEE ICASSP* 2023, pp. 1–5.

Daniluk, P., M. Gozdziewski, S. Kapka, and M. Kosmider (2020, July). Ensemble of autoencoder based systems for anomaly detection. *Technical Report*, DCASE2020 Challenge.

Foggia, P., N. Petkov, A. Saggese, N. Strisciuglio, and M. Vento (2015). Audio surveillance of roads: A system for detecting anomalous sounds. *IEEE Transactions on Intelligent Transportation Systems* 17(1), 279–288.

Giri, R., S. V. Tenneti, K. Helwani, F. Cheng, U. Isik, and A. Krishnaswamy (2020, July). Unsupervised anomalous sound detection using self-supervised classification and group masked autoencoder for density estimation. *Technical Report*, DCASE2020 Challenge.

Guan, J., F. Xiao, Y. Liu, Q. Zhu, and W. Wang (2023). Anomalous sound detection using audio representation with machine id based contrastive learning pretraining. *In IEEE ICASSP* 2023, pp. 1–5.

Kawaguchi, Y. and T. Endo (2017). How can we detect anomalies from subsampled audio signals? In *2017 IEEE 27th International Workshop on Machine Learning for Signal Processing (MLSP)*, pp. 1–6. IEEE.

Koizumi, Y., Y. Kawaguchi, K. Imoto, T. Nakamura, Y. Nikaido, R. Tanabe, H. Purohit, K. Suefusa, T. Endo, M. Yasuda, and N. Harada (2020, November). Description and discussion on DCASE2020 challenge task2: Unsupervised anomalous sound detection for machine condition monitoring. In *Proceedings of the Detection and Classification of Acoustic Scenes and Events 2020 Workshop* (DCASE2020), pp. 81–85.

Koizumi, Y., S. Saito, H. Uematsu, N. Harada, and K. Imoto (2019, November). ToyADMOS: A dataset of miniature-machine operating sounds for anomalous sound detection. In *Proceedings of IEEE Workshop on Applications of Signal Processing to Audio and Acoustics (WASPAA)*, pp. 308–312.

Li, Y., X. Li, Y. Zhang, M. Liu, and W. Wang (2018). Anomalous sound detection using deep audio representation and a blstm network for audio surveillance of roads. *IEEE Access* 6, 58043–58055.

Liu, Y., J. Guan, Q. Zhu, and W. Wang (2022). Anomalous sound detection using spectraltemporal information fusion. In *IEEE ICASSP* 2022, pp. 816–820. IEEE.

Mai, K. T., T. Davies, L. D. Griffin, and E. Benetos (2022). Explaining the decisions of anomalous sound detectors. *In 7th Workshop on Detection and Classification of Acoustic Scenes and Events (DCASE)*, pp. 1–5.

Primus, P. (2020, July). Reframing unsupervised machine condition monitoring as a supervised classification task with outlier-exposed classifiers. *Technical Report*, DCASE2020 Challenge.

Purohit, H., R. Tanabe, T. Ichige, T. Endo, Y. Nikaido, K. Suefusa, and Y. Kawaguchi (2019, November). MIMII Dataset: Sound dataset for malfunctioning industrial machine investigation and inspection. In *Proceedings of the Detection and Classification of Acoustic Scenes and Events 2019 Workshop* (DCASE2019), pp. 209–213.

Tan, M. and Q. Le (2019). Efficientnet: Rethinking model scaling for convolutional neural networks. In *International Conference on Machine Learning*, pp. 6105–6114. PMLR.

Wilkinghoff, K. (2021). Sub-cluster adacos: Learning representations for anomalous sound detection. *In 2021 International Joint Conference on Neural Networks (IJCNN)*, pp. 1–8.

Recent Advances in Reliability and Maintenance Modeling – Okamura, Inoue & Xiao (Eds)
© 2025 the Author(s), ISBN: 978-1-032-79098-5

Resilience of trade network in uncertain environment based on trade data

Changchun Lv, Jiansong Bao, Cong Yao, Yizhuo Sun & Dongli Duan
School of Information and Control Engineering, Xian University of Architecture and Technology, Xian, China

ABSTRACT: Due to the trade system is closed related to the national economy, interruption of trade would cause the economy to stagnate and throw the economy into a tailspin for the development of the trade globalization. Hence how to evaluate the resilience of the trade system is critical. The topology property of the trade system has drawn great attention, but the flow of products between countries is a dynamic process, and the interaction between countries should be considered. In this paper, we propose a trade network model to establish the trade network based on the trade data. Then a resilience evaluation model that considers the interaction between countries is introduced. The average activity is calculated to evaluate the resilience of the trade network as time changes. The results show the trade network becomes more and more homogeneous as the year increases. For the resilience of the trade network, the topology property is positive to the average activity of the trade network. More generally, the proposed model demonstrates how trade system resilience is and opens an innovative resilience evaluation model that considers the dynamics between nodes in the trade network.

Keywords: Trade system, Complex network, Dynamics, Resilience

1 INTRODUCTION

International trade and globalization are the cornerstones of the present society, it means that any item could be bought all over the world [1,2]. The globalization of trade brings convenience to people and reduces the cost of goods. However, the trade system faces many challenges, such as trade protectionism, and the COVID-19 pandemic [3,4]. These challenges would cause trade interruption which could impact the economy and society and even cause economic stagnation. A model to evaluate the resilience of the trade system is urgently needed.

The trade system exits the importer and exporter, and the product flows from the exporter to the importer. It can be described as networks in which the node represents the country and the link represents the relation of countries. The topology of the trade system has drawn great attention [5,6]. Sartori *et al.* [7] analyzed the links between virtual water trade and vulnerability to external shocks from the perspective of network analysis. Fair *et al.* [8] proposed a preferential dependency network model of the global wheat trade network and used the model to predict the response of wheat trade network indicators to shocks of different lengths and severity. Dong *et al.* [9] constructed the wheat trade competition network and analyzed the overall characteristics, core-edge structure and the evolution of the network over time. Torreggiani *et al.* [10] analyzed the closely linked trading groups in each specific crop layer from the perspective of multi-networks. Gutierrez-Moya *et al.* [11] studied the international wheat trade network and its evolution characteristics from 2009 to 2013 based on a complex network method.

However, the above research focuses on the topology analysis of the trade system, the product of a country in the trade system should flow to other countries. Hence the interaction between countries due to trade is a dynamic process, the dynamics should be considered

 DOI: 10.1201/9781003491309-14

in the trade system. For the resilience of a complex network with dynamical behavior, Gao *et al.* [12] revealed universal resilience patterns in complex networks and applied the model to the mutualistic networks and gene regulatory networks. In this paper, we first establish the trade network model, and then the birth-death model is introduced to represent the interaction between countries. Finally, based on the proposed model, the resilience evaluation model is proposed.

The rest of this paper is structured as follows. Section 2 describes the methodology of this paper. The trade network model and the resilience evaluation model are proposed. The results including the trade network in different years and the numerical algorithm to evaluate the resilience of the trade network are shown in section 3. The conclusion and future research works are discussed in section 4.

2 METHODOLOGY

The overview of the trade network resilience evaluation model includes three phases: data collection, trade network model, and resilience evaluation model.

2.1 *Data collection*

The data in this paper is from the CEPII-BACI database, and its data source is the United Nations Commodity Trade Statistics Database (UN Comtrade), which has high authority and accuracy. The data is divided by year and includes the exporter, the importer, the product code, the product value, and the product quantity. In this paper, we choose the trade data in 1998, 2005, and 2020 for the trade network model and resilience evaluation model.

2.2 *Trade network model*

Before evaluating the resilience of the trade system, we need to establish the topology of the system from the trade data. Figure 1 shows the process of how to establish the trade network from the trade database. The step is as follows: *i)* For a given trade data in a specific year, data cleaning is needed to delete the data that is missing or reduplicative. *ii)* The node in the trade network should be determined which includes exporters and importers. *iii)* For the value or quantity from exporters to importers has different dimensions, we need to do the process of data normalization. In this paper, we choose the value to do the data normalization. Repeating *i)* to *iii)* based on the CEPII-BACI database, the trade network for each year can be got.

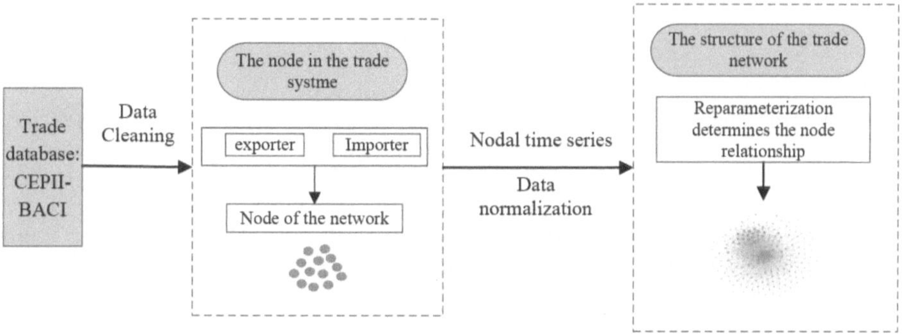

Figure 1. The process of how to establish the trade network from trade database.

From the step in Figure 1, the trade network is directed for the product flows from exporters to importers. Meanwhile, the trade network would be weighted. That is because the demand for each product is different, and regardless of the weight of each link would make the evaluation of the resilience inaccurate. Specifically, if country i and country j appear in a row at the same time, we consider there is a link between country i and country j. The weight between these countries would be different. For example, the resource countries, such as Australia and Russia, there are more exports of products. For the country, such as America and England, there are more imports of products. It is necessary to consider the weight between countries, we use the standardized value between countries which is to be the weight of links. The parameters in the trade network G are defined as follows:

- i represents the country in the CEPII-BACI database and it also is the node in the trade network.
- w_{ij} represents the weight between country i and country j.
- A_{ij} is the network adjacency matrix. If $A_{ij}> 0$ represents there is a link from country j to country i, otherwise, $A_{ij}= 0$.

Obviously, the trade network is a directed weighted network, and the topological property is needed to analyze. The out-strength and in-strength are classical indices in directed weighted networks[13]. The sum of weight from i to other nodes is used to determine the out-strength of node i, it can be calculated through $w_{out} = \sum_{j=G} A_{ji}$. The sum of weight from other nodes to i is used to determine the in-strength of node i, it can be calculated through $w_{in} = \sum_{j=G} A_{ij}$. The sum of in-strength and out-strength of node i is defined as the total weight of node i. These indicators which consider static properties of the trade network can be used to represent the topology property.

2.3 *Resilience evaluation model*

For the product flowing between countries, the node in the trade network has out-strength or in-strength which represents the export and import of product. The interaction between nodes should be a dynamic process. The birth-death model has an application in population dynamics. In the model, the node in the network represents the site, and each site has a population, the people would flow from site i and site j [14]. As a result, the flow of the product includes outflow and inflow, analogous to the flow process of people in the birth-death model. Therefore, it is more appropriate to replicate the flow process of people associated with the trade system using the birth-death model. The birth-death model also has the advantage of being able to dynamically analyze the interaction between nodes in the trade network. Here we consider the dynamical models with the birth-death model which can be written as

$$dx_i dt = -x_i^2 + \sum_{j=1}^{N} A_{ij} x_j \tag{1}$$

Within the framework of the birth-death model, all nodes in the network are labeled with one of two states: inflow, and outflow. The dynamics include an inflow model an interaction by one of its nearest neighbors, and the product in a node outflow at a rate. $x_i(t)$ represents the trade volume of node i at time t. In this paper, the trade value of node i is used to assess the performance of each country in the trade system, indicating the throughput of country i.

When the dynamics reaches the steady state, the Equation (1) can be written as

$$-x_i^{*2} + \sum_{j=1}^{N} A_{ij} x_j^{*} = 0 \tag{2}$$

Through Equation (2), we could get the steady-state trade value of each node to assess the trade level of each node. Meanwhile, the average trade value $\langle x^* \rangle$ of the overall network can be got by taking the mean of the steady-state values of all nodes. According to Ref.[12], here we choose $\langle x^* \rangle$ to represent the resilience of the trade system in a specific year.

3 RESULTS

Based on the proposed model in Section 2, we first establish the trade network from the CEPII-BACI database in a year. The topology and the weighted distribution are shown. Then the algorithm for evaluating the resilience of the trade network is proposed. Finally, the resilience is calculated based on the proposed algorithm for the trade network in 1996, 2005, and 2020.

Using the process in Figure 1, three years of trade data including 1996, 2005, and 2020 is chosen. The topology of each trade network is shown in Figure 2(a-c). The size of nodes represents the strength. It means that the larger the size of the node is, the higher the strength of the node is. Figure 2(d-f) shows the strength distribution of each trade network. The results show that the network becomes more homogeneous as the year increases. It is because the development of international trade tends to the integration of the international market and the globalization of trade.

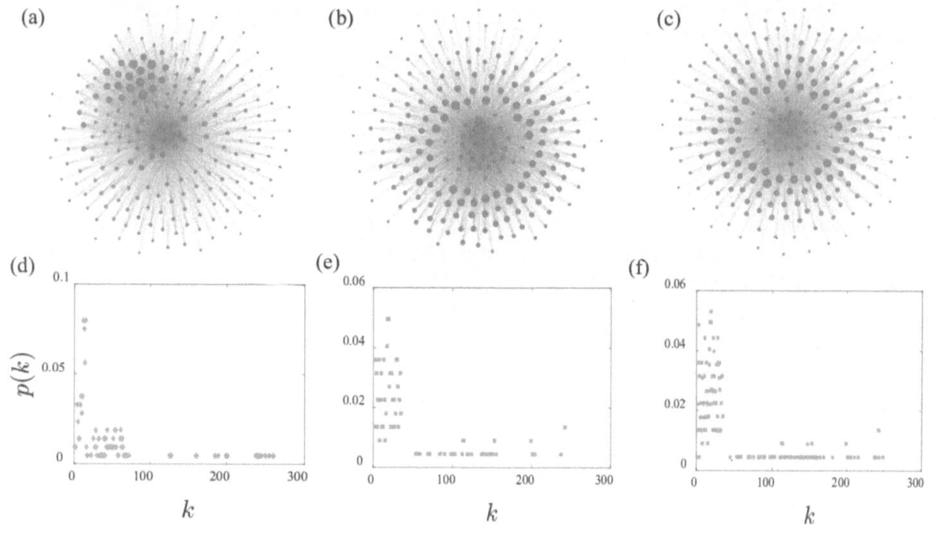

Figure 2. (a-c) The topology of the trade system. (d-f) The strength distribution of the trade network.

To evaluate the resilience of the trade network, the steady-state state of each node should be got. Here we integrated numerically the dynamics of Equation (1) using a fourth-order Runge-Kutta method. Through the method, the node in the trade network would reach the steady state. The pseudocode of the numerical algorithm for the steady-state additivity of each node is presented in Algorithm 1. The numerical results of each node in 1996, 2005, and 2020 are shown in Figure 3. The results show that the activity of the node would reach the steady state over time through the initialization. It is found that the steady-state activity of the node in 1996 is higher than the activity in 2020. In other words, as the year increases, the steady-state activity of the node becomes smaller.

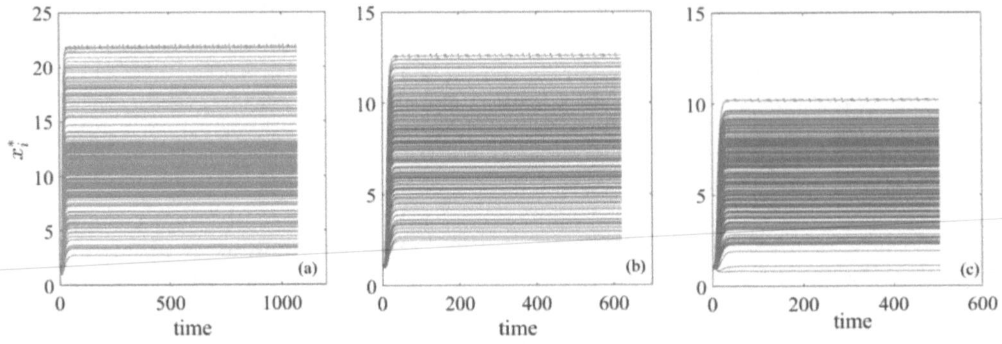

Figure 3. (a-c) The process of each node from the inial state to the steady-state state.

Algorithm 1. Numerical algorithm for the steady-state activity of each node

Require: Network G = (V, E), the initial value x_0, the temporal interval $[t_0, t_{end}]$;
1: Initialization $(x_{i0}^1, \cdots, x_0^N)$;
2: $[i, j, s] = \mathbf{find}(A)$;
3: $M\_system : dx_i dt = -x_i^2 + \sum_{j=1}^{N} A_{ij} x_j$;
4: $options = []$;
5: $[t, xss] = \mathbf{ode}45(M\_system([,],, options, A))$;
6: $x = xs(\mathbf{end}, :)$;
7: return the steady-state activity of each node $(x_i^*, \cdots, x_N^*)$.

According to the above analysis, the topology of the trade network and the steady state of each node can be got. To evaluate the resilience of the overall network, the steady-state activity of all nodes is combined by averaging. Therefore, the average activity of the trade network is calculated to evaluate the resilience of the network. The resilience of the network in 1996, 2005, and 2020 is shown in Table 1. Meanwhile, the average degree and average strength of the network are also shown. The results show as the year increases, the average degree, the average strength, and the average activity decrease. It means the trade system becomes less resilient as the year increases. In particular, there is a positive correlation between the structure property and the dynamic property.

Table 1. The resilience of trade network in 1996, 2005, and 2020.

The specific year	Average degree	Average strength	Resilience
1996	39.24	18.69	11.54
2005	36.79	15.12	7.86
2020	31.03	11.98	6.13

*A specific year corresponds to a trade network.

4 CONCLUSIONS

Our study aimed to evaluate the resilience of trade network in uncertain environments based on trade data. Firstly, a trade network model based on the trade data is proposed. Through the process, the node in the trade network is determined which includes importers and exporters, and the weight between nodes is calculated through normalizing the values between the nodes. The trade network should be a directed weighted network. Based on the trade network, we propose the resilience evaluation model. The birth-death dynamics is introduced to represent the interaction between nodes which includes outflow and inflow of the product. Then the numerical algorithm is proposed to get the steady-state state of each node, and the resilience of the trade network is evaluated by averaging the steady-state activity of all nodes. Intriguingly, as the year increases, the trade network becomes more and more homogeneous, and less resilient.

Our work suggests several avenues for future research. The globalization of trade would cause the trade system to become less resilient, but the reason should be analyzed in the future. In addition, the model in this study focuses on the trade system. The resilience of trade networks for specific products also should be analyzed.

ACKNOWLEDGMENTS

This work was supported by the National Natural Science Foundation of China (Grant Nos.72301206), in part by the National Natural Science Foundation of Shaanxi province (Grant Nos.2023-JC-QN-0802), in part by Students Scientific Research Training (Grant Nos.202310703064).

REFERENCES

[1] Borchert, I. and Yotov, Y. V. (2017). Distance, globalization, and international trade. *Economics Letters* 153, 32–38.

[2] Mukherjee, I. (2008). Impact of globalization on international trade. *ICFAI Journal of International Business* 3(1).

[3] Bhagwati, J., Krishna, P. and Panagariya, A. (2014). *The World Trade System: Trends and Challenges.* Columbia University.

[4] Zurek, M., Ingram, J., Sanderson Bellamy, A. *et al.* (2022). Food system resilience: concepts, issues, and challenges. *Annual Review of Environment and Resources* 47, 511–534.

[5] De Benedictis, L. and Tajoli, L. (2011). The world trade network. *The World Economy* 34(8), 1417–1454.

[6] Ji, G., Zhong, H., Nzudie, H.L.F. *et al.* (2024). The structure, dynamics, and vulnerability of the global food trade network. *Journal of Cleaner Production* 434, 140439.

[7] Sartori, M. and Schiavo, S. (2015). Connected we stand: A network perspective on trade and global food security. *Food Policy* 57, 114–127.

[8] Fair, K. R., Bauch, C. T. and Anand, M. (2017). Dynamics of the global wheat trade network and resilience to shocks. *Scientific Reports* 7(1), 7177.

[9] Dong, C., Yin, Q., Lane, K. J. *et al.* (2018). Competition and transmission evolution of global food trade: A case study of wheat. *Physica A: Statistical Mechanics and Its Applications* 509, 998–1008.

[10] Torreggiani, S., Mangioni, G., Puma, M. J. *et al.* (2018). Identifying the community structure of the food-trade international multi-network. *Environmental Research Letters* 13(5), 054026.

[11] Gutierrez-Moya, E., Adenso-Dłaz, B. and Lozano, S. (2021). Analysis and vulnerability of the international wheat trade network. *Food Security* 13, 113–128.

[12] Gao, J., Barzel, B., Barabsi, A. L. (2016). Universal resilience patterns in complex networks. *Nature* 530 (7590), 307–312.

[13] Li, M., Fan, Y., Chen, J. *et al.* (2005). Weighted networks of scientific communication: the measurement and topological role of weight. *Physica A: Statistical Mechanics and its Applications* 350(2–4), 643–656.

[14] Lv, C., Yuan, Z., Chen, J. *et al.* (2022). Cascading failure in networks with dynamical behavior against multi-node removal. *Chaos, Solitons and Fractals* 160, 112270.

Recent Advances in Reliability and Maintenance Modeling – Okamura, Inoue & Xiao (Eds)
© 2025 the Author(s), ISBN: 978-1-032-79098-5

Reliability and failure mechanisms of thermal imaging modules undergone highly accelerated stress tests

Nikolay Lyapunov, Mikita Marus, Him Ting Wong & Tak Lam Chan
Centre for Advances in Reliability and Safety (CAiRS), New Territories, Hong Kong

Stanislav Markov
Meridian Innovation Ltd, New Territories, Hong Kong

Haibo Hu
Department of Electrical and Electronic Engineering, Hong Kong Polytechnic University, Hong Kong
Centre for Advances in Reliability and Safety (CAiRS), New Territories, Hong Kong

ABSTRACT: In this work, reliability of thermal imaging modules undergone highly accelerated stress tests was investigated and the following failure mechanisms such as lens degradation, die detachment, loss of vacuum and electrical failure were observed and extensively analyzed. The module, being a complex assembly, can experience individual or co-occurrent failure mechanisms; some of such co-occurrences of the failure mechanisms and their interdependencies have also been tackled. Besides, the necessity of an annealing procedure introduced between the application to the module of a saturated highly accelerated stress test and the following health check of the module is explained. Essentially, such annealing procedure helps prevent undesired moisture-related electrical failure of the module as well as differentiate intrinsic degradation of the module from that caused by the residual moisture trapped inside the module during the application of the stress test.

1 INTRODUCTION

Applications of thermal imaging modules (the modules) that identify the temperature of infrared radiating objects vary from such that are to some extent tolerant to incorrect identification of the temperature to such that have zero tolerance (Perić *et al.* 2019; Shaikh *et al.* 2019). With degradation of the modules, they might identify the temperature incorrectly, and it is important to continuously monitor their degradation to judge if the module is still reliable. While there is a lot of literature on reliability of other electronic devices (Liu *et al.* 2020) including those whose reliability has been investigated by means of thermal imaging, surprisingly there is no much research conducted on reliability of the thermal imaging modules themselves with only a few literatures found (Elßner 2014; Elßner and Vogt 2015; Premachandran *et al.* 2004).

Although thermal imaging modules are complex electronic devices, they can be simplified to a combination of a lens and a sensor. Degradation of either element leads to incorrect temperature readouts and a drop in sensitivity of the module. Thus, under harsh conditions the antireflective coating of the lens tends to degrade and the sensor which is a die created by wafer level vacuum packaging (WLVP) (Hilton and Temple 2016) tends to lose vacuum due to the bonding ring degradation. To talk about the reliability of the module it is important to understand its overall structure.

Printed circuit board (PCB) is the foundation of the module. The die is attached to the PCB by thermal paste and wire bonded. The lens holder is also attached to the PCB, and the lens is attached to the lens holder. There is a pair of removable lenses (referred to as stressed and unstressed lenses). Stressed lens undergoes the stress test together with the module while

 DOI: 10.1201/9781003491309-15

unstressed lens does not. This will be helpful in identifying lens degradation factor from overall degradation of the module.

The die fabricated through WLVP consists of the bottom part of the 'active' wafer and the top part of the 'passive' wafer. The top part of the die contains an infrared radiation transparent window through which infrared radiation is transmitted towards the matrix of infrared radiation sensitive elements (or pixels) and which is surrounded by a getter frame (or getter) preventing outgassing. The bottom part of the die contains a matrix of pixels of the size 62 by 80 and the supporting periphery. The bottom part is bonded to the top part via a bonding ring under vacuum. There are different ways of bonding techniques among which are diffusion bonding (Chidambaram and Wickramanayaka 2015), thermocompression bonding (Al Farisi *et al.* 2017; Malik *et al.* 2015) and eutectic bonding (Chidambaram *et al.* 2012, n.d.; Xu *et al.* 2017).

The pixels are microbolometers that are membranes suspended in vacuum on the two legs and have the following working principle. When infrared radiation is absorbed by the membrane of the microbolometer, its temperature changes leading to the resistance change. Through this resistance change the temperature of the infrared radiating object is identified. Typically, microbolometers are organized into matrices that are placed in vacuum to prevent thermal loss through gas convection. As the performance of the module depends on the vacuum level inside the vacuum packaging, reliability of the bonding ring through which the vacuum packaging is sealed is of the top priority.

2 METHODOLOGY

2.1 *Stress test*

It is important to know the lifetime of the module or, in other words, when it will fail under the operating conditions. As the operating conditions are mild, harsher conditions or environmental stress tests can be applied to make the module fail in a shorter time. With the reliability of modern electronic devices constantly growing, traditional temperature-humidity stress tests become insufficient and highly accelerated stress tests gradually come to their replacement. While they can still be referred as temperature-humidity stress tests, highly accelerated stress tests are performed in hermetic stress test chambers with the pressure equal to or elevated above atmospheric and lead to accelerated occurrence of moisture-related failures in the stressed devices.

There are a variety of environmental stress tests that are performed on such and analogous electronic devices to accelerate their moisture-related degradation. When traditional temperature-humidity stress tests are not sufficient to cause degradation (or it takes extremely long time), an application of unsaturated / saturated water vapor referred as unsaturated / saturated highly accelerated stress tests (UHAST / SHAST) is resorted to (Danielson *et al.* 1989; Gunn *et al.* 1983, 1981; Shirley and De Guzman 1993; Walker *et al.* 1994). UHAST is a combination of high temperature (110 degrees), high relative humidity (85 percent) and high water vapor pressure (121.8 kPa). SHAST is a combination of high temperature (121 degrees), high relative humidity (100 percent) and high water vapor pressure (204.9 kPa). In this work, 15 modules were split into two groups of 8 and 7 and passed through UHAST / SHAST, respectively.

2.2 *Post-stress annealing*

While some of the stress tests (such as UHAST) do not cause accumulation of moisture inside the module, some of them (such as SHAST) do. In such cases, the moisture must be dried by post-stress annealing introduced between application of the stress test and the following measurement, thus the description of such procedure also referred to as accelerated recovery is provided in comparison to natural recovery. Besides affecting the performance of the module, the moisture inside the module if not dried can lead to moisture-related electrical failure of the module when it is powered up.

Figures 1(a)-(b) show, respectively, the readout temperature data collected from the module and the corresponding sensitivity data for natural and accelerated recovery at 100 degrees for 24 hours. As can be seen for accelerated recovery, sensitivity right after the stress test with the moisture inside the module is about 10 percent. After accelerated recovery it is above 90 percent. For natural recovery, sensitivity right after the stress test with the moisture inside the module is about 15 percent. After natural recovery it is above 90 percent. It should be noted that natural recovery takes around 15 days which is unacceptable. Thus, after each stress test that causes accumulation of moisture, accelerated recovery is performed, i.e. post-stress annealing.

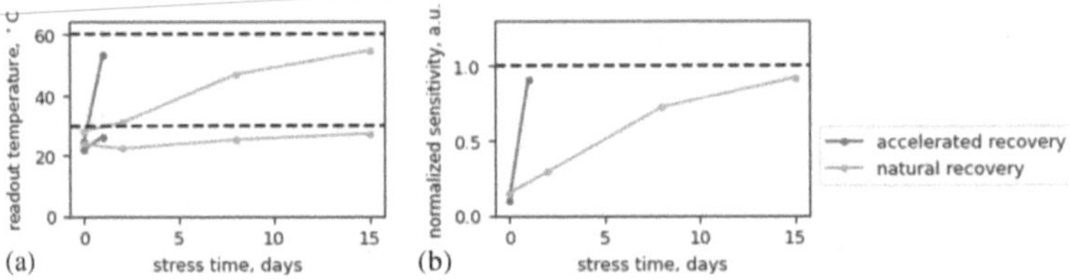

Figure 1. (a) Representation of accelerated and natural recovery through readout temperature data. (b) Representation of accelerated and natural recovery through corresponding sensitivity data.

2.3 *Sensitivity measurement*

The modules undergo environmental stress tests, and their performance is checked before and after application of the stress test in order to monitor their degradation. Sensitivity measurement is performed in the following measurement setup. The module is placed on a stand facing the reference infrared radiating object (the black body) of the size 20 cm by 20 cm. The stand is placed on the rail so that the distance between the module and the black body can be adjusted, but is typically set to 50 cm. The temperature of the black body is set to 30°C (referred as low temperature) and 60°C (referred as high temperature). Among the readout data obtained from the module are readout temperatures, die temperature and V_{dd} collected at the two temperatures of the black body. Based on the readout temperatures, sensitivity of the module is calculated.

3 RESULTS AND DISCUSSION

3.1 *Measurement data*

From sensitivity measurement V_{dd}, die temperature, readout temperature for 30 degrees black body temperature and readout temperature for 60 degrees black body temperature are obtained. Based on the readout temperatures sensitivity is calculated. In the ideal world, readout temperature for 30 degrees black body temperature would be 30 degrees, readout temperature for 60 degrees black body temperature would be 60 degrees, corresponding sensitivity would be equal to 1. Thus, on the readout temperature graphs horizontal dashed lines are provided for reference at 30 and 60 degrees and on the sensitivity graphs horizontal dashed lines are provided for reference at 1. While V_{dd} and die temperature are single values, readout temperatures are matrices of the size 62 by 80. Image segmentation is performed on the readout temperature data to identify the center of the black body. Then 9 values are taken from this area for both low temperature and high temperature, averaged (so that 2 matrices are scaled down to 2 single values) and used to calculate sensitivity as their delta divided by 30 (as a result being a single value as well).

Sensitivity measurement is performed on the module with stressed lens and with unstressed lens. Readout temperature for 30 and 60 degrees as well as sensitivity depend on the lens – the values might vary for unstressed lens and stressed lens as can be seen from the experimental data in the top row and in the bottom row, respectively, Figures 2(a)-(d). As mentioned, 15 modules were split into two groups of 8 and 7 and passed through UHAST / SHAST, respectively. These two groups were further split into 2 groups of 3–4 modules in each and here is an example of the measurement data obtained for one of them that had gone through SHAST.

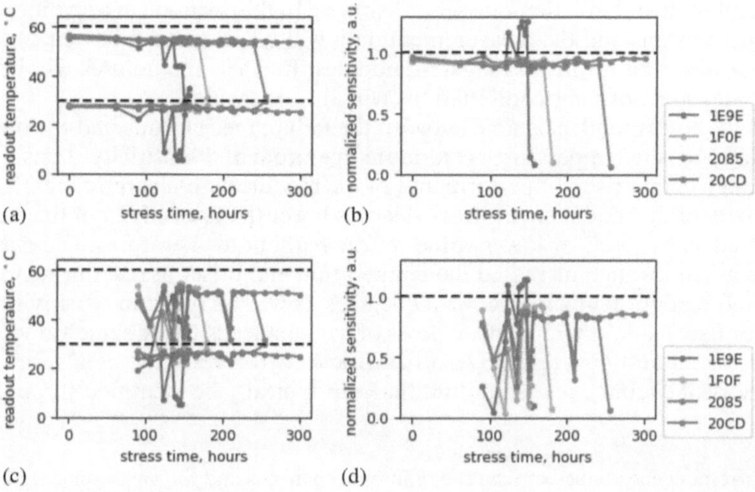

Figure 2. (a) Readout temperature data with the unstressed lens. (b) Corresponding sensitivity data with the unstressed lens. (c) Readout temperature data with the stressed lens. (d) Corresponding sensitivity data with the stressed lens.

Such readout temperature and sensitivity ups and downs for stressed lens could be attributed to no post-stress annealing being established yet. As can be seen, after around 150 hours readout temperature and sensitivity fluctuations are gone (since post-stress annealing was established) except one more appearance at around 210 hours (was needed for an additional investigation). Die temperature and V_{dd} do not depend on the lens. Thus, the difference between these data for stressed and unstressed lenses is just measurement to measurement fluctuations and only one data is presented for die temperature and V_{dd}, Figures 3(a)-(b).

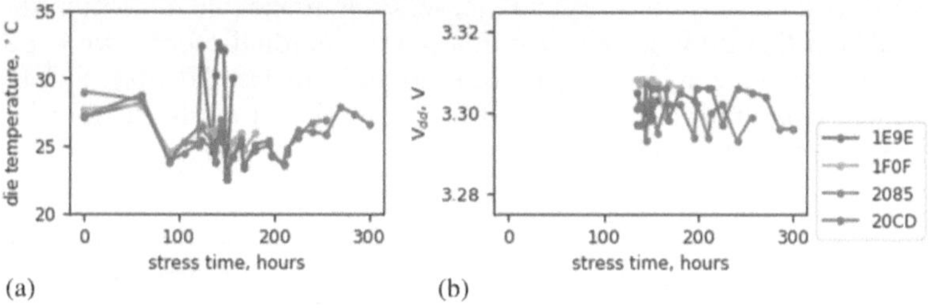

Figure 3. (a) Die temperature data. (b) V_{dd} data.

All the graphs, Figures 2(a)-(d) and 3(a)-(b) are scaled from 0 hours to 300 hours for easier comparison of the data obtained for unstressed lens (originally 0 hours to 300 hours) and stressed lens (originally 90 hours to 300 hours). Such discrepancy is because the original plan did not include comparative analysis of the performance of the module with 2 different lenses. The plan was modified after 90 hours of stress test. It should also be noted that V_{dd} monitoring was introduced even later – after 120 hours of stress test due to the software modification that allowed it.

3.2 *Failure features*

Through sensitivity measurement data including V_{dd}, die temperature, readout temperature and sensitivity lens degradation, die detachment, loss of vacuum, electrical failure can be identified as all of them have a unique set of features. In this section correspondence between the failure mechanisms and the measurement data will be provided followed by a summary of the failures observed in the investigated modules. Further, the failures will be addressed individually and some of their combinations will also be tackled.

Table 1 is the correspondence table between the failure mechanisms and the measurement data including V_{dd}, die temperature, readout temperature and sensitivity. Lens degradation is found through comparison of performance of the module with unstressed lens and stressed lens – sensitivity of the module with stressed lens is lower than sensitivity of the same module with unstressed lens caused by lower values of the readout temperatures. Die detachment is typically identified through increased die temperature and offset in readout temperature for 30 degrees and readout temperature for 60 degrees as well as offset in sensitivity calculated based on the offset readout temperature. Loss of vacuum leads to the same values of readout temperature for 30 and 60 degrees (zero delta) and zero sensitivity, as a result. Electrical failure is electrical circuitry related failure that can typically be identified through V_{dd}.

Table 1. Correspondence table between the failure mechanisms and the measurement data.

Failure mechanism	V_{dd}	Die temperature	Readout temperature	Sensitivity
lens degradation	no	no	lower	lower
die detachment	no	yes	higher	higher
vacuum loss	no	no	low = high	zero
electrical failure	yes	no	no	no

Table 2 is a summary of the failures observed in the investigated modules. Parts of the serial numbers of the modules are used for identification purposes. Thus, in the group of 15 modules, 4 cases of die detachment, 4 cases of vacuum loss and 4 cases of electrical failure were observed. While in all UHAST modules there is no such problem, in all SHAST modules stressed lenses degraded. The time metric is not applicable to them in the same way as it is applicable to other failures (lens degradation is a gradual process while other failures happen at a particular moment in time) and thus not indicated in the table. Speculatively one can claim that lens degradation is the earliest failure, followed by die detachment, followed by vacuum loss, followed by electrical failure.

Table 2. Summary of the failures observed in the 15 investigated modules.

Stress test	Reference name	Lens degradation	Die detach	Vacuum loss	Electrical failure
SHAST	1E9E	yes	120 hours	150 hours	no
	1F0F	yes	no	180 hours	no
	2085	yes	no	255 hours	no
	20CD	yes	no	no	315 hours
	2008	yes	no	120 hours	no
	20D9	yes	120 hours	no	165 hours
	20F0	yes	135 hours	no	180 hours
UHAST	1EDC	no	120 hours	no	315 hours
	1F03	no	no	no	no
	2079	no	no	no	no
	207E	no	no	no	no
	1E9A	no	no	no	no
	1F68	no	no	no	no
	1F8D	no	no	no	no
	20F8	no	no	no	no

3.3 *Lens degradation*

No lens degradation was observed in the modules exposed to UHAST. Their performance with unstressed lens and with stressed lens is identical, and no difference can be seen between the microscopy images of unstressed and stressed lenses. Contrarily, lens degradation was observed in the modules exposed to SHAST – by comparing performance of the same module with unstressed lens at the top and stressed lens at the bottom Figures 2(a)-(d). The difference in the lens quality is clearly seen in the microscopy images of the lenses, Figures 4 (a)-(b). The module with unstressed lens performed 10 percent better than the same module with the stressed lens. That is the direct proof of lens degradation in case of having a pair of removable lenses.

(a) (b) (c)

Figure 4. (a) Microscopy image of the unstressed lens. (b) Microscopy image of the stressed lens. (c) Microscopy image of a detached die.

3.4 *Die detachment*

Die detachment is typically identified through the offset in readout temperature for 30 degrees and readout temperature for 60 degrees as well as through the offset in sensitivity based on the readout temperature and through increased die temperature. The offset happens because the module identifies readout temperature through its own temperature (die temperature) which after die detachment differs from the module calibration data.

4 cases of die detachment were observed and 1 of them can be seen on the graphs, Figures 2 (a)-(d) and 3(a). As can be seen for 1E9E module, after 120 hours of stress test its die temperature abruptly increased and, on average, is 5 degrees higher than die temperature of the other modules. At the same time readout temperature for 30 degrees and readout temperature for 60 degrees both have negative offset (the values are lower than for other modules). It is important to note that the value of the offset is different for low temperature and for high temperature. Thus, for low temperature the readout value can be 20 degrees lower than it is expected to be and for high temperature the readout value can be 10 degrees lower than it is expected to be. In other words, the offsets are two-fold different which might originate from the two-fold difference between the black body temperatures set to 30 and 60 degrees, respectively. It is important to note that the same values of offset for both low temperature and high temperature would be a sign of poor calibration rather than die detachment. It follows then that sensitivity calculated based on the offset readout values is also offset (positively) – giving the values of sensitivity higher than 100 percent which carries no physical sense any longer. An example of die detachment is presented in Figure 4(c). As can be seen, there is an angle between the die and the PCB caused by degradation of the thermal paste between the PCB and the die. Such detached die is only held by the wire bonding.

It should also be noted that there are a few factors affecting the die temperature. One of them is the room temperature. If the room temperature is 25 degrees, the die temperature starts from 25 degrees as well. Then the die temperature grows with the operation time of the module till it reaches its temperature saturation point. Thus, another factor is the exact moment when the measurement is taken – either at the beginning of the operation of the module, or when the module reached its temperature saturation point, or at any point of time in between. These 2 factors are addressed intrinsically.

3.5 *Loss of vacuum*

4 cases of vacuum loss were observed and 3 of them can be seen on the graphs, Figures 2(a)-(d). 3 cases of vacuum loss are just vacuum loss and 1 case is vacuum loss which happened to the same module as die detachment. However, it should be noted that these two failures are not related. All cases of vacuum loss have in common zero delta between low temperature and high temperature (for both temperatures of the black body readout temperature values are around 30 degrees) and sensitivity close to zero. This happens because the bonding ring is degraded under the applied stress test, leakage occurs and the vacuum in the microcavity is filled with the ambient air. It should be noted that the getter that prevents the outgassing phenomena cannot help with the leakage.

In such cases a soldering iron is placed in front of the module with a temperature exceeding the high temperature of the black body. The appearance of the soldering iron in the thermal image proofs vacuum loss – the module can only see extremely high temperatures.

3.6 *Electrical failure*

Electrical failure is electrical circuitry related failure that can typically be identified through V_{dd} of the module. As can be seen from the plots, there were no issues with V_{dd} since the recording of its value was started. That means no electrical failure is observed when the module is still responsive. However, electrical failures when the module is operated normally before the stress test and not responsive after the stress test have been observed. That is what can be classified as an ultimate electrical failure or simply an ultimate failure of the module.

Such electrical failure was observed for 4 modules. The rest of the modules might have had such electrical failure as well, but they were either suspended from further stress test due to the long time with no failures or withdrawn for detailed failure analysis after having vacuum loss issue which was of the primary interest of the current investigation.

Among the 4 modules mentioned above, 3 of them happened to the same modules as die detachment. While the co-occurrence of die detachment and vacuum loss mentioned above is not related, die detachment and electrical failure might be related and might be attributed to wire bonding, Figure 4(c). As can be seen, after die detachment, the die, originally held by thermal paste and not by the wires, is now held by the wires and not by the thermal paste. That increases the chances of the wires to be broken thus disconnecting the die from the rest of the module and leading to no more response from the module.

4 CONCLUSIONS

In this work, reliability of thermal imaging modules that have gone through unsaturated and saturated highly accelerated stress tests was investigated. Performance of the modules was checked before and after application of the stress test in order to monitor their degradation. Through extensive analysis of the measurement data collected from the modules and including V_{dd}, die temperature, readout temperature and sensitivity, a number of failure mechanisms such as electrical failure, loss of vacuum, die detachment and lens degradation were identified. It was demonstrated that all these failure mechanisms have a unique set of features. As a result, these failures can be identified non-destructively through the performance of the modules instead of destructive and costly failure analyses.

As the module is a complex assembly, some failures can happen at the same time and it is important to be able to identify not only individual failures, but also their combinations.

Some cases of co-occurrent failure mechanisms have been investigated. As it was found, some of them, such as die detachment and vacuum loss, might not be related and some of them, such as die detachment and electrical failure, might be related – die detachment increases the chances of a following electrical failure.

A recommendation on handling highly accelerated stress tests is given, i.e. introduction of a post-stress annealing. Such procedure introduced between application of the stress test and the following measurement is equally important for correct analysis of the performance of the modules that underperform in case of residual moisture inside and for preventing moisture-related electrical failure of the module.

ACKNOWLEDGMENTS

The work presented in this article is supported by the Centre for Advances in Reliability and Safety (CAiRS) admitted under AIR@InnoHK Research Cluster.

REFERENCES

Al Farisi, M. S., Hirano, H., Frömel, J. and Tanaka, S., (2017). Wafer-level hermetic thermo-compression bonding using electroplated gold sealing frame planarized by fly-cutting. *J. Micromechanics Microengineering* 27, 015029. https://doi.org/10.1088/1361-6439/27/1/015029

Chidambaram, V. and Wickramanayaka, S., (2015). Al-Ge diffusion bonding for hermetic sealing application. *J. Electron. Mater.* 44, 2387–2395. https://doi.org/10.1007/s11664-015-3683-y

Chidambaram, V., Yeung, H.B. and Shan, G., (2012). Development of metallic hermetic sealing for MEMS packaging for harsh environment applications. *J. Electron. Mater.* 41, 2256–2266. https://doi.org/10.1007/s11664-012-2107-5

Chidambaram, V., Yeung, H.B. and Shan, G., n.d. Development of CMOS Compatible Bonding Material and Process for Wafer Level MEMS Packaging Application under Harsh Environment.

Danielson, D., Marcyk, G., Babb, E. and Kudva, S., (1989). HAST applications: acceleration factors and results for VLSI components, in: *27th Annual Proceedings., International Reliability Physics Symposium. Presented at the 27th Annual Proceedings., International Reliability Physics Symposium, IEEE*, Phoenix, AZ, USA, pp. 114–121. https://doi.org/10.1109/RELPHY.1989.36331

Elßner, M., (2014). Vacuum quality evaluation for uncooled micro bolometer thermal imager sensors. *Microelectron. Reliab.* 54, 1758–1763. https://doi.org/10.1016/j.microrel.2014.07.094

Elßner, M. and Vogt, H., (2015). Failure mechanisms of microbolometer thermal imager sensors using chip-scale packaging. *Microelectron. Reliab.* 55, 1901–1905. https://doi.org/10.1016/j.microrel.2015.07.040

Gunn, J.E., Camenga, R.E. and Malik, S.K., (1983). Rapid Assessment of the Humidity Dependence of IC Failure Modes by Use of HAST, in: *21st International Reliability Physics Symposium. Presented at the 21st International Reliability Physics Symposium, IEEE*, Phoenix, AZ, USA, pp. 66–72. https://doi.org/10.1109/IRPS.1983.361963

Gunn, J.E., Malik, S.K. and Mazumdar, P.M., (1981). Highly Accelerated Temperature and Humidity Stress Test Technique (HAST), in: *19th International Reliability Physics Symposium. Presented at the 19th International Reliability Physics Symposium, IEEE*, Las Vegas, NV, USA, pp. 48–51. https://doi.org/10.1109/IRPS.1981.362972

Hilton, A. and Temple, D., (2016). Wafer-level vacuum packaging of smart sensors. *Sensors* 16, 1819. https://doi.org/10.3390/s16111819

Liu, Y., Wang, Y., Fan, Z., Hou, Z., Zhang, S. and Chen, X., (2020). Lifetime prediction method for MEMS gyroscope based on accelerated degradation test and acceleration factor model. Eksploat. *Niezawodn. – Maint. Reliab.* 22, 221–231. https://doi.org/10.17531/ein.2020.2.5

Malik, N., Tofteberg, H.R., Poppe, E., Finstad, T.G. and Schjølberg-Henriksen, K., (2015). Environmental stress testing of wafer-level au-au thermocompression bonds realized at low temperature: strength and hermeticity. *ECS J. Solid State Sci. Technol.* 4, P236–P241. https://doi.org/10.1149/2.0201507jss

Perić, D., Livada, B., Perić, M., Vujić, S., (2019). Thermal imager range: predictions, expectations, and reality. *Sensors* 19, 3313. https://doi.org/10.3390/s19153313

Premachandran, C.S., Chong, S.C., Chai, T.C. and Iyer, M., (2004). Vacuum packaging development and testing for an uncooled IR bolometer device, in: *2004 Proceedings. 54th Electronic Components and Technology Conference (IEEE Cat. No.04CH37546). Presented at the 2004 54th Electronic Components and Technology Conference, IEEE*, Las Vegas, NV, USA, pp. 951–955. https://doi.org/10.1109/ECTC.2004.1319453

Shaikh, S., Akhter, N. and Manza, R., (2019). Current Trends in the Application of Thermal Imaging in Medical Condition Analysis 8.

Shirley, C.G. and De Guzman, M.S., (1993). Moisture-induced gold ball bond degradation of polyimide-passivated devices in plastic packages, *in: 31st Annual Proceedings Reliability Physics 1993. Presented at the Proceedings of IEEE International Reliability Physics Symposium*, IEEE, Atlanta, GA, USA, pp. 217–226. https://doi.org/10.1109/RELPHY.1993.283321

Walker, B.J., Dordi, Y.N., Kitchin, J. and Steele, J.A., 1994. A comparison of HAST to conventional THB testing on commodity SRAM's, *in: 1994 Proceedings. 44th Electronic Components and Technology Conference. Presented at the 1994 Proceedings. 44th Electronic Components and Technology Conference, IEEE*, Washington, DC, USA, pp. 877–882. https://doi.org/10.1109/ECTC.1994.367525

Xu, J., Ding, Z., Chidambaram, V., Ji, H. and Gu, Y., (2017). High Vacuum and High Robustness Al-Ge Bonding for Wafer Level Chip Scale Packaging of MEMS Sensors, in: *2017 IEEE 67th Electronic Components and Technology Conference (ECTC). Presented at the 2017 IEEE 67th Electronic Components and Technology Conference (ECTC), IEEE*, Orlando, FL, USA, pp. 956–960. https://doi.org/10.1109/ECTC.2017.228

Recent Advances in Reliability and Maintenance Modeling – Okamura, Inoue & Xiao (Eds)
© 2025 the Author(s), ISBN: 978-1-032-79098-5

Generating synthetic warranty data

Sarah E. Marshall
Department of Information Systems and Operations Management, Business School, University of Auckland, Auckland, New Zealand

Richard Arnold & Stefanka Chukova
School of Mathematics and Statistics, Victoria University of Wellington, Wellington, New Zealand

Yu Hayakawa
School of International Liberal Studies, Waseda University, Tokyo, Japan

ABSTRACT: Good warranty data is scarce, and even data that has been collected reliably is subject to censoring. This can limit the amount of data available for conducting warranty cost analysis, or can require the use of specialised statistical methods for censored data. An alternative method is to identify products that have been fully-observed for the warranty coverage period, and then generate synthetic data based on this subset. This provides a larger set of data, which can be used for further analysis. This paper presents an approach for synthetic data generation. The approach will be demonstrated and evaluated using an example with automotive warranty data. Opportunities for future work will be discussed.

1 INTRODUCTION

Product warranties are typically offered for a set period of time, or usage, or both (Blischke 1995; Blischke *et al.* 2011). For many household appliances, the warranty specifies a set period of time since purchase. In the case of vehicles, the warranty is issued in terms of age (time since purchase) and usage (mileage since purchase). This work focuses on two-dimensional warranties, such as those typically offered with the purchase of a new vehicle. Each time a vehicle owner makes a claim, the age and mileage are recorded. Additional information may also be recorded, such as the cost or reason for the claim (Blischke *et al.* 2011). Whilst the manufacturer can accurately determine the age of all vehicles, using sales information, the usage is typically only observed when a claim is made. This means that manufacturers do not know whether or not vehicles leave the warranty coverage due to their age or mileage, nor if the vehicles have been put out of use, nor if the owner has chosen not claim for failures. Warranty data is thus subject to censoring (Blischke *et al.* 2011). It is also the case that warranty data is rarely shared outside of the manufacturer due to concerns about commercial confidentiality.

It is for these reasons and others, that good warranty data is scarce, and even data that has been collected reliably is subject to censoring. This can limit the amount of data available for conducting warranty cost analysis, and can require the use of specialised statistical methods for censored data. This paper proposes an alternative method – identification of vehicles that have been fully-observed for the warranty coverage period, and then generation of synthetic data based on this subset. This provides a larger dataset for further analysis.

Synthetic data generation is a growing field with many papers published on the subject in recent years. Recent reviews published indicate the relevance of synthetic data across many disciplines areas, including healthcare (Gonzales *et al.* 2023; Hernandez *et al.* 2022), finance (Assefa *et al.* 2021), government (Bonnày *et al.* 2019) and manufacturing (Buggineni *et al.*

DOI: 10.1201/9781003491309-16

2024; Libes *et al.* 2017). Methods typically employed include synthetic minority oversampling technique (SMOTE) and generative adversarial networks (GANS) (Buggineni *et al.* 2024). The nature of warranty data differs from data used in many existing synthetic data generation methods, in that it is multi-dimensional time-series data. Many existing synthetic data generation methods for multivariate datasets do not account for the dependence among multiple related observations over time, and those that do are generally domain specific. Another challenge is that many methods result in a 'black box' approach, which may contain invisible and unmodifiable assumptions. To the best of our knowledge, synthetic data generation has not been applied in the field of warranty analysis.

Our work contributes to this research gap by providing a simple transparent method for synthetic data generation that takes advantage of the specific nature of warranty data. We wish to create a synthetic dataset that contains the same structure as the original, without first having to construct an explicit model for the correlation and dependence that always exists in a warranty trajectory. The work presented here is an initial step towards generating synthetic data that can be used for warranty cost analysis.

The remainder of this paper is structured as follows. An approach for generating synthetic data is presented in Section 3. Methods for comparing sets of warranty data, and thus evaluating the approach are presented in Section 4. Section 5 contains an example that demonstrates the synthetic data generation approach and the techniques for evaluation. Section 6 concludes the paper.

2 PRELIMINARIES

In this work we assume that a two-dimensional warranty is offered based on age and usage, under which the product is covered until it reaches age A or usage U, whichever occurs first. The following notation is used to describe a single product:

A	maximum age covered within warranty period
U	maximum usage covered within warranty period
n	number of claims made during the warranty period
t_i	age of product at time of claim i, $i = 0,\ldots,n$, $t_0 = 0$, $t_i < t_{i+1}$, and $t_n < A$
u_i	usage of product at time of claim i, $i = 0,\ldots,n$, $u_0 = 0$, $u_i < u_{i+1}$, and $u_n < U$
x_i	ith interclaim time, $x_i = t_i - t_{i-1}$, $i = 1,\ldots,n$
y_i	ith interclaim usage, $y_i = u_i - u_{i-1}$, $i = 1,\ldots,n$
z	trajectory $z = \{(t_0, u_0), (t_1, u_1), (t_2, u_2), \ldots, (t_n, u_n)\}$

When plotted in two-dimensional space, the linear interpolation of points at which age and usage are known is called a 'trajectory'. It is assumed that age and usage begin at 0, i.e. $(t_0, u_0) = (0, 0)$. The slope of the trajectories varies according to the usage rate. For example, if a product is heavily used then it will leave the warranty coverage because usage limit is exceeded, though the age is still less than the maximum age. Another product may have low usage and thus leave the warranty coverage due to age. Figure 1 depicts examples of these types of usage patterns.

Definition: A *full observation* is defined as one in which the product has a claim within a specified interval of a time units prior to the maximum warranty age (i.e. $t_n \in (A - a)$), or a specified usage of u usage units before the end of the warranty usage period $u_n \in (U - u)$.

The values of a and u must be chosen by the analyst, and may vary in different applications.

3 SWANN: A METHOD FOR GENERATING SYNTHETIC WARRANTY DATA

In this section we present our method for generating *S*ynthetic *WA*rranty data using *N*earest *N*eighbours (SWANN). SWANN is inspired by the Synthetic Minority Oversampling Technique (SMOTE) (Chawla *et al.* 2002), which is common in dealing with unbalanced

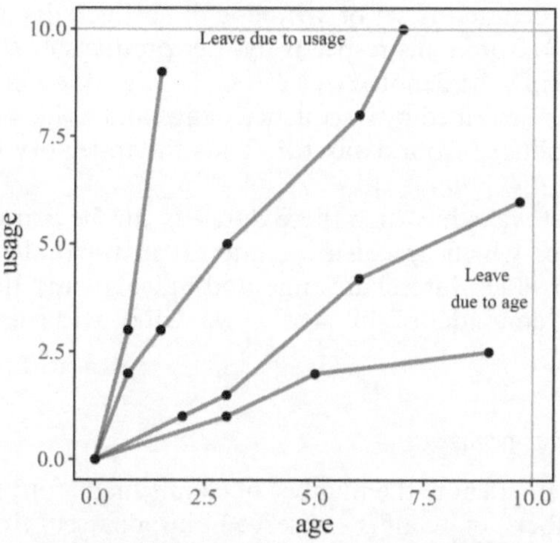

Figure 1. Example trajectories for 4 products with a warranty with a maximum age of $A = 10$ time units and a maximum usage of $U = 10$ units.

datasets in data science. It uses the idea of generating new data by combining values from a selected data point and one of its nearest neighbours. SWANN extends this idea by applying to warranty trajectories. The steps involved in SWANN are described below.

From a dataset containing only 'full observations', and for some chosen value k:

(1) Randomly select a product, v
(2) Identify all other products that have the same number of claims n_v as product v
(3) From these selected products, identify the k nearest neighbours of v based on the desired metric (e.g. the Euclidean distance between the age and usage of the last claim, with usage scaled by a typical usage rate λ to convert it into time units). If v does not have at least k nearest neighbours, return to step 1.
In our application the distance measure is: For products $i \in N^{n_v}$ (set of all products with n_v claims) $= \sqrt{(t_{i,n_v} - t_{v,n_v})^2 + ((u_{i,n_v} - u_{v,n_v})/\lambda)^2}$
(4) Randomly select one of the k nearest neighbours and denote this product as w
(5) Generate a new trajectory $v^* = \{(t_{v^*,1}, u_{v^*,1}), (t_{v^*,2}, u_{v^*,2}), \ldots, (t_{v^*,n_v}, u_{v^*,n_v})\}$ based on v and w as follows:
(i) Generate r from a Uniform (0,1) distribution
(ii) For each claim $j = 1, \ldots, n_v$, compute the vector between the age and usage of $(t_{v,j}, u_{v,j})$ and $(t_{w,j}, u_{w,j})$, and multiply by r to get the new age and usage, i.e.

$$t_{v^*,j} = t_{v,j} + r(t_{w,j} - t_{v,j})$$

$$u_{v^*,j} = u_{v,j} + r(u_{w,j} - u_{v,j})$$

(6) Repeat from step 1 until the required number of synthetic trajectories have been generated

4 EVALUATION – COMPARISON OF ORIGINAL AND SYNTHETIC DATASETS

To assess the performance of SWANN, we compare the properties of the synthetic trajectories with the original data from which they were generated. Let D_O denote the set of original data and let D_S denote the set of synthetic data. Each dataset consists of warranty trajectories for M_O and M_S products respectively. Let products in D_O be denoted by v_{Oi} for $i = 1, \ldots, M_O$, products in D_S be denoted by v_{Si} for $i = 1, \ldots, M_S$. As in previous sections the trajectory for a product is described by a sequence of age and usage points. Let n_{dj} denote the number of claims by product j from dataset d. Thus the trajectory for product j in dataset D_O, is $\{(t_{Oj,1}, u_{Oj,1}), \ldots, (t_{Oj,n_{Oj}}, u_{Oj,n_{Oj}})\}$.

There are a number of ways in which these datasets can be compared. In this work, we will focus on the metrics which are of most interest in warranty cost analysis, namely: number of claims per product, interclaim time and mileage, and the number of claims by time t or usage u. In the remainder of this section we will provide details of how to compare two datasets.

4.1 *Number of claims per product*

As discussed above, Let $n_{d,j}$ denote the number of claims made for product j from dataset d. Then the mean number of claims observed in dataset d can be denoted by $\bar{N}_d = (\sum_{j=1}^{M_d} n_{dj})/M_d$. Let μ_N^d denote the population mean number of claims per product, for dataset d. Then, to compare the mean number of claims per product, a t-test can be used, with the following hypotheses.

$$H_0 : \mu_N^O = \mu_N^S$$

$$H_1 : \mu_N^O \neq \mu_N^S$$

The p-value resulting from this t-test is denoted p_N. The distribution of the number of claims per product can be assessed graphically, by comparing bar charts depicting the number of products with each number of claim.

4.2 *Interclaim time and mileage*

The behaviour of system can vary with the number of claims, especially if the system is ageing. Thus when considering the time between claims, we consider the interclaim time and mileage by *claim number*. For claim i, the mean interclaim time for dataset d is $\bar{X}_d^i = (\sum_{j=1}^{M_d} x_{dj,i})/M_d$ Let $\mu_{X_i}^d$ denote the population mean of the interclaim time for claim i, for dataset d. Then, to compare the mean interclaim time for claim i, a t-test can be used, with the following hypotheses.

$$H_0 : \mu_{X_i}^O = \mu_{X_i}^S$$

$$H_1 : \mu_{X_i}^O \neq \mu_{X_i}^S$$

The p-value resulting from this t-test is denoted p_{X_i}. The time between claims can also be compared graphically, for example, using boxplots for each dataset and claim number, or ECDF plots by claim number.

Similarly, the interclaim mileage can be evaluated by replacing x with y in the expressions above.

4.3 *Number of claims by age t or mileage u*

The number of claims by age t (mileage u) can be explored by examining the distribution of the age (mileage) at which claims occur. This can be done graphically using an ECDF plot. Across multiple replications, the mean ECDF can be computed and compared.

5 EXAMPLE

In this section the generation and comparison procedures from section 3 and 4, respectively, are demonstrated for a small example. The example used here relates to an automotive warranty.

5.1 *Parameters*

For this example, the "original" data is simulated and the times between claims (interclaim time) is exponentially distributed with a rate of 0.5 per year (1/24 per month) and the mileage between claims (interclaim mileage) is exponentially distributed with a rate of 1 per 10000 miles. For simplicity, age and mileage are independent in this example, however, in practice this is not usually the case. It is assumed that the warranty coverage is 36 months and 36000 miles. For this example, a trajectory is considered a "full trajectory" if the age and mileage of the last claim is at least 90% of the warranty coverage, and $k = 5$ neighbours will be used. A typical usage rate of $\lambda = 1000$ miles/month is used.

5.2 *Application of SWANN*

Figure 2 shows the age and mileage of the last (3rd) claim for all vehicles with full trajectories and with 3 claims. In this figure the red point corresponds to the selected vehicle v. The blue points are the 5 nearest neighbours with a blue square indicating the one that was selected. A random number was generated, in this case $r = 0.202$. For each claim, a vector was generated between the corresponding red and blue points and this was multiplied by r to get the purple trajectory. This process was repeated for each claim. Notice that since r is close to 0, the generated and original trajectories are very similar. In a similar way, the required number of trajectories can be generated.

5.3 *Evaluation of SWANN – a single replications*

To demonstrate the process of evaluating SWANN, 200 synthetic trajectories were generated from an original dataset of 200 trajectories.

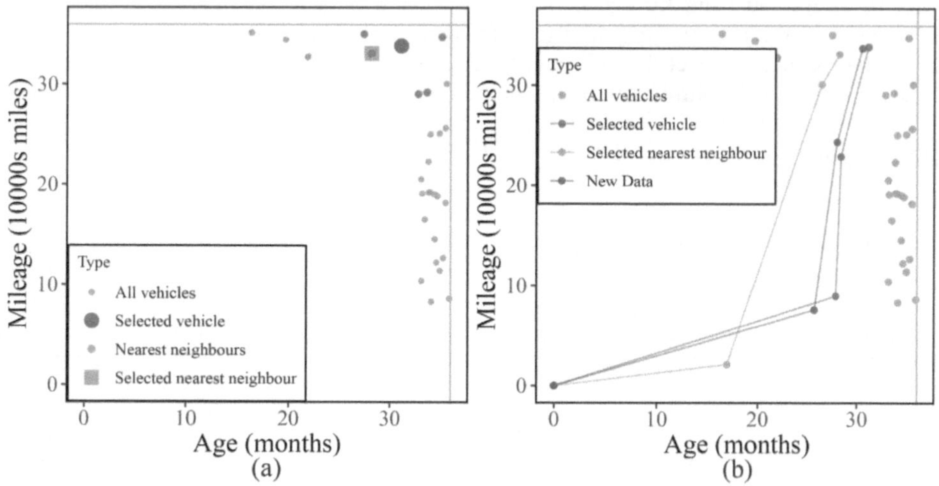

Figure 2. Example of a single iteration of SWANN. (a) k nearest neighbours, and selected neighbour for a vehicle with 3 claims, $k = 5$ (b) Synthetic trajectory for a vehicle with 3 claims.

5.3.1 *Number of claims per vehicle*

First, the number of claims per vehicle is compared across the two datasets. As shown in Figure 3, in the original dataset, vehicles made up to 6 claims, whereas in the synthetic data a maximum of 4 claims was observed. However the numbers of vehicles making between 1 and 3 claims is similar across the two datasets. The mean number of claims per vehicle in the original dataset was 2.02 and the mean number of claims per vehicle in the generated dataset was 1.985. The p-value from a t-test comparing the means was 0.717. Therefore there is insufficient evidence to reject the null hypothesis that two distributions are the same. This provides evidence for the similarity of the original and synthetic datasets.

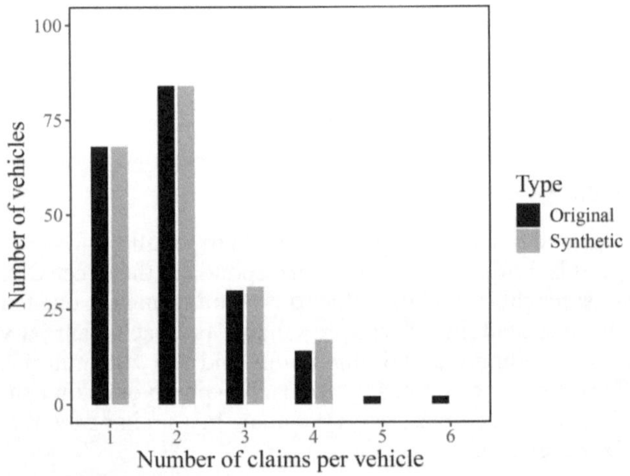

Figure 3. Distribution of number of claims per vehicle from original and synthetic datasets, $M_O = M_S = 200$ vehicles.

5.3.2 *Interclaim time and mileage*

The mean interclaim time by claim number for each dataset is shown in Table 1. The table also shows the p-values for the t-tests for claim numbers 1 to 4. In each case, there is insufficient evidence to reject the null hypothesis. Similar results are shown for the interclaim mileage in Table 2. The similarity of the interclaim time can also be seen in Figure 4 shows the distribution of the interclaim time and mileage by claim number for each dataset. These results provide evidence for the similarity of the two datasets.

Table 1. Comparison of interclaim time by claim number for the original and synthetic datasets.

	N trajectories		Mean Interclaim Time		
Claim Number	Synthetic	Original	Synthetic	Original	p-value
1	200	200	19.11	19.61	0.658
2	132	132	13.87	12.97	0.390
3	48	48	7.49	8.18	0.567
4	17	18	6.60	7.43	0.595
5		4		2.97	
6		2		4.10	

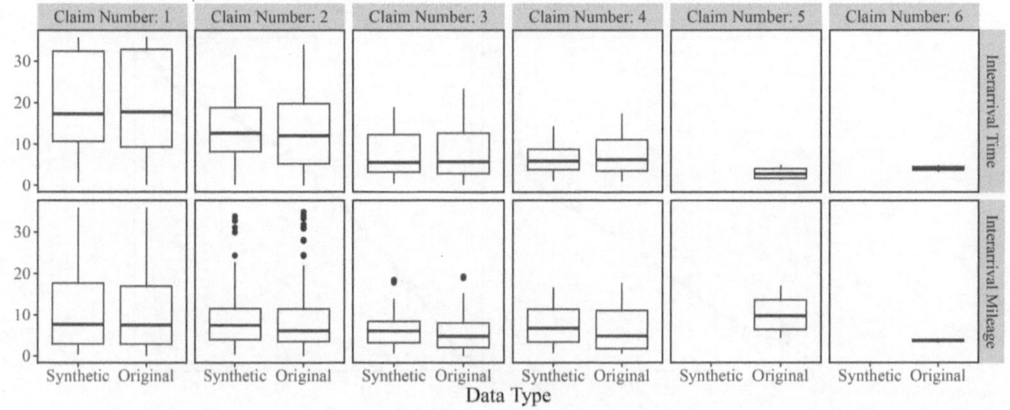

Figure 4. Comparison of interclaim distributions for age and mileage by claim number.

Table 2. Comparison of interclaim mileage by claim number for the original and synthetic datasets.

	N trajectories		Mean Interclaim Mileage		
Claim Number	Synthetic	Original	Synthetic	Original	p-value
1	200	200	11.15	11.04	0.914
2	132	132	8.77	8.85	0.934
3	48	48	6.45	5.98	0.610
4	17	18	7.42	6.43	0.569
5		4		10.28	
6		2		3.80	

5.3.3 *Number of claims by age t or mileage u*

The two datasets can also be compared by examining the distribution of the age at the time of the claims. This can be done by claim number, as shown for the interclaim time and mileage and it was found that the results are consistent with what is shown above, but are omitted here due to brevity. The distribution of claims can also be compared across all claim numbers. This allows for comparisons such as the proportion of claims that have been observed by a certain age or mileage. Figure 5 presents the empirical distribution of the age and mileage for both datasets, and shows strong similarity between the two datasets. Statistical comparisons of the datasets (e.g. KS-test) will be explored in future work.

The example presented above demonstrates a single application of SWANN for a single set of original data. This is, of course, insufficient for making judgments about the quality of the method. This will be addressed in the next section.

5.4 *Evaluation of SWANN – multiple replications*

In this section preliminary results relating to the evaluation of multiple replications of SWANN are presented. SWANN was repeated a large number of times to capture the variability. For this example two types of replication were considered. First, for a single original dataset, multiple synthetic datasets were generated. Secondly for the same input parameters, multiple original datasets were simulated and from each of these multiple synthetic datasets were generated.

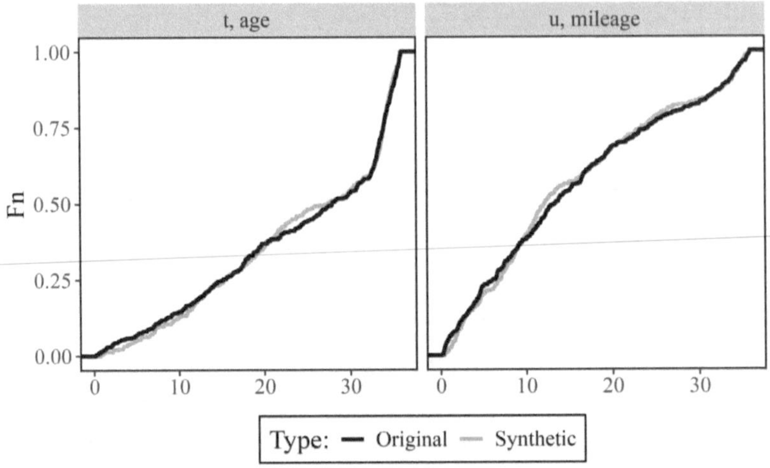

Figure 5. Empirical cumulative distribution function for age $F_n(t)$ and mileage $F_n(u)$ at the time of claims for the original and synthetic datasets.

Figure 6 shows the ECDF of one original dataset and the 30 synthetic datasets generated from it. The mean ECDF of the synthetic datasets is also shown. These plots indicate a similarity between number of claims by each age and mileage between the two datasets.

Figure 7 shows results from 30 synthetic datasets generated from each of the 30 original datasets. In particular, the figure shows the ECDF of the 30 original datasets (black), the mean ECDF of the 30 original datasets (red), the mean ECDFs of the 30 synthetics for each original (blue – this corresponds to the blue in Figure 6), and the mean of these means across all original datasets (light blue – this is effectively the mean of all the blue lines). What is of particular interest is the similarity between the light blue and red lines.

These results suggest that across multiple replications there is consistency between the original and synthetic datasets. It is possible to compare multiple instances on the other metric, using both graphical and statistical methods. This will be explored in future work.

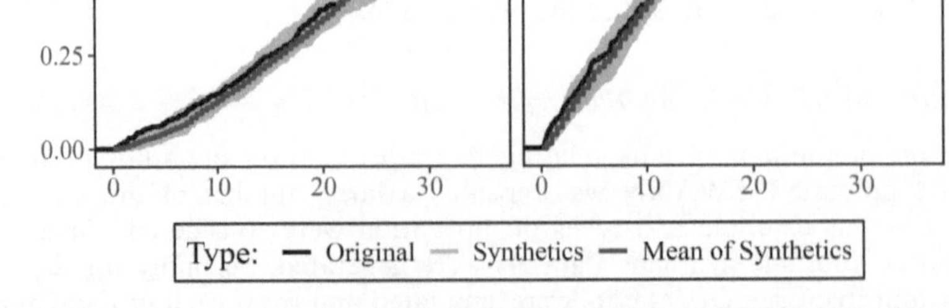

Figure 6. Empirical cumulative distribution function for age $F_n(t)$ and mileage $F_n(u)$ at the time of claims for 1 original and 30 synthetic datasets.

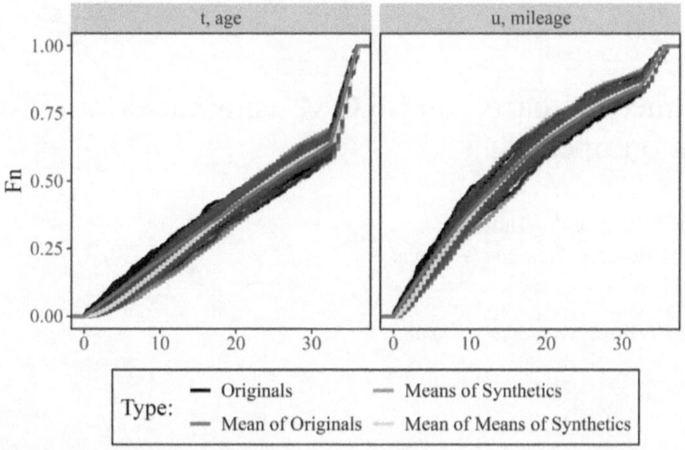

Figure 7. Empirical cumulative distribution function for age $F_n(t)$ and mileage $F_n(u)$ at the time of claims for 30 original and 30 synthetic datasets.

6 CONCLUSION

Synthetic data generation is a new topic in warranty analysis. This paper provides a first step towards generating synthetic data. In addition to expanding the evaluation methods for multiple replications, as mentioned above, in future work SWANN needs to be evaluated for different original parameters. The selection of the parameter k also requires investigation. Alternative generation methods such as jackknifing or bootstrapping could also be compared. The use of synthetic data generated using SWANN for warranty cost analysis is also an area for future work.

REFERENCES

Assefa, S. A., Dervovic, D., Mahfouz, M., Tillman, R. E., Reddy, P., and Veloso, M. (2021). Generating synthetic data in finance: Opportunities, challenges and pitfalls. *In Proceedings of the First ACM International Conference on AI in Finance, ICAIF '20*, pages 1–8, New York, NY, USA. Association for Computing Machinery.

Blischke, W. (1995). *Product Warranty Handbook*. CRC Press.

Blischke, W. R., Karim, M. R., and Murthy, D. N. P. (2011). *Warranty Data Collection and Analysis. Springer Series in Reliability Engineering*. Springer-Verlag, London.

Bonnéry, D., Feng, Y., Henneberger, A. K., Johnson, T. L., Lachowicz, M., Rose, B. A., Shaw, T., Stapleton, L. M., Woolley, M. E., and Zheng, Y. (2019). The promise and limitations of synthetic data as a strategy to expand access to state-level multi-agency longitudinal data. *Journal of Research on Educational Effectiveness*, 12(4):616–647.

Buggineni, V., Chen, C., and Camelio, J. (2024). Enhancing manufacturing operations with synthetic data: A systematic framework for data generation, accuracy, and utility. *Frontiers in Manufacturing Technology*, 4.

Chawla, N. V., Bowyer, K. W., Hall, L. O., and Kegelmeyer, W. P. (2002). SMOTE: Synthetic Minority Over-sampling Technique. *Journal of Artificial Intelligence Research*, 16:321–357.

Gonzales, A., Guruswamy, G., and Smith, S. R. (2023). Synthetic data in health care: A narrative review. *PLOS Digital Health*, 2(1):e0000082.

Hernandez, M., Epelde, G., Alberdi, A., Cilla, R., and Rankin, D. (2022). Synthetic data generation for tabular health records: A systematic review. *Neurocomputing*, 493:28–45.

Libes, D., Lechevalier, D., and Jain, S. (2017). Issues in synthetic data generation for advanced manufacturing. In *2017 IEEE International Conference on Big Data (Big Data)*, pages 1746–1754.

Recent Advances in Reliability and Maintenance Modeling – Okamura, Inoue & Xiao (Eds)

Estimation method based on SRGM using uncertain fault information on open BTS

S. Miyamoto, L. Zhou & Y. Tamura
Yamaguchi University, Ube, Yamaguchi, Japan

S. Yamada
Tottori University, Tottori, Yamaguchi, Japan

ABSTRACT: Bug Tracking System (BTS) is used for debug to manage reported faults in software development. BTS has been used in the reliability evaluation since it records detail information that can analyze. It also uses for Open Source Software (OSS) reliability evaluation. However, OSS project BTS rejects more faults than commercial project BTS. Furthermore, it takes long time for the reported fault to be confirmed. Therefore, it cannot apply the latest Software Reliability Growth Model (SRGM) when it is constructed by confirmed fault only. In this paper, we analyze the relationship between the reported faults and the fixed faults using the Hyper Text Transfer Protocol (HTTP) of client OSS fault data. Moreover, we propose the evaluation method of software reliability even in situations where it is unclear whether the reported fault is correct. As the result, we find the trend toward the correlation between the confirmed fault and the rejected fault. Moreover, we estimate the reliability growth curves of corrected faults with high accuracy using only reported faults.

1 INTRODUCTION

In software development, it is important to prevent making faults in the source code. However, it is practically difficult to prevent because the software is artifacts of human activity. Therefore, many developers try to detect and removed fault by unit test and integration test after cording. However, it is nearly impossible for the predefined test case to cover all faults in the testing process. Furthermore, the development costs are excessive when the tests are conducted for long periods of time. It is necessary to set an appropriate testing time by evaluating software reliability. In software reliability evaluation, The approach has been based on probability and deep learning theory (Lyu 1996; San *et al.* 2021; Tamura *et al.* 2023). The SRGM based on non-homogeneous Poisson process (NHPP) can easily grasp the convergence of faults. It is known that the software faults tends to follows NHPP. Several researchers has proposed various types of SRGM considering the situation of software development (Kapur *et al.* 2009; Li and Pham 2017; Musa *et al.* 1987). Many fault reports are used for the software reliability assessment. In commercial software development, the fault report is usually managed by closed BTS in the software development project. On the other hand, OSS fault report is managed by the open BTS by OSS development project. In the BTS, there is fault confirmation process for check the detail fault information. In the open BTS, more reports rejected and compared to closed BTS. Moreover, it tends to take longer for modifications in order to be confirmed. It is possible to make the SRGM based on NHPP more reflective of the actual state of the software by constructing only the faults that have been approved. However, the time is long until fault is confirmed. Then, it only can be shown old state SRGM based on NHPP. In this paper, we focus on the fault report until

 DOI: 10.1201/9781003491309-17

confirm. Moreover, we compare the SRGM based on NHPP that constructed from the reported fault data and the confirmed fault data.

2 ANALYSIS OF FAULT REPORTED ON THE OPEN BTS

2.1 *Dataset collection*

We collected the fault dataset for analyze the difference in trends of the reported faults and the confirmed fault from the open BTS. In this paper, we focused on the HTTP client OSS that is technically mature and does not have numerous functions. Moreover, we narrow the popular HTTP client OSS on python package index (PyPI) to collect small noise dataset (Python Software Foundation). Then, we select OSS based on the following criteria:

(1) Including HTTP word in the OSS description on PyPI.
(2) Include HTTP Client function in the OSS.
(3) Top 4 libraries in terms of downloads that meet the above two criteria on PyPI from March 24th to 31st, 2024.

As the result, we selected HTTP client OSS as shown in Table 1. There were 11 OSS with HTTP in the tag among the top 100 software downloads in PyPI. Seven of them were non-HTTP client OSS such as HTTP Server. Then we selected urllib3, requests, aiohttp, and h11 (urllib3, Python Software Foundation, aio-libs, Hyper).

Table 1. Top 4 most downloaded HTTP clients on PyPI from March 24th to 31st, 2024.

Total rank	OSS	Number of downloads
3	urllib3	128802654
5	requests	110197755
63	aiohttp	27269399
99	h11	18254159

For the selected OSS, we conducted an analysis using fault data reported in the issue data on GitHub and the number of downloads data on BigQuery (GitHub, Google).

2.2 *Relationship the confirmed fault and the rejected one*

We consider the SRGM based on NHPP waveform. The SRGM based on NHPP is calculated to measure the degree of fault convergence. In popular OSS, the number of reported and confirmed faults is enormous. Therefore, if the tendency of each occurrence is the same, the SRGM based on NHPP waveforms are expected to be similar. We can compare the two waveforms using normalization. In this paper, we normalized the x-axis and y-axis of the SRGM based on NHPP by dividing by each maximum value. Here, it is known that the SRGM based on NHPP in OSS is greatly affected by the number of users (Miyamoto *et al.* 2023). Therefore, we use the execution time of OSS considering the number of users. The execution time at i times is as follows:

$$t_i = \Delta T_i n_i, \tag{1}$$

where, n_i is the number of downloads at i times, T_i is the duration of observing n_i, respectively. Figure 1 shows the normalized SRGM based on NHPP. Also, we consider the fitting of the curve from the mean absolute error (MAE). Given the coordinates (x_1, y_1) and (x_2, y_2)

of two points, (x_3, y_3) between them is calculated as:

$$y_3 = \frac{y_2 - y_1}{x_2 - x_1}(x_3 - x_1) + y_1. \tag{2}$$

From the linearly interpolated function, we obtain the average error when the execution time is divided into 100 equal parts. Table 2 shows the MAE of normalized SRGM based on NHPP.

From Table 2, it can be seen that there is no significant difference between the waveforms of reported fault SRGM based on NHPP and confirmed fault SRGM based on NHPP. Also, it can be seen that there is no significant difference in each error rate depending on the number of downloads. These results show that the waveforms of reported and confirmed fault SRGM based on NHPP are similar. It is also shown that there is the possibility to estimate the waveform of certified fault SRGM based on NHPP from reported fault SRGM based on NHPP.

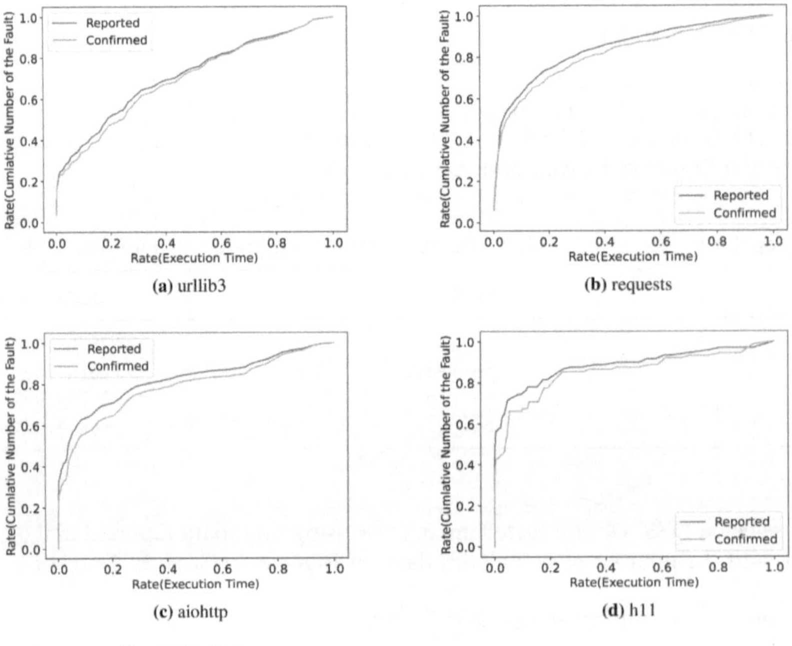

Figure 1. The normalized SRGM.

Table 2. The evaluation result by MAE for normalized SRGM based on NHPP.

Total rank	OSS	MAE
3	urllib3	1.702×10^{-2}
5	requests	2.728×10^{-2}
63	aiohttp	3.339×10^{-2}
99	h11	1.057×10^{-2}

2.3 *Relationship between confirmed fault and rejected fault*

We investigated the relationship between confirmed fault and rejected fault for waveform of SRGM based on NHPP. The number of reported fault consists of the number of confirmed fault and the number of rejected fault. Then, we analyzed the relationship between confirmed fault and rejected fault. Figure 2 shows the scatter plot and regression line for confirmed fault and rejected fault. The regression line was derived using the following equation as follows:

$$\min_{w} ||Xw - y||_2^2, \tag{3}$$

where w is the coefficients of liner function. We also performed an uncorrelation test on each OSS. The significance level is set at 5% and the population correlation coefficient is assumed to be zero. Table 3 shows the correlation coefficients and p-values for confirmed and rejected faults for each OSS. From Table 3, the p-value is well below the significance level and the hypothesis is rejected for each OSS. Therefore, we found that each OSS is not uncorrelated. Moreover, correlation coefficients exceeding 0.3 are observed for urllib3, requests, and aiohttp. We found that when the number of reported faults is large, the correlation coefficient between rejected faults and confirmed ones tends to be proportional. On the other hand, when the number of reported faults is small, the results of h11 show that there is no proportional relationship.

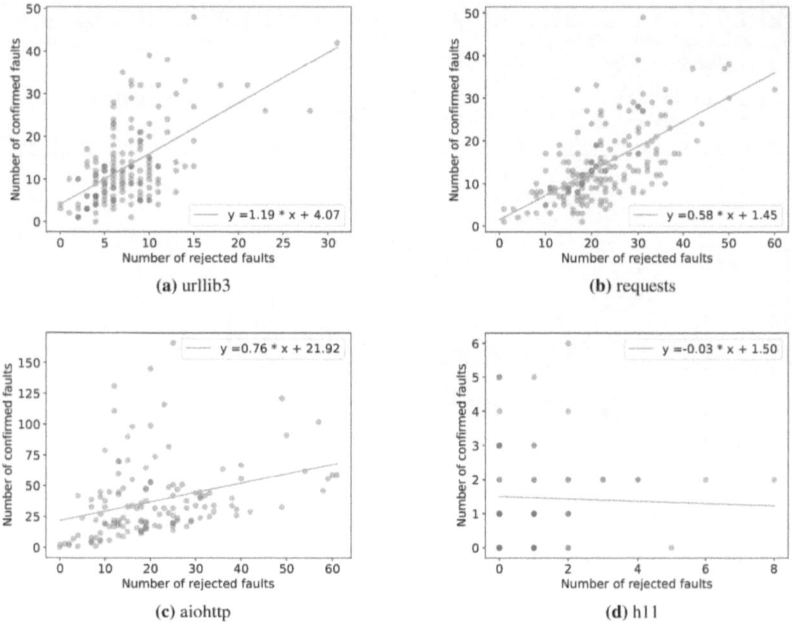

Figure 2. The scatter plot of the number of rejected faults and the number of confirmed ones.

2.4 *Distribution of each type of the faults*

It is known that the distribution of the number of fault occurrences follows Poisson distribution. Since the software fault detection process is NHPP, it does not show the perfect Poisson distribution, although it does show close form. On the other hand, reported fault includes rejected fault that didn't confirm. Therefore, we analyze the distribution of each type of fault. Figure 3 shows the distribution of fault types in each OSS.

Figure 3 shows that the confirmed and rejected fault results are closer to the strong Poisson distribution than the reported result for each OSS. In addition, the reported faults do not show the distribution that deviates significantly from the Poisson distribution. Therefore,

Table 3. The evaluation result by correlation coefficient.

OSS	Correlation	P-value
urllib3	5.680×10^{-1}	2.599×10^{-15}
requests	6.405×10^{-1}	2.988×10^{-21}
aiohttp	3.415×10^{-1}	4.416×10^{-5}
h11	2.177×10^{-1}	2.713×10^{-2}

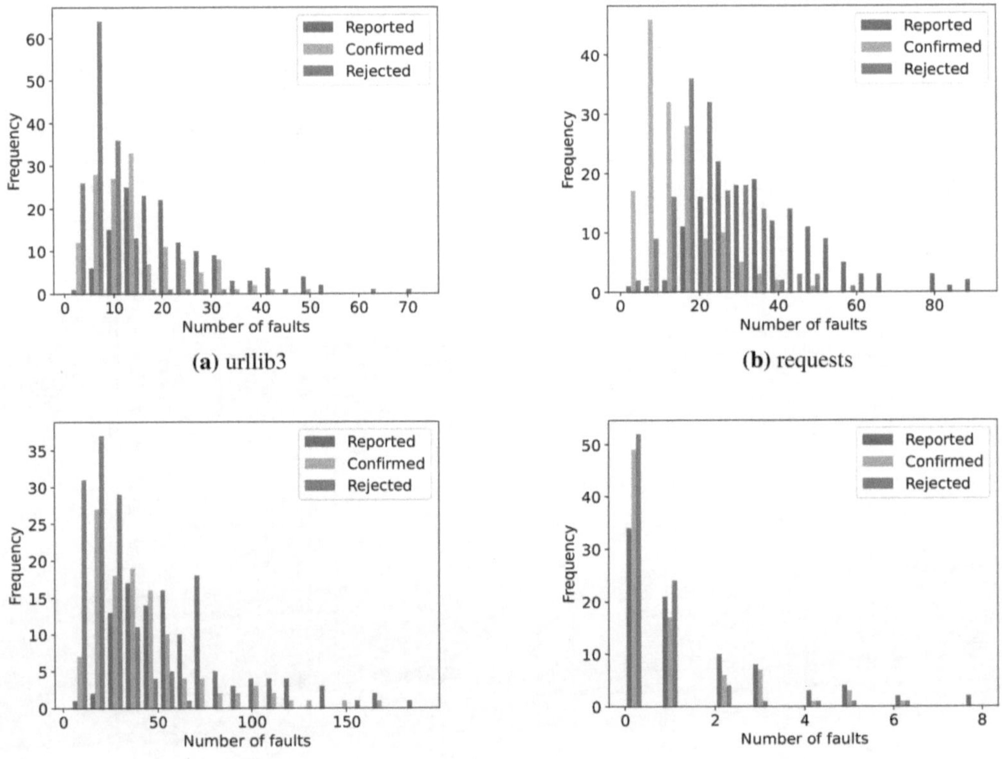

(a) urllib3 (b) requests (c) aiohttp (d) h11

Figure 3. The distribution for each types of faults.

the number of reported faults can be treated as an NHPP as well as the number of confirmed faults. In NHPP, the probability of observing m fault at time t is as follows:

$$\Pr\{N(t) = m\} = \frac{\{F(t)\}^m}{m!}\exp(-F(t)), \tag{4}$$

where $F(t)$ is the distribution function. In SRGM based on NHPP, the mean value function is $F(t)$. Furthermore, it fluctuates continuously. Therefore, if the original measurements are data from NHPP, the trend is removed and the data show Poisson distribution centered at about zero. Then, moving average is subtracted from the measured values to check whether each data is NHPP. Figure 4 shows the distribution of the number of fault detrended by moving average for each OSS. Figure 4 shows the normal distribution for all fault types. Therefore, it can be seen that each fault follows NHPP.

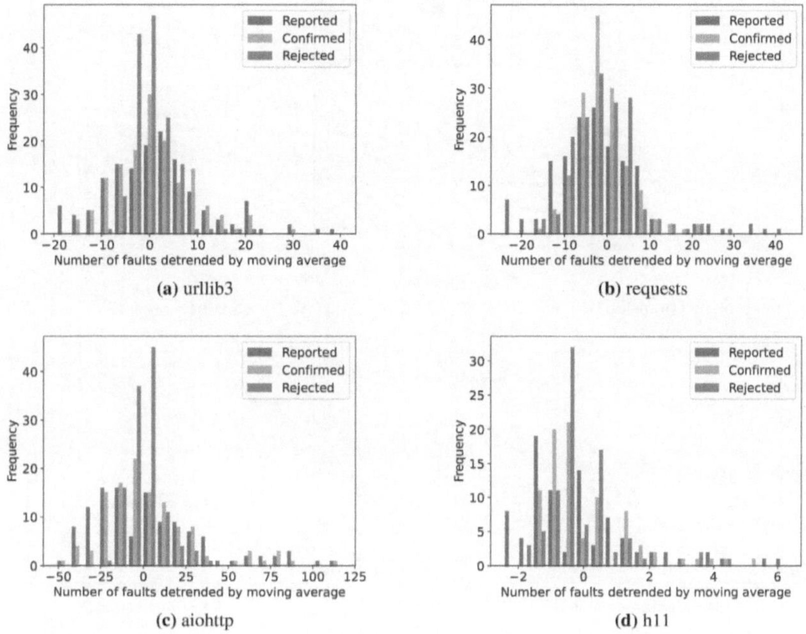

(a) urllib3 (b) requests (c) aiohttp (d) h11

Figure 4. The fault distribution based on the moving average for each types of faults.

3 NUMERICAL EXAMPLES

From Section 2, there was no significant difference between the waveforms of SRGM based on NHPP constructed with reported faults and SRGM based on NHPP constructed with confirmed faults. The analysis also showed that the proportionality relationship where the number of rejected faults increases as the number of confirmed faults increases, and that SRGM based on NHPP waveforms equivalent to those of confirmed faults can be obtained even if only reported faults are constructed. Therefore, we estimated each SRGM based on NHPP and compared the errors. Here, it is necessary to correct the SRGM based on NHPP for the number of reported faults and the waveforms for Confirmed faults. Therefore, for each SRGM based on NHPP, we normalized each SRGM based on NHPP. In this paper, exponential model, inflection S-shaped model, and logarithmic Poison execution time model proficient model, which are frequently utilized, are applied as mean value functions (Goel A. L. *et al.* 1979; Musa J. D. *et al.* 1987; Ohba M. 1984). We performed estimation based on the maximum likelihood estimation method using each mean value function and compared the differences in waveforms. Figure 5 shows the results of SRGM based on NHPP estimation in each OSS. Table 4 shows evaluation results by mean absolute percentage error (MAPE).

Table 4. The evaluation results by MAPE of the reported fault and the measured SRGM based on NHPP in each OSS.

OSS	Exp	Inf	Log
urllib3	4.297×10^{-2}	1.782×10^{-1}	4.141×10^{-5}
requests	4.749×10^{-2}	5.554×10^{-2}	3.190×10^{-4}
aiohttp	6.617×10^{-2}	6.848×10^{-2}	9.447×10^{-4}
h11	5.148×10^{-2}	5.355×10^{-2}	3.496×10^{-3}

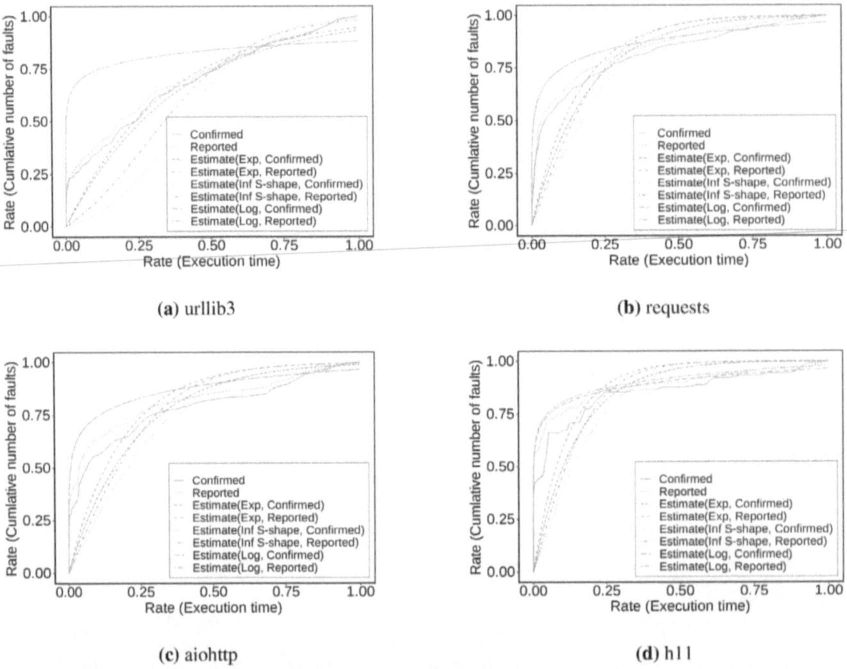

Figure 5. The estimated SRGM.

From Figure 5, except for the results for the learned S-Shape SRGM based on NHPP in urllib3, the MAPE of the SRGM based on NHPP waveforms is within 7%. In particular, the logarithmic type has the MAPE below 1% in all cases. It can be seen that the reported fault SRGM based on NHPP and confirmed SRGM based on NHPP waveforms are very similar. Furthermore, the results for h11, which has the lowest number of reported faults among the four OSS's selected for this paper, are compared with the other models in terms of MAPE. It can be concluded that the number of reported faults can be used to predict the number of confirmed fault SRGM based on NHPP even when the number of reported faults is small.

4 CONCLUDING REMARKS

In this paper, we have analyzed the relationship between the reported faults and the confirmed fault on the open BTS for the popular HTTP libraries on PyPI. Furthermore, we analyzed the distribution of each type of the fault for OSS. Then, we considered the applicability of the reported fault SRGM based on NHPP for estimating the confirmed fault SRGM based on NHPP waveform by comparing the waveform of estimated mean value function. As the result, we found that the waveforms of SRGM based on NHPP constructed by confirmed and reported fault are very similar. We also found that the confirmed and the rejected faults are not uncorrelated. Furthermore, we have shown that each type of fault follows NHPP. We have found that the errors in the SRGM based on NHPP waveforms estimated by the reported faults were small. In the future, we will further examine the applicability of this research by analyzing OSS with the few number of downloads.

ACKNOWLEDGMENTS

This work was supported in part by the JSPS KAKENHI Grant No. 23K11066 in Japan.

REFERENCES

aio-libs., "aiohttp," https://github.com/aio-libs/aiohttp.

GitHub., "GitHub," https://github.com/.

Goel, A. L., and Okumoto, K. (1979). Time-dependent error-detection rate model for software reliability and other performance measures, *IEEE transactions on Reliability*, vol. 28, no. 3, pp. 206–211.

Google., "BigQuery," https://cloud.google.com/bigquery.

Hyper., "h11," https://github.com/python-hyper/h11.

Kapur, P. K., Shatnawi, O., Aggarwal, A. G., and Kumar, R. (2009), Unified framework for developing testing effort dependent software reliability growth models, *WSEAS Transactions on Systems*, vol. 8, no. 4, pp. 521–531.

Li, Q., and Pham, H. (2017). NHPP software reliability model considering the uncertainty of operating environments with imperfect debugging and testing coverage, *Applied Mathematical Modelling*, vol. 51, pp. 68–85.

Lyu, M. R. (1996), *Handbook of Software Reliability Engineering*, IEEE computer society press Los Alamitos, vol. 222.

Miyamoto, S., Tamura, Y., and Yamada, S. (2023). A method of reliability assessment based on stochastic model for open source software, *Proceedings of the 28th ISSAT International Conference on Reliability and Quality in Design*, pp. 274–278.

Musa, J. D., Iannino, A., & Okumoto, K. (1987). *Software reliability: measurement, prediction, application*, McGraw-Hill, Inc.

Musa, J. D., Iannino, A., and Okumoto, K. (1987). *Software Reliability: Measurement, Prediction, Application*, McGraw-Hill, Inc.

Ohba, M. (1984). Inflection S-shaped software reliability growth model, *Stochastic Models in Reliability Theory: Proceedings of a Symposium Held in Nagoya, Japan, April 23–24, 1984*, pp. 144–162, Springer.

Python Software Foundation., "PyPI," https://PyPI.org/.

Python Software Foundation., "requests," https://github.com/psf/requests.

San, K. K., Washizaki, H., Fukazawa, Y., Honda, K., Taga, M., and Matsuzaki, A. (2021), Deep cross-project software reliability growth model using project similarity-based clustering, *Mathematics*, vol. 9, no. 22, p. 2945.

Tamura, Y., Miyamoto, S., Zhou, L., Anand, A., Kapur, P. K., and Yamada, S. (2024). OSS reliability assessment method based on deep learning and independent Wiener data preprocessing, *Journal International Journal of System Assurance Engineering and Management*, Vol. ahead-of-print No. ahead-of-print, Springer, https://doi.org/10.1007/s13198-024-02288-w, pp. 1–9.

urllib3., "urllib3," https://github.com/urllib3/urllib3.

Recent Advances in Reliability and Maintenance Modeling – Okamura, Inoue & Xiao (Eds)

New schemes of mean deviation, redundant x units and scalability in reliability theory

T. Nakagawa & S. Mizutani
Department of Business Administration, Aichi Institute of Technology, Toyta, Japan

ABSTRACT: To protect the environment of Earth and maintenance of all living things, we have to make the development of reliability theory, as information technologies have advanced rapidly. This chapter proposes three new schemes of mean deviation, redundant x units, and scalability. Mean deviation shows good examples of statistical applications to reliability engineering, redundant x units is a mathematical extension of a standard redundant system with n units, and scalability is a property in the context of system design and structure to handle a growing of work. Three schemes would be interested in some researchers and be much useful for theoretical development and practical application in fields of reliability, statistics and others.

1 INTRODUCTION

As information technologies advance rapidly and environment conservation progresses very first, it has become necessary to continue discussing more important subjects from the viewpoints of reliability. That is, reliability should be a requirement in the maintenance of all things and environmental protection. From the above viewpoints, we propose three new schemes of *Mean deviation*, *Redundant with x units* and *Scalability*.

Mean deviation is a measure in statistics which makes the differences between observed data and their mean. This is applied to data analysis in quality control, financial analysis, education assessment, and market research. However, this is less commonly used than other measures such as standard deviation and variance, because the calculation of absolute differences is required [1]. When the unit is replaced preventively at time T, we obtain the mean deviation between time T and failure, and the deviation rate per mean replacement time, and derive optimal time T^* to minimize them.

High system reliability can be achieved by redundancy and maintenance. The most typical model is a standard parallel system, which consists of n identical units in parallel. The reliabilities of many redundant systems were computed and summarized [2]. A variety of random parallel systems [3] and hierarchical structures [4] were taken up, and their optimization problems were discussed. We propose redundant systems with x units and investigate their reliability measures such as the mean time to failure and failure rate, and consider their replacement models.

Scalability is the measure of a system's ability to increase or decrease in performance and cost in response to changes in application and system processing demands in information technology such as software and hadware engineering [5]. For example, scalability is important in designing software systems that can handle increasing loads of users or data [6], and in designing hardware systems that can be expanded or upgraded to meat growing demands [7]. We apply scalability to reliability models, and compute reliability, mean time to failure (MTTF), failure rate and median of redundant systems as the total number of n units increases by one. A series-parallel system is given as one example.

 DOI: 10.1201/9781003491309-18

2 MEAN DEVIATION

Mean deviation is a measure in statistics which represents the differences between observed data x_i $(i = 1, 2, \ldots, n)$ and their mean μ [1] as mean deviation = $(1/n)\sum_{i=1}^{n}|x_i - \mu|$.

Most important problem in maintenance model is how to maintain preventively an operating unit just before its failure. A variety of maintenance policies has been considered, their optimal policies have been discussed extensively [8–11], and compared as Which-Is-Better problems [12].

It is assumed in Section 2 that the failure time X of an operating unit has a general distribution $F(t) \equiv \Pr\{X \leq t\}$, a density function $f(t) \equiv \mathrm{d}F(t)/\mathrm{d}t$ and its finite $\mu \equiv E\{X\} = \int_0^\infty \overline{F}(t)\mathrm{d}t$, where $\overline{\Phi}(t) \equiv 1 - \Phi(t)$ for any function $\Phi(t)$. These notations are used through this paper.

Preventive maintenance aims to perform some necessary maintenances of the unit before its failure. For this purpose, we take up a standard age replacement in which the unit is replaced at time T $(0 \leq T < \infty)$ or at failure, whichever occurs first. Then, the mean deviation between time T and failure is [12,13]

$$D_1(T) = \int_0^\infty |T - t|\mathrm{d}F(t) = \int_0^T F(t)\mathrm{d}t + \int_T^\infty \overline{F}(t)\mathrm{d}t. \tag{1}$$

Clearly, $D_1(0) \equiv \mu$ and $D_1(\infty) \equiv \lim_{T\to\infty} D_1(T) = \infty$.

We find optimal T_1^* to minimize $D_1(T)$ and call T_1^* *deviation time*. Differentiating $D_1(T)$ with respect to T and setting it equal to zero, $F(T) = 1/2$, which is a *median* T_M of a failure distribution $F(t)$, i.e., optimal deviation time T_1^* is a median of $F(t)$.

2.1 *Deviation rate*

We want to use an operating unit as long as possible. So that, we define as Deviation rate ≡ (Mean deviation)/(Mean time to maintenance). When the unit is replaced preventively at time T $(0 < T \leq \infty)$, the mean time to replacement is

$$T\overline{F}(T) + \int_0^T t\mathrm{d}F(t) = \int_0^T \overline{F}(t)\mathrm{d}t. \tag{2}$$

Thus, the deviation rate is

$$D_2(T) = \frac{\int_0^T F(t)\mathrm{d}t + \int_T^\infty \overline{F}(t)\mathrm{d}t}{\int_0^T \overline{F}(t)\mathrm{d}t}, \qquad \lim_{T\to 0} D_2(T) = \lim_{T\to\infty} D_2(T) = \infty. \tag{3}$$

We find optimal T_2^* to minimize $D_2(T)$. Differentiating $D_2(T)$ with respect to T and setting it equal to zero,

$$\frac{F(T)}{\overline{F}(T)}\int_0^T \overline{F}(t)\mathrm{d}t - \int_0^T F(t)\mathrm{d}t = \mu, \tag{4}$$

whose left-hand side increases strictly with T from 0 to ∞. Thus, there exists a finite and unique T_2^* $(0 < T_2^* < \infty)$ which satisfies (4), and the resulting rate is

$$D_2(T_2^*) = \frac{F(T_2^*)}{\overline{F}(T_2^*)} - 1. \tag{5}$$

It is easily shown that $T_2^* > T_M$, because $F(T_2^*) > 1/2$.

Table 1. Optimal T_1^* and T_2^* when $F(t) = 1 - e^{-(t/10)^m}$.

	$m = 1.0$		$m = 1.5$		$m = 2.0$		$m = 2.5$	
α	T_1^*	T_2^*	T_1^*	T_2^*	T_1^*	T_2^*	T_1^*	T_2^*
0.1	0.1054	0.4371	0.2231	0.4656	0.3246	0.5102	0.4065	0.5525
0.2	0.2231	0.6327	0.3679	0.6259	0.4724	0.6530	0.5488	0.6828
0.3	0.3567	0.8025	0.5029	0.7560	0.5972	0.7639	0.6621	0.7809
0.4	0.5108	0.9693	0.6390	0.8773	0.7147	0.8639	0.7644	0.8673
0.5	0.6931	1.1462	0.7832	1.0002	0.8326	0.9622	0.8636	0.9504
μ	1.0000		0.9027		0.8862		0.8873	

2.2 *Deviation factor*

We introduce the following two kinds of deviations: If the unit fails at time t after time T $(0 \le T < \infty)$, this causes a deviation time $\alpha(t - T)$, and if it fails before time T, this causes a deviation time $\beta(T - t)$. In most replacement models, load or overhead after failure would be higher than that before failure. So that, it is assumed that $0 < \alpha \le \beta < 1$ and $\alpha + \beta = 1$.

The mean deviation between time T and failure is

$$D_1(T) = \beta \int_0^T (T - t)\mathrm{d}F(t) + \alpha \int_T^\infty (t - T)\mathrm{d}F(t) = \beta \int_0^T F(t)\mathrm{d}t + \alpha \int_T^\infty \overline{F}(t)\mathrm{d}t. \tag{6}$$

Clearly, $D_1(0) = \alpha\mu$ and $D_1(\infty) \equiv \lim_{T\to\infty} D_1(T) = \infty$.

We find optimal T_1^* to minimize $D_1(T)$. Differentiating $D_1(T)$ with respect to T and setting it equal to zero, $F(T) = \alpha/(\alpha + \beta) = \alpha$, which is called *Deviation factor*. In particular, when $\alpha = \beta = 1/2$, $F(T) = 1/2$, i.e., $T_1^* = T_M$, which is a median of a failure distribution $F(t)$.

Furthermore, from (2) and (6), the deviation rate is

$$D_2(T) = \frac{\beta \int_0^T F(t)\mathrm{d}t + \alpha \int_T^\infty \overline{F}(t)\mathrm{d}t}{\int_0^T \overline{F}(t)\mathrm{d}t}, \quad \lim_{T\to 0} D_2(T) = \lim_{T\to\infty} D_2(T) = \infty. \tag{7}$$

We find optimal T_2^* to minimize $D_2(T)$. Differentiating $D_2(T)$ with respect to T and setting it equal to zero,

$$\frac{1 - \alpha}{\overline{F}(T)} \int_0^T [F(T) - F(t)]\mathrm{d}t = \alpha\mu, \tag{8}$$

whose left-hand side increases strictly with T from 0 to ∞. Thus, there exists a finite and unique T_2^* $(0 < T_2^* < \infty)$ which satisfies (8), and the resulting rate is

$$D_2(T_2^*) = \frac{F(T_2^*) - \alpha}{\overline{F}(T_2^*)}, \tag{9}$$

and $T_2^* > T_1^*$, because $F(T_2^*) > \alpha$. In addition, T_2^* increases strictly with α from T_1^* to ∞.

Example 1. When $F(t) = 1 - e^{-t^m}$ $(m \ge 1)$, T_1^* satisfies $1 - e^{-T^m} = \alpha$, and from (8), T_2^* satisfies

$$\frac{1}{\Gamma(1 + 1/m)} \int_0^T \left[e^{-t^m + T^m} - 1\right]\mathrm{d}t = \frac{\alpha}{1 - \alpha}.$$

Table 1 presents optimal T_1^* and T_2^* when $F(t) = 1 - e^{-t^m}$. This indicates that both T_1^* and T_2^* increase strictly with α and m, and $T_2^* > T_1^*$ and $\mu > T_1^* = T_M$ in case of $\alpha = 0.5$, T_1^* increase strictly with m, however, T_2^* has no nonotone property of m.

3 REDUNDANT SYSTEM WITH x UNITS

Consider redundant systems with n $(n = 1, 2, \ldots)$ identical units where each unit has the same reliability $q \equiv 1 - p$ $(0 < q \le 1)$. Then, a parallel system and a series system have the respective reliabilities [2]:

$$R_P(n) = 1 - p^n, \quad R_S(n) = q^n \quad (n = 1, 2, \ldots). \tag{10}$$

3.1 *Parallel system with x units*

Suppose that a parallel system consists of a positive real number x $(0 < x < \infty)$ units. Then, replacing n with x in (10), reliability of the system is $R_P(x) = 1 - p^x$, which increases strictly with x from 0 to 1.

It is assumed that each unit has a failure distribution $F(t)$, a density function $f(t) \equiv \mathrm{d}F(t)/\mathrm{d}t$, and failure rate $h(t) \equiv f(t)/\overline{F}(t)$. Denoting that $p = F(t)$ formally, reliability and MTTF are, respectively,

$$R_P(t; x) = 1 - F(t)^x, \quad \mu_P(x) = \int_0^\infty [1 - F(t)^x]\mathrm{d}t. \tag{11}$$

3.2 *Failure rate*

Failure rate is, from (11),

$$h_P(t; x) = \frac{xh(t)[F(t)^{x-1} - F(t)^x]}{1 - F(t)^x}. \tag{12}$$

We investigate the reliability properties of failure rate $h_P(t; x)$ when $h(t)$ increases with t from $h(0)$ to $h(\infty)$. Denoting for $0 < x < \infty$ and $0 < y < 1$,

$$L(y) \equiv \frac{y^{x-1} - y^x}{1 - y^x}, \quad \lim_{y \to 1} L(y) = \frac{1}{x}, \quad \lim_{y \to 0} L(y) = \begin{cases} \infty & \text{for } 0 < x < 1, \\ 1 & \text{for } x = 1, \\ 0 & \text{for } x > 1, \end{cases}$$

$$L(y) = \begin{cases} \text{decreases with } y \text{ from } \infty \text{ to } 1/x & \text{for } 0 < x < 1, \\ 1 & \text{for } x = 1, \\ \text{increases with } y \text{ from } 0 \text{ to } 1/x & \text{for } x > 1. \end{cases}$$

Thus, we have

$$h_P(t; x) = \begin{cases} \text{unknown} & \text{for } 0 < x < 1, \\ h(t) & \text{for } x = 1, \\ \text{increases witht from } h(0) \text{ to } h(\infty) & \text{for } x > 1. \end{cases} \tag{13}$$

In particular, when $F(t) = 1 - \mathrm{e}^{-\lambda t}$,

$$R_P(t;x) = 1 - (1 - \mathrm{e}^{-\lambda t})^x, \ \mu_P(x) = \frac{1}{\lambda}\sum_{j=1}^{\infty}\left(\frac{1}{j} - \frac{1}{j+x}\right).$$

$$h_P(t;x) = \frac{\lambda x \mathrm{e}^{-\lambda t}(1 - \mathrm{e}^{-\lambda t})^{x-1}}{1 - (1 - \mathrm{e}^{-\lambda t})^x}, \qquad \lim_{t\to\infty} h(t;x) = \lambda. \tag{14}$$

Thus, we have

$$h_P(t;x) = \begin{cases} \text{decreases with } t \text{ from } \infty \text{ to } \lambda & \text{for } 0 < x < 1, \\ \lambda & \text{for } x = 1, \\ \text{increases with } t \text{ from } 0 \text{ to } \lambda & \text{for } x > 1. \end{cases}$$

We cannot know whether $h_P(t;x)$ increases or decreases in the case where $h(t)$ increases strictly with t from $h(0)$ to $h(\infty)$ for $0 < x < 1$. It is assumed that the failure time has a Weibull distribution $F(t) = 1 - \mathrm{e}^{-(\lambda t)^m}$ $(m > 1)$. Then, failure rate is

$$h_P(t;x) = \frac{mx\lambda(\lambda t)^{m-1}\mathrm{e}^{-(\lambda t)^m}[1 - \mathrm{e}^{-(\lambda t)^m}]^{x-1}}{1 - [1 - \mathrm{e}^{-(\lambda t)^m}]^x}. \tag{15}$$

We analyze the properties of $h_P(t;x)$ for $0 < x < 1$ and $m > 1$. Letting

$$L_1(t) \equiv \frac{t^{m-1}}{(1 - \mathrm{e}^{-t^m})^{1-x}},$$

we have $\lim_{t\to\infty} L_1(t) = \infty$,

$$\lim_{t\to 0} L_1(t) = \frac{1}{1-x}\frac{m-1}{m}\lim_{t\to 0}\frac{(1 - \mathrm{e}^{-t^m})^x}{t} = \begin{cases} 0 & \text{for } mx > 1, \\ 1 & \text{for } mx = 1, \\ \infty & \text{for } mx < 1. \end{cases}$$

Therefore, we may easily expect the following result:

- If $mx > 1$, then $h_P(t;x)$ increases strictly with t from 0 to ∞.
- If $mx = 1$, then $h_P(t;x)$ increases strictly with t from λ to ∞.
- If $mx < 1$, then $h_P(t;x)$ decreases strictly with t from ∞ firstly, and after that, increases strictly to ∞.

When $F(t) = 1 - \mathrm{e}^{-t^2}$, failure rate is

$$h_P(t;x) = \frac{2xt\mathrm{e}^{-t^2}(1 - \mathrm{e}^{-t^2})^{x-1}}{1 - (1 - \mathrm{e}^{-t^2})^x}.$$

Figure 1 presents failure rate $h_P(t;x)$ for $x = 0.3$, 0.5, 0.8. As expected above, $h_P(t;x)$ increases strictly from 0 to ∞ for $x = 0.8$, increases strictly from $\lambda = 1$ to ∞ for $x = 0.5$, and decreases first, and after that, increases strictly to ∞ for $x = 0.3$, and $h_P(t, 0.3) > h_P(t, 0.5) > h_P(t, 0.8)$.

The mean deviation between time T and system failure is, from (1),

$$D_P(T) = \int_0^T F(t)^x \mathrm{d}t + \int_T^\infty [1 - F(t)^x]\mathrm{d}t, \tag{16}$$

and optimal T_P^* to minimize $D_P(T)$ is given by $F(T)^x = 1/2$.

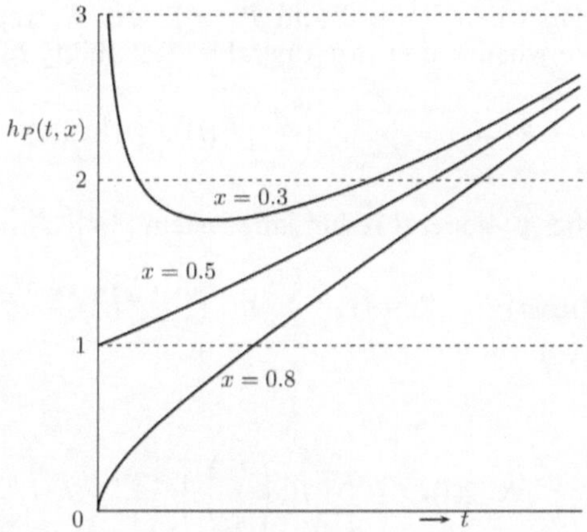

Figure 1. Failure rate $h_P(t;x)$ for $x = 0.3, 0.5, 0.8$, when $F(t) = 1 - e^{-t^2}$.

4 SCABILITY

We consider redundant systems with n identical units, and compute their reliabilities, MTTFs, failure rates and medians, as the total number n increases by one.

4.1 *K-out-of-n system*

We consider a K-out-of-n system in which the system operates if and only if K $(K = 1, 2,\ldots, n)$ units of the total n $(n = 1, 2,\ldots)$ units are operable, where each unit has a failure distribution $F(t)$, mean $\mu \equiv \int_0^\infty \overline{F}(t)\mathrm{d}t$, and failure rate $h(t) \equiv f(t)/\overline{F}(t)$. Then, reliability is [2,8]

$$R_{K,n}(t) = \sum_{j=K}^{\infty} \binom{n}{j} \overline{F}(t)^j F(t)^{n-j}. \tag{17}$$

We obtain the following scabilities of the system: For reliability $R_{K,n}(t)$,

$$\begin{aligned} \widetilde{R}_{K,n}(t) &\equiv R_{K,n+1}(t) - R_{K,n}(t) = \binom{n}{k-1} \overline{F}(t)^K F(t)^{n+1-K} > 0, \\ r_{K,n}(t) &\equiv \widetilde{R}_{K,n+1}(t) - \widetilde{R}_{K,n}(t) = \binom{n+1}{K-1} \overline{F}(t)^K F(t)^{n+1-K} \left[\frac{K-1}{n+1} - \overline{F}(t)\right]. \end{aligned} \tag{18}$$

Thus, we have

(1) If $n + 1 < (K - 1)/\overline{F}(t)$, then $r_{K,n}(t)$ decreases with n.
(2) If $n + 1 \geq (K - 1)/\overline{F}(t)$, then $r_{K,n}(t)$ increases with n.

Letting n^* be a minimum which satisfies $n^* + 1 \geq (K - 1)/\overline{F}(t)$, the difference $r_{K,n}(t)$ decreases with n, takes a minimum at $n = n^* + 1$, and after that, increases with n.

4.2 *General redundant system*

Suppose that K is a random variable with a probability function $p_k \equiv \Pr\{K = k\}$ $(k = 0, 1,\ldots, n)$ for a specified n $(n = 1, 2,\ldots)$, where $p_0 = 0$. It is noted that

$P_k \equiv \Pr\{K \le k\} = \sum_{j=0}^{k} p_j$, where $P_0 \equiv 0$ and $P_n = 1$, and P_k represents the probability that the system is operable when k units are operable. Reliability of the system at time t is

$$R_{P,n}(t) = \sum_{k=0}^{n} P_k \binom{n}{k} \overline{F}(t)^k F(t)^{n-k}, \tag{19}$$

which represents reliabilities of general redundant system [3,4]. Thus,

$$\widetilde{R}_{P,n}(t) \equiv R_{P,n+1}(t) - R_{P,n}(t) = \sum_{k=0}^{n} p_{k+1} \binom{n}{k} \overline{F}(t)^{k+1} F(t)^{n-k} > 0. \tag{20}$$

MTTF of the system is

$$\begin{aligned} \mu_{P,n} &\equiv \int_0^\infty R_{P,n}(t)\mathrm{d}t = \sum_{k=0}^{n} P_k \binom{n}{k} \int_0^\infty \overline{F}(t)^k F(t)^{n-k} \mathrm{d}t, \\ \widetilde{\mu}_{P,n} &\equiv \mu_{P,n+1} - \mu_{P,n} = \sum_{k=0}^{n} p_{k+1} \binom{n}{k} \int_0^\infty \overline{F}(t)^{k+1} F(t)^{n-k} \mathrm{d}t. \end{aligned} \tag{21}$$

In particular, when $F(t) = 1 - \mathrm{e}^{-\lambda t}$,

$$\begin{aligned} \mu_{P,n} &= \sum_{k=1}^{n} P_k \binom{n}{k} \int_0^\infty \mathrm{e}^{-k\lambda t}(1 - \mathrm{e}^{-\lambda t})^{n-k} \mathrm{d}t = \frac{1}{k} \sum_{k=1}^{n} \frac{1}{k} p_k, \\ \widetilde{\mu}_{P,n} &= \frac{1}{(n+1)\lambda} P_{n+1} \le \frac{1}{(n+1)\lambda}. \end{aligned}$$

Failure rate of the system is

$$\begin{aligned} h_{P,n}(t) &= -\frac{R'_{P,n}(t)}{R_{P,n}(t)} = \frac{nh(t)\sum_{k=1}^{n} p_k \binom{n-1}{k-1} \overline{F}(t)^k F(t)^{n-k}}{\sum_{k=1}^{n} P_k \binom{n}{k} \overline{F}(t)^k F(t)^{n-k}}, \\ \widetilde{h}_{P,n}(t) &\equiv h_{P,n+1}(t) - h_{P,n}(t) = \sum_{k=1}^{n+1} \sum_{j=k}^{n+1} P_k P_j \left(\frac{kp_k}{P_k} - \frac{jp_j}{P_j} \right) \left[\frac{n!(n+1)!(k-j)}{k!j!(n+1-k)!(n+1-j)!} \right]. \end{aligned} \tag{22}$$

Thus, if kp_k/P_k decreases with k, then $\widetilde{h}_{P,n}(t)$ decreases with n,

Finally, median $M_{P,n}$ of the system is

$$\sum_{k=1}^{n} P_k \binom{n}{k} \overline{F}(t)^k F(t)^{n-k} = \frac{1}{2}. \tag{23}$$

Thus, letting $M_{P,n+1}$ be a solution of

$$\sum_{k=1}^{n+1} P_k \binom{n+1}{k} \overline{F}(t)^k F(t)^{n+1-k} = \frac{1}{2},$$

we easily have $M_{P,n} \le M_{P,n+1}$, i.e., median $M_{P,n}$ decreases with n.

Example 2. Consider a parallel system with identical n units. Then, reliability is

$$R_n(t) = 1 - F(t)^n, \quad \widetilde{R}_n(t) = F(t)^n\overline{F}(t) > 0, \quad r_n(t) = -F(t)^n\overline{F}(t)^2 < 0,$$

which mean that reliaility $R_n(t)$ increases with n, and its increase rate decreases with n to 0.

MTTF is

$$\mu_n = \int_0^\infty [1 - F(t)^n]\mathrm{d}t, \quad \widetilde{\mu}_n = \int_0^\infty F(t)^n\overline{F}(t)\mathrm{d}t,$$

When $F(t) = 1 - \mathrm{e}^{-\lambda t}$, $\mu_n = (1/\lambda)\sum_{j=1}^{n}(1/j)$, which increases with n to ∞, and $\widetilde{\mu}_n = \frac{1}{(n+1)\lambda}$ which decreases with n from $1/\lambda$ to 0 where $\widetilde{\mu}_0 \equiv 0$.

Failure rate is

$$h_n(t) = \frac{nh(t)\overline{F}(t)F(t)^{n-1}}{1 - F(t)^n},$$

which increases with n from $h(t)$, where $h_0(t) \equiv 0$, and

$$\widetilde{h}_n(t) = \frac{h(t)F(t)^{n-1}}{\sum_{j=0}^{n} F(t)^j \sum_{j=0}^{n-1} F(t)^j}\left[-\sum_{j=0}^{n}[1 - F(t)^j]\right] < 0,$$

which decreases with n from $h(t)$ to 0. This means that failure rate $h_n(t)$ increases with n, and its increase rate decreases with n to 0.

Median M_n is given by $F(t)^n = 1/2$, which increases with n, and $\widetilde{M}_n = M_{n+1} - M_n$, where M_{n+1} is given by $F(t)^{n+1} = 1/2$. When $F(t) = 1 - \mathrm{e}^{-\lambda t}$,

$$\widetilde{M}_n = -\frac{1}{\lambda}\left[\ln\left(1 - \frac{1}{2^{n+1}}\right) - \ln\left(1 - \frac{1}{2^n}\right)\right] = -\frac{1}{\lambda}\ln\frac{1 - (1/2)^{n+1}}{1 - (1/2)^n},$$

which decreases with n from $(1/\lambda)\ln 2$ to 0.

Example 3. We consider a series-parallel system which consists of n subsystems in series, where each subsystem has identical m units in parallel. It is assumed that each unit has an identical reliability $q \equiv 1 - p$ $(0 < q \leq 1)$. Then, system reliability is [2,8]

$$R_{n,m}(p) = (1 - p^m)^n. \tag{24}$$

In particular, when $n = m = x$ $(0 < x < \infty)$, reliability in (24) is

$$R_x(p) = (1 - p^x)^x, \quad \lim_{x\to 0}(1 - p^x)^x = \lim_{x\to\infty}(1 - p^x)^x = 1. \tag{25}$$

We find a minimum x^* to minimize $R_x(p)$ in (25). Differentiating $\ln R_x(p)$ with x and setting it equal to zero,

$$p^x \ln p^x = (1 - p^x)\ln(1 - p^x).$$

Noting that $a \ln a$ for $0 < a < 1$ increases strictly from $-\infty$ to 0, x^* satisfies

$$p^x = (1 - p^x), \quad \text{i.e.,} \quad p^x = \frac{1}{2}. \tag{26}$$

Thus, x^* agrees with meddian of a parallel system with x units, and for $0 < p < 1$, $x^* = -\ln 2/\ln p$. This means that $R_x(p)$ decreases with x from 1 to $R_{x^*}(p)$ for $x \leq x^*$, and increases from $R_{x^*}(p)$ to 1 for $x > x^*$.

5 CONCLUSIONS

We have obtained optimal replacement policies to minimize the mean deviation and deviation rate. It has been shown the optimal times can be given even if the failure time is exponential and any maintenance cost is not estimated. Especially, the model with deviation factor would be applied to built up effective maintenances of redundant systems [12] and damage models [14,15], and to form good plans of inspections [4].

Reliability theory should advance like the development of mathematics. Until now, every researchers have studied in a lot of effort redundant systems with a natural number n of units. A redundant system with a positive real number x of units is the first model in the area of reliability theory. As technologies have advanced greatly and everything become opaque, the existence of such systems will go on increasing.

Scalability has been already applied to various fields in technology but also in business and other field [5,6]. However, there are little theoretical papers. There is a close connection between reliability and scalability. At the beginning of this chapter, scalability would be studied theoretically and developed extensively as a research partner of reliability.

ACKNOWLEDGMENT

This work is supported by JSPS KAKENHI Grant Number 24K07937.

REFERENCES

[1] Pham, H. (2022). *Statistical Reliability Engineering*. Springer Nature, Switzerland.
[2] Nakagawa, T. (2008). *Advanced reliability models and maintenance policies*. Springer, London.
[3] Nakagawa, T. (2014). *Random Maintenance Policies*. Springer, London.
[4] Ito, K. and Nakagawa, T. (2023). *Optimal Inspection Models with Their Applications*. Springer Nature, Switzerland.
[5] Gorton, I. (2022). *Foundations of Stable System: Designing Distributed Architectures*, O'Reilly Media.
[6] Lie, H. H. (2009). *Software Performance and Scalability: A Quantitative Approach*. Wiley-Blackwell
[7] Imran, M., Bashir. F., Jafri, A. R., Rashid, M., and Islam, M. N. (2021). A systematic review of scalable handware archtectures for pattern matching in network security. *Comp Elect Eng* 92: 107169.
[8] Barlow, R. E. and Proschan F. (1965). *Mathematical Theory of Reliability*. Wiley, New York.
[9] Rausand, M. and Hoyland, A. (2004). *System Reliability Theory*. Wiley, New York.
[10] Pham, H. (ed.) (2003). *Handbook of Reliability Engineering*. Springer, London.
[11] Nakagawa, T. (2005). *Maintenance Theory of Reliability*. Springer, London.
[12] Mizutani, S., Zhao, X. and Nakagawa, T. (2023). *Which-Is-Better (WIB): Problems in Reliability Theory*. Springer Nature, Switzerland.
[13] Zhao, X., Li, B., Mizutani, S. and Nakagawa, T. (2022). A revisit of age-based replacement models with exponential failure distributions. *IEEE Trans Reliab* 71: 1477–1487.
[14] Nakagawa, T. (2007). *Shock and Damage Models in Reliability Theory*. Springer, London.
[15] Zhao, X. and Nakagawa, T. (2018). *Advanced Maintenance Policies for Shock and Damage Models*. Springer, London.

Recent Advances in Reliability and Maintenance Modeling – Okamura, Inoue & Xiao (Eds)
© 2025 the Author(s), ISBN: 978-1-032-79098-5

Measures to minimize sales support costs for financial service providers to continue doing business with corporate clients

T. Nakamura
Department of Social Management Engineering, Graduate School of Engineering, Tottori University, Tottori, Japan

J. Koyanagi
Department of Information Science, Shunan University, Yamaguchi, Japan

K. Ito
Department of Science and Engineering, Yamato University, Osaka, Japan

ABSTRACT: Financial service providers such as banks, trust companies, and securities firms are focused on understanding the environment in which their corporate clients operate, and on identifying changes in their needs related to the securities finance business in a timely and qualified manner. Financial service providers are exploring efficient sales support measures while implementing effective sales support by their headquarters (hereinafter referred to as sales support), rather than relying solely on corporate sales representatives to understand the changing needs of corporate clients. Sales support can be broadly classified into two categories: relatively high-cost sales support, such as accompanying corporate sales representatives on visits, and relatively low-cost sales support, such as providing information using some information communication environments. The purpose of this study is to explore effective sales support measures to sustain contact relationships between corporate clients of financial service providers and corporate sales representatives of financial service providers. Specifically, Sales support has a set annual budget, and measures to minimize sales support costs are to be selected while taking this annual budget into account. In this paper, we consider the tactics of selecting sales support measures according to the rate of change in the effect of contacts with corporate clients.

1 INTRODUCTION

Financial service providers and corporate clients are closely related in the area of financial activities of corporate clients. Financial service providers with a large number of corporate clients face the challenge of how to set up effective corporate client contact hypotheses and ensure that corporate sales representatives have efficient opportunities to contact corporate clients. Therefore, a financial service provider could consider a method in which the head office is partially responsible for providing information on the environment surrounding corporate clients, while providing sales support to corporate sales representatives as needed. Sales support by the headquarters is expected to be provided in two ways. The first is relatively low-cost support, such as training provided to corporate sales representatives via the Web and social networking services. The second type of support is more expensive, such as attending and responding to corporate clients when they contact the company. If the contact status of corporate sales representatives indicates a high potential for securities financing business, head office sales support measures can collaborate with relevant head office departments, taking into account priorities and other factors.

DOI: 10.1201/9781003491309-19

There are various research examples of models for determining business priorities. For example, there is an urgency/importance matrix (Eisenhower Matrix) that categorizes and prioritizes business tasks on two axes: urgency and importance. If this is to be used, it is necessary to determine the key tasks regarding the potential for securities finance business based on the contact situation with corporate clients. Another model prioritizes tasks with higher ICE scores, which are evaluated by measuring Impact, Confidence, and Ease of Realization (EASE) of business ideas and project proposals. Since scores tend to be subjective in this model, methods such as multiple scoring and evaluation are considered effective.

Richard *et al.* [1] present a performance pyramid that measures performance in nine dimensions. The nine dimensions are as follows. market satisfaction, financial measures, customer satisfaction, Flexibility, Productivity, Quality, Delivery, Process time, cost. There are multiple factors that control operations. A number of models have been studied that consider multiple evaluation factors and narrow down candidates based on correlation coefficients.

Lin *et al.* [2] explain that this performance pyramid links organizational strategy and operations by replacing top-down with bottom-up performance measures by capturing organizational levels horizontally and client priority-based goals vertically.

Chui [3] describes that the results of corporate activities can be improved by applying corporate business capabilities to business activities. We believe that corporate sales activities to financial service providers can measure the effectiveness of support by applying sales support to sales activities. It would also be possible to consider what kind of criteria could be used to make the decision more effective and efficient.

Jihyun [4] describes that business value in corporate management is a value that guides corporate decision-making and helps determine the health of the corporation. He also describes that in order to effectively assist corporate decision makers, it is necessary to measure their business value and identify problem areas that need immediate improvement. This suggests that sales support should have a standard to promptly respond to changes in contact situations with corporate clients.

Chun [5] discusses the detection of temporal outliers and methods for their analysis, focusing on the impact of temporal changes on the execution of business processes. The environment surrounding corporate clients is likely to change over time. These changes can be ascertained when the corporate sales representative contacts a corporate client, and changes in the number of contacts may indicate that the corporate sales representative may be acquiring some information from the corporate client. Therefore, it is expected that the sales support will timely and qualifiedly identify the timing of changes in the number of contacts and provide support that leads to securities financing business. We believe that capturing this timing will also help to prioritise business.

Nakamura *et al.* [6] and [7] have introduced the model that there are some strategies for sales support based on the number of contacts made by the sales representative. Two types of corporate clients are assumed: continuing clients and defected clients. Continuing clients are corporate clients that are handled by the corporate sales representative from the beginning of the fiscal year to the end of the fiscal year. On the other hand, a defected client is a corporate client who is handled by a corporate sales representative at the beginning of the fiscal year but who has requested to cease transactions until the end of the fiscal year. We believe that financial service providers will select strategies to decrease the number of defected clients in order to secure business opportunities with their clients.

Therefore, They assume a model with a term during which Financial service providers determine whether a client is defected (the defect determination period) and whether sales support is necessary for the organization to respond to the client (the support determination period).

In these models, there is one type of sales support cost by the headquarters. This paper assumes two types of sales support costs and carries out a comparison with a model under conditions where there is an annual budget constraint on sales support costs.

2 MODEL ASSUMPTIONS

Sales support measures are implemented according to the contact effects with corporate clients. The model is set up as follows.

1) We consider the planning period T. For the term t (1,..., T), the sales representative tries to contact with the client and contact effect e_t is obtained for term t. It is assumed e_t is the integer number $0, 1, \ldots \bar{k}$, such as visit number of term t. The probability of $e_t = k$ is denoted by $w(k)$ for all terms. The expected number λ of e_t is calculated by $\lambda = \sum_{k=0}^{\bar{k}} kw(k)$
2) Two types of sales support are considered: the first is low-level support, such as utilizing web conferencing, while the second is high-level support, such as accompanying visits. Each support cost is denoted by C_1, C_2 $(0 \leq C_1 \leq C_2)$, respectively.
3) At the beginning of term t, the support decision is made among three options: no support, low-level support and high-level support. If low-level support is implemented, the contact effect increases by α_1, and by high-level support the contact effect increases by α_2. It is assumed that α_1 and α_2 are integer numbers and $0 \leq \alpha_1 \leq \alpha_2$. It indicates that the contact effect becomes $k + \alpha_1$ $(k + \alpha_2)$ with probability $w(k)$ by low-level (high-level) support.
4) At the end of term t (after the contact effect for term t is determined), the decision on whether or not the client continues to trade in the next $t + 1$ term is determined by the contact effects over the previous three terms (e_{t-2}, e_{t-1}, e_t). Let denote the probability of continuing to trade in the term $t + 1$ by $q(i, j, k)$ $(i = e_{t-2},\ j = e_{t-1},\ k = e_t)$. That is, a client has a probability of $1 - q(i, j, k)$ of being a defected client and remains defected until the end of the T term.
5) Let $x = k + y_1 j + y_2 i$ $(1 \geq y_1 \geq y_2)$, then it is assume that $q(i, j, k) = r(x)$. This indicates that the past contact effect is gradually diminished by y_1 and y_2. In this research, the probability of continuing a transaction $r(x)$ is assumed to be a sigmoid function (1) parallel to the x-axis, since the more contacts a client has, the more likely it is that the client will continue trading.

$$r(x) = \frac{1}{1 + \exp(-\beta(x - \gamma))} \tag{1}$$

The β and γ are constants, β is called "gain", similar to the setting in Nakamura [7].

6) The support decision at the beginning of term t is made if the client continues the trade at the end of term $t - 1$. Since the contact effects before term $t - 3$ have no effect for the future, the decision is made by the contact effect e_{t-2}, e_{t-1}, which are denoted by i, j respectively. Then the state of the system is expressed by (i, j). Our objective is to determine the optimal action for state (i, j) which satisfies some constraints and minimize the specified cost.

3 MARKOV DECISION PROCESS (MDP) MODEL

If managers know the defect cost and support cost for each corporate client, they can consider strategies concerning client targets for sales support. MDP is an effective method for this purpose. By MDP, managers can determine the sales strategy to minimize the total expected cost. In this section, we assume one client and one staff. If the client defects, we assume the defect cost C_3 is incurred. Then the optimality equation to minimize the total expected cost becomes

$$V^t(i, j) = \min\{W_0^t(i, j), W_1^t(i, j), W_2^t(i, j)\} \tag{2}$$

$$W_0^t(i,j) = \sum_{k=0}^{\overline{k}} w(k)[q(i,j,k)V^{t+1}(j,k) + C_3\{1 - q(i,j,k)\}] \quad (3)$$

$$W_1^t(i,j) = \sum_{k=0}^{\overline{k}} w(k)[q(i,j,k+\alpha_1)V^{t+1}(j,k+\alpha_1) + C_3\{1 - q(i,j,k+\alpha_1)\}] + C_1 \quad (4)$$

$$W_2^t(i,j) = \sum_{k=0}^{\overline{k}} w(k)[q(i,j,k+\alpha_2)V^{t+1}(j,k+\alpha_2) + C_3\{1 - q(i,j,k+\alpha_2)\}] + C_2 \quad (5)$$

This recurrence relation is calculated by $V^{T+1}(i,j) \equiv 0$ for all (i,j), (the system ends without cost at $T+1$) and $V^1(i,j)$ shows the minimum total expected cost if the system starts in state (i,j). In this paper, the system starts in state $(\overline{k},\overline{k})$ which means the system starts in the best contact effect without support. The optimal action for state (i,j) at term t is determined to be 'no support' if $V^t(i,j) = W_0^t(i,j)$, to be 'low-level support' if $V^t(i,j) = W_1^t(i,j)$, and to be 'high-level support' if $V^t(i,j) = W_2^t(i,j)$, respectively. Let $A_M^t(i,j)$ be the optimal action for state (i,j) at term t.

$$A_M^t(i,j) = \begin{cases} 0 & \textit{if optimal action is 'no support'}, \\ 1 & \textit{if optimal action is 'low-level support'}, \\ 2 & \textit{if optimal action is 'high-level support'}. \end{cases} \quad (6)$$

After the optimal action for every state and term is determined, the expected number of low-level and high-level support and the defect probability of the client can be calculated as follows. Let $M_1^t(i,j)$, $M_2^t(i,j)$ and $M_3^t(i,j)$ be the expected number of low and high support, and defect probability for state (i,j) at term t. Then with the terminal value $M_1^{T+1}(i,j) = M_2^{T+1}(i,j) = M_3^{T+1}(i,j) = 0$, the values for each term t are calculated as follows.

$$M_1^t(i,j) = \begin{cases} \sum_{k=0}^{\overline{k}} w(k)q(i,j,k)M_1^{t+1}(j,k) & \textit{if } A_M^t(i,j) = 0 \\ 1 + \sum_{k=0}^{\overline{k}} w(k)q(i,j,k+\alpha_1)M_1^{t+1}(j,k+\alpha_1) & \textit{if } A_M^t(i,j) = 1 \\ \sum_{k=0}^{\overline{k}} w(k)q(i,j,k+\alpha_2)M_1^{t+1}(j,k+\alpha_2) & \textit{if } A_M^t(i,j) = 2 \end{cases} \quad (7)$$

$$M_2^t(i,j) = \begin{cases} \sum_{k=0}^{\overline{k}} w(k)q(i,j,k)M_2^{t+1}(j,k) & \textit{if } A_M^t(i,j) = 0 \\ \sum_{k=0}^{\overline{k}} w(k)q(i,j,k+\alpha_1)M_2^{t+1}(j,k+\alpha_1) & \textit{if } A_M^t(i,j) = 1 \\ 1 + \sum_{k=0}^{\overline{k}} w(k)q(i,j,k+\alpha_2)M_2^{t+1}(j,k+\alpha_2) & \textit{if } A_M^t(i,j) = 2 \end{cases} \quad (8)$$

$$
M_3^t(i,j) = \begin{cases} \sum_{k=0}^{\overline{k}} w(k)[q(i,j,k)M_3^{t+1}(j,k) + 1 - q(i,j,k)] & \textit{if } A_M^t(i,j) = 0 \\ \sum_{k=0}^{\overline{k}} w(k)[q(i,j,k+\alpha_1)M_3^{t+1}(j,k+\alpha_1) + 1 - q(i,j,k+\alpha_1)] & \textit{if } A_M^t(i,j) = 1 \\ \sum_{k=0}^{\overline{k}} w(k)[q(i,j,k+\alpha_2)M_3^{t+1}(j,k+\alpha_2) + 1 - q(i,j,k+\alpha_1)] & \textit{if } A_M^t(i,j) = 2 \end{cases} \tag{9}
$$

4 STATIONARY POLICY MODEL

MDP model can be applied if both defect and support costs are estimated. However, the defect cost is usually hard to be estimated. We believe that it would be effective if measures could be selected to minimize the total cost of sales support within a certain defect probability under the assumption that defection will occur to some extent. In this section, we assume the defect cost C_3 is unknown and estimate the defect probability under the stationary policy. The expected total support costs is also calculated. Then the manager can select one stationary policy, according to the defect probability and the expected total support costs. For example, the manager can select the minimum expected total support costs with the constraint that the defect probability is less than the given value, for example, less than 5%.

Stationary policy means that the same action is taken for state (i,j), irrespective to term t. The optimal (non-stationary) policy by MDP model takes different action for state (i,j) according to term t, therefore it may be difficult to execute the policy. It takes long time to search the best stationary policy among all stationary policies, then we consider two types of stationary policies, based on the index assigned to each state. Contact effect is expressed by the number $0,\ldots,\overline{k}+\alpha_2$ by considering high-level support, then all states are $(0,0),\ldots,(\overline{k}+\alpha_2,\overline{k}+\alpha_2)$. For each state, we set the index from 0 to $(\overline{k}+\alpha_2+1)^2-1$. Consider the stationary policy (h,l) that means the following policy.

- The high-level support is implemented for the state (i,j) whose index u satisfies $0 \le u \le h$.
- The low-level support is implemented for the state (i,j) whose index u satisfies $h < u \le l$.
- No support is implemented for the state (i,j) whose index u satisfies $l < u$.

We calculate the values for all combination (h,l) $(-1 \le h \le \overline{k}+\alpha_2,\ h \le l \le \overline{k}+\alpha_2)$, $h=-1$ means no high-level support, $h=l$ means no low-level support, and find the best stationary policy.

We consider following two types of stationary policies.

4.1 *State index model (SI model)*

For the state (i,j), the index is given according to the value $jy_1 + iy_2$. The smaller the value is, the smaller the index is. If two states have same value, the state with smaller j has smaller index. If the value is small, the probability to continue the trade becomes small. Thus, the manager should support the sales representative to prevent defect, taking the above into consideration.

For example, consider that the contact effect without support is numbered from 0 to $\overline{k}=4$, and $\alpha_2=2$. Let $y_1=0.8$, $y_2=0.5$, then the values to assign the index are calculated, shown in Table 1. By these values the order of support is shown in the Table 2.

Table 1. Values of $jy_1 + iy_2$ for SI model.

-	$j = 0$	1	2	3	4	5	6
$i = 0$	0	0.8	1.6	2.4	3.2	4	4.8
1	0.5	1.3	2.1	2.9	3.7	4.5	5.3
2	1	1.8	2.6	3.4	4.2	5	5.8
3	1.5	2.3	3.1	3.9	4.7	5.5	6.3
4	2	2.8	3.6	4.4	5.2	6	6.8
5	2.5	3.3	4.1	4.9	5.7	6.5	7.3
6	3	3.8	4.6	5.4	6.2	7	7.8

Table 2. Support order matrix for SI model.

-	$j = 0$	1	2	3	4	5	6
$i = 0$	0	2	6	11	18	25	32
1	1	4	9	15	22	29	36
2	3	7	13	20	27	34	40
3	5	10	17	24	31	38	43
4	8	14	21	28	35	41	45
5	12	19	26	33	39	44	47
6	16	23	30	37	42	46	48

For example, The stationary policy $h = 3, l = 6$ means that in the states with index $0, 1, 2, 3$, i.e., states $(0, 0)$, $(1, 0)$, $(0, 1)$, $(2, 0)$, high-level support is implemented, in the states with index $4, 5, 6$, i.e., states $(1, 1)$, $(3, 0)$, $(0, 3)$, low-level support is implemented, in the other states, no support is implemented.

4.2 *Derivative index model (DI model)*

For the state (i,j), the index is given according to the value $r'(x)$, where $x = \lambda + jy_1 + iy_2$. The value λ is the expected value of the contact effect without support. The merit of sigmoid function is that $r'(x)$ is easily calculated by $r'(x) = \beta r(x)[1 - r(x)]$ The larger the value is, the smaller the index is. If the value is large, the increase of probability to continue the trade becomes large. Thus the effect of support is large with smaller index, for example, see Tables 3 and 4. For this example, the values are $\overline{k} = 4$, $\alpha_2 = 2$, $y_1 = 0.8$, $y_2 = 0.5$, $\lambda = 2.4$, $\beta = 2$, $\gamma = 3$.

Table 3. Values of support for DI model.

-	$j = 0$	1	2	3	4	5	6
$i = 0$	356	481	210	52	11	2	0.45
1	495	317	90	20	4	1	0.17
2	428	153	35	7	1	0.3	0.06
3	243	63	13	3	1	0.11	0.02
4	108	24	5	1	0.2	0.04	0.01
5	43	9	2	0.37	0.7	0.02	0.003
6	16	3	1	0.14	0.3	0.01	0.001

* All numbers in the table are 1000 times the actual values.

Table 4. Support order matrix for DI model.

-	$j=0$	1	2	3	4	5	6
$i=0$	3	1	6	11	18	25	32
1	0	4	9	15	22	29	36
2	2	7	13	20	27	34	40
3	5	10	17	24	31	38	43
4	8	14	21	28	35	41	45
5	12	19	26	33	39	44	47
6	16	23	30	37	42	46	48

The calculation is similar to MDP model. First, set set the action for each state (i,j) by h,l. In SI model, let action for state (i,j) be $A_S(i,j)$. Then the values $M_1^t(i,j)$, $M_2^t(i,j)$, $M_3^t(i,j)$ are calculated by Equation (7), (8), (9) by replacing $A_M^t(i,j)$ by $A_S(i,j)$ (similar in DI model).

5 NUMERICAL RESULTS

We compare MDP model, SI model and DI model with the following values $C_1 = 5$, $C_3 = 200$, $\alpha_1 = 2$, $\alpha_2 = 4$, $\overline{k} = 4$, $w(0) = 0.1$, $w(1) = 0.1$, $w(2) = 0.3$, $w(3) = 0.3$, $w(4) = 0.2$, $y_1 = 0.8$, $y_2 = 0.5$, $\gamma = 3$. For the numerical examples, the values β (gain of sigmoid function) and C_2 (high-level support cost) are changed. The results are calculated for the following 12 cases in Table 5.

Table 5. Case table.

Case	β	C_2	D_{cmax}
1	1	10	5%
2	1	20	5%
3	1	50	5%
4	2	10	5%
5	2	20	5%
6	2	50	5%
7	3	10	5%
8	3	20	5%
9	3	50	5%
10	1	10	10%
11	2	10	10%
12	3	10	10%

Each variable is defined as follows. E_L be the expected number of low-level support, E_H be the expected number of high-level support, S_{cost} be the expected support cost, D_{prob} be the defect probability, (D_{cmax} is the maximum allowable D_{prob} in the SI and DI models.ï¼‰ T_{cost} be the expected total cost for T terms.

The results of MDP model are shown in Table 6 under the conditions of different β and C_2 (Case 1-9). For Case 10-12, MDP results are the same as Case 1, 4, 7, because the value of D_{cmax} is not used in MDP model.

Comparing Cases 4, 5, and 6 in Table 6, C_2 is larger, S_{cost} is smaller, but T_{cost} is larger. As C_2 increases, high-level support (E_H) decreases. Then, S_{cost} decreases, but D_{prob} and T_{cost}

Table 6. MDP model result for Case 1-9.

Case (β, C_2)	E_L	E_H	S_{cost}	D_{prob}	T_{cost}
1 (1,10)	6.392	1.056	42.521	7.4%	57.307
2 (1,20)	8.327	0.006	41.762	8.8%	59.411
3 (1,50)	8.336	0.000	41.679	8.9%	59.419
4 (2,10)	3.921	0.470	24.310	3.5%	31.266
5 (2,20)	4.632	0.002	23.207	4.5%	32.123
6 (2,50)	4.635	0.000	23.176	4.5%	32.123
7 (3,10)	3.669	0.224	20.590	1.9%	24.478
8 (3,20)	3.998	0.000	19.990	2.5%	24.965
9 (3,50)	3.998	0.000	19.990	2.5%	24.965

increases because of less support. Also, comparing cases 1, 4, and 7, as β increases, defect probability decreases for all states. Then, D_{prob} decreases, therefore S_{cost} and T_{cost} also decrease. We compare MDP model with SI model and DI model in Tables 8 and 9, concerning to S_{cost}, D_{prob} and T_{cost}.

Firstly, we searched for SI model that S_{cost} minimize under the condition that D_{prob} is less than or equal to 5%. The results of SI model are shown in Table 7 under the conditions of different β and C_2 (Case 1-9). Although C_3 is not used in SI model and DI model, we include C_3 as a reference value for the same C_3 as in MDP model for comparison with MDP model.

Table 7. SI model results for Case 1-9.

Case (β, C_2, D_{cmax})	h	l	E_L	E_H	S_{cost}	D_{prob}	T_{cost} *
1 (1,10,5%)	28	47	5.995	1.942	49.398	4.9%	59.298
2 (1,20,5%)	15	56	10.020	0.165	53.399	5.0%	63.396
3 (1,50,5%)	9	58	10.668	0.015	54.106	5.0%	64.083
4 (2,10,5%)	3	25	4.402	0.046	22.475	4.8%	32.147
5 (2,20,5%)	3	25	4.402	0.046	22.937	4.8%	32.610
6 (2,50,5%)	−1	27	4.593	0.000	22.963	4.8%	32.646
7 (3,10,5%)	3	19	3.022	0.114	16.255	4.9%	26.142
8 (3,20,5%)	3	19	3.022	0.114	17.397	4.9%	27.285
9 (3,50,5%)	−1	20	3.501	0.000	17.504	4.3%	26.095

*T_{cost} as a reference value for comparison with MDP model, same as Tables 8, 9 and 10.

Comparing Cases 1-3 ($\beta = 1$) and Cases 4-6 ($\beta = 2$) in Table 7, C_2 is larger, S_{cost} and T_{cost} are larger, but they are different features when $\beta = 3$. Also, comparing cases 1, 4, and 7, as β increases, S_{cost} decreases.

Comparing MDP model with SI model in Cases 1-3, D_{prob} in SI model is less than that of MDP model, because D_{cmax} is 5%, which is less than D_{prob} in MDP model. However, S_{cost} increases to decrease D_{prob}. In Case 4-6, D_{prob} and S_{cost} are almost same values. In Case 7-9, opposite results of Case 1-3 are obtained.

Secondly, we searched for DI model that S_{cost} minimize under the condition that D_{prob} is less than or equal to 5%. The results of DI model are shown in Table 8 under the conditions of different β and C_2 (Case 1-9).

The results in Table 8 are similar to Table 7.

In Tables 7 and 8, we searched optimal policy by D_{cmax} = 5%. Moreover, we compared the effect of D_{cmax} = 10% of SI and DI model in case 10-12 in Tables 9 and 10.

Table 8. DI model results for Case 1-9.

Case (β, C_2, D_{cmax})	h	l	E_L	E_H	S_{cost}	D_{prob}	T_{cost}*
1 (1,10,5%)	28	47	5.995	1.942	49.398	4.9%	59.298
2 (1,20,5%)	15	56	10.020	0.165	53.399	5.0%	63.396
3 (1,50,5%)	9	58	10.668	0.015	54.106	5.0%	64.083
4 (2,10,5%)	2	25	4.402	0.046	22.474	4.8%	32.149
5 (2,20,5%)	2	25	4.402	0.046	22.937	4.8%	32.611
6 (2,50,5%)	−1	26	4.593	0.000	22.963	4.8%	32.646
7 (3,10,5%)	2	19	3.023	0.114	16.253	4.9%	26.152
8 (3,20,5%)	2	19	3.023	0.114	17.392	4.9%	27.291
9 (3,50,5%)	−1	20	3.501	0.000	17.504	4.3%	26.095

Table 9. Effect of D_{cmax} and β of SI model.

Case (β, C_2, D_{cmax})	h	l	E_L	E_H	S_{cost}	D_{prob}	T_{cost}*
1 (1,10,5%)	28	47	5.995	1.942	49.398	4.9%	59.298
4 (2,10,5%)	3	25	4.402	0.046	22.475	4.8%	32.149
7 (3,10, 5%)	3	19	3.023	0.114	16.255	4.9%	26.152
10 (1,10,10%)	12	41	7.222	0.305	39.158	10.0%	59.066
11 (2,10,10%)	−1	20	3.369	0.000	16.843	10.0%	36.800
12 (3,10,10%)	3	15	2.224	0.153	12.653	9.8%	32.183

Table 10. Effect of D_{cmax} and β DI model.

Case (β, C_2, D_{cmax})	h	l	E_L	E_H	S_{cost}	D_{prob}	T_{cost}*
1 (1,10,5%)	28	47	5.995	1.942	49.398	4.9%	59.298
4 (2,10,5%)	2	25	4.402	0.046	22.474	4.8%	32.149
7 (3,10,5%)	2	19	3.023	0.114	16.253	4.9%	26.152
10 (1,10,10%)	12	41	7.222	0.305	39.158	10.0%	59.066
11 (2,10,10%)	−1	20	3.369	0.000	16.843	10.0%	36.800
12 (3,10,10%)	2	15	2.227	0.151	12.641	9.8%	32.243

In Cases 10-12, D_{prob} increases and S_{cost} decreases as D_{cmax} increases aiming to reduce support costs. It can also be observed that β increases, S_{cost} decreases. In other word, it can be assumed that a large β and a large D_{cmax} can be used as a strategy to reduce S_{cost}. Each condition of the optimal policy has almost the same result by Tables 9 and 10.

The difference of the values are produced the difference of the stationary policy shown in Table 11.

The optimal policy ($h = 3, l = 25$) in the SI model in Table 11 denotes the high-level support (action 2) to the 4 cases in support order 0 to 3, low-level support (action 1) to the 22 cases in support order 4 to 25, and no support above support order 26. Comparing Tables 9 and 10, there is a difference in the optimal action for state (i, j) of $(0, 0)$. We think the SI model selects high-level support because the state is the lowest, but the DI model selects low-level support because the high-level for (0,0) would not be expected to have a high support effect.

Table 11. Comparison of optimal policies for SI model and DI model for Case 4.

Optimal Policy ($h = 3, l = 25$) in SI model

-	$j = 0$	1	2	3	4	5	6	7	8
$i = 0$	2	2	1	1	1	0	0	0	0
1	2	1	1	1	1	0	0	0	0
2	2	1	1	1	0	0	0	0	0
3	1	1	1	1	0	0	0	0	0
4	1	1	1	0	0	0	0	0	0
5	1	1	0	0	0	0	0	0	0
6	1	1	0	0	0	0	0	0	0
7	1	0	0	0	0	0	0	0	0
8	0	0	0	0	0	0	0	0	0

Optimal Policy ($h = 2, l = 25$) in DI model

-	$j = 0$	1	2	3	4	5	6	7	8
$i = 0$	1	2	1	1	1	0	0	0	0
1	2	1	1	1	1	0	0	0	0
2	2	1	1	1	0	0	0	0	0
3	1	1	1	1	0	0	0	0	0
4	1	1	1	0	0	0	0	0	0
5	1	1	0	0	0	0	0	0	0
6	1	1	0	0	0	0	0	0	0
7	1	0	0	0	0	0	0	0	0
8	0	0	0	0	0	0	0	0	0

6 CONCLUSION

We considered the optimal support strategy for financial service providers to continue doing business with their corporate clients through two types of sales support costs.

Firstly, we derived the optimal conditions under which the total expected cost is minimized as a Markov decision process model (MDP model) under the conditions that sales support cost and defect cost can be assumed. In MDP model, we obtained the following results. For the same gain of the probability of continuing to trade according to the sigmoid function, the higher the support cost, the higher the expected defect probability and the higher the total expected optimal cost. As the gain in the sigmoid function increases, the expected defect probability and the total expected optimal cost tend to decrease. It is considered possible to reduce the expected defect probability when the gain is large.

Secondly, we estimated the defect probability under the stationary policy since the defect cost is usually hard to be estimated. Under the assumption that the cost of defection is unknown, an acceptable defect probability was set and within that range, a stationary policy with a small expected cost of support was estimated. The differences were estimated using SI model that preferentially supports smaller values representing contact status and DI model that preferentially supports higher rates of change in values representing contact status.

The conditions for an optimal policy that minimizes expected support costs have roughly the same result. However, in cases of low contact status, we could observe the possibility of cases in which SI and DI models provide different support. If there is a significant change in contact status in practice, we believe this suggests the possibility of different results in SI and DI models.

REFERENCES

[1] Lynch, R. L. and Cross, K. F., (1995). *Measure Up! How to Measure Corporate Performance*, Blackwell, *Second edition*. ISBNl-55786-718-6.

[2] Lin. F. and Philip, M., (1997). The influence of service process on performance measurement. *The Journal of Management Accounting, Japan*, VOL.5, NO.2 :7–27.

[3] Chui, Y. Y., (2021). Measurement of enterprise smart business performance on a smart business management. *The Institute of Electronics, Information and Communication Engineers, TRANS.INF. &SYST.*, VOL.E104-D, NO.1 :56–62.

[4] Jihyun. L. and Sungwon. K., (2015). Techniques for measuring business process based on business values. *The Institute of Electronics, Information and Communication Engineers,TRANS.INF. &SYST.*, VOL. E98-D, NO.4 :911–921.

[5] Chun, G. P. and Hyuun, A., (2019). Temporal outlier detection and correlation analysis of business process executions. *The Institute of Electronics, Information and Communication Engineers,TRANS. INF. &SYST.*, VOL.E102-D, NO.7 :1412–1416.

[6] Takahiro N., Junji K. and Kodo I., (2021). The optimal exercise policy through the effective marketing support under the uncertainty. *Proc. of the 26th ISSAT International Conference Reliability and Quality in Design*, 210–214.

[7] Takahiro N., Junji K. and Kodo I., (2021). The effect of selective marketing support under the uncertainty. *Proc. of the Reliability and Maintenance Engineering Summit* 2021, 162–169.

Recent Advances in Reliability and Maintenance Modeling – Okamura, Inoue & Xiao (Eds)
© 2025 the Author(s), ISBN: 978-1-032-79098-5

Optimal execution times with forward and backward recoveries

Kenichiro Naruse
Josai International University, Togane, Chiba, Japan

Toshio Nakagawa
Aichi Institute of Technology, Yakusa, Toyota, Aichi, Japan

ABSTRACT: We take up a computer system with forward and backward recoveries to keep on running normally. Most computer systems are made of several modules. Even if some modules cause errors, they are fixed by re-executing. This paper considers two kinds of modules such as M-L module and M-S module: M-L module is that its error rate is large and a process executes forward recovery on error, and M-S module is that its error rate is small and a process executes backward recovery on error. Furthermore, a computer system works for a job with constant processing time S, which is divided into N tasks with equal intervals $T \equiv S/N$, and checkpoints are placed at periodic time kT. Three kinds of checkpoints such as self-checking, dual-checking and majority-checking are adopted. We obtain the total execution times to complete N tasks successfully and compute optimal number N^* to minimize them numerically.

1 INTRODUCTION

If some failures occur in computer system, we can not continue to work, and especially, IoT and some measuring devices have been used for various scenes and places. To keep computers with high-reliability, some useful fault-tolerant computing techniques have been adopted (Abd-El-Barr2007; Lee and Anderson 1990; Naruse and Nakagawa 2020, 2022 Siewiorek and Swarz 1982). The performances and reliabilities of redundant modular structures (Pradhan and Vaidya 1992), and checkpoint schemes with task duplications (Ziv and Bruck 1997, 1998) were evaluated. The optimal instruction-retry period to minimize the probability of a dynamic failure by a triple modular controller was derived (Kim and Shin 1996]). The evaluation models with finite checkpoints and bounded rollback were discussed (Ohara *et al.* 2006). Furthermore, checkpointing schemes for a set of multiple tasks in real-time systems were investigated (Zhang and Chakrabarty 2004), and optimal scheme models with two kinds of errors were derived (Naruse and Nakagawa 2024). To keep computer systems with high-reliability, one of techniques is forward and backward recovery. Forward recovery is that the process keeps running forward even in a fault state without backward (Naruse and Nakagawa 2023). Such recoveries are applied actually to weather satellite pictures, Voice of IP telephone systems and so on (Pradhan and Vaidya 1994) and backward recovery method is that the process returns to the first one or repeats the same one again. Such methods are applied to database systems and so on (Gray *et al.* 1981; Jeon *et al.* 2012).

We consider a computer system built by two types of modules such as M-L and M-S modules:

(a) M-L module works as a terminal module like a camera element, and when some errors occur in it, the system needs to recover using a forward recovery tool.

 DOI: 10.1201/9781003491309-20

(b) M-S module works as a control unit like a camera controller, and when some errors occur in it, the system needs to recover using backward recovery tools.

We obtain the execution times of a computer system with the above two modules and derive optimal policies to minimize them numerically.

2 THREE CHECKING STRUCTURES

We consider a computer system with self-checking, dual-checking and majority-checking (Nakagawa *et al.* 2003).

2.1 *Self-checking*

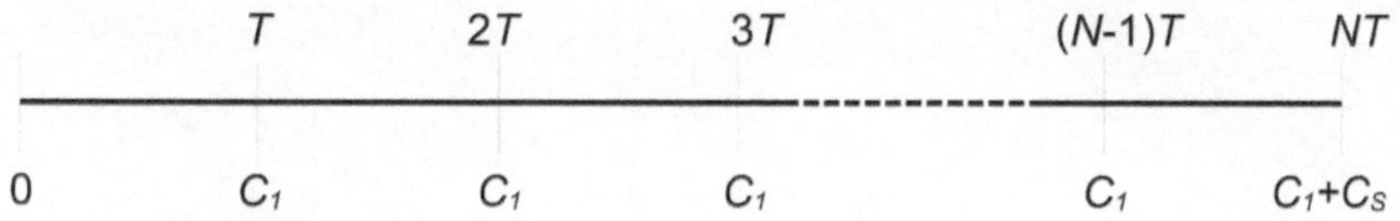

Figure 1. Checkpoint scheme for self-checking model.

Suppose that a computer system works for a job with a native execution time $S(0 < S < \infty)$ which does not include any overheads. Then, we divide S equally into $N(N = 1, 2, \cdots)$ tasks interval T, and make the following assumptions in Figure 1:

(1) Checkpoints are placed at periodic times $kT(k = 1, 2, \cdots, N)$ where $NT = S$ to see if the process is normal nor not.
(2) The system has self-checking at each checkpoint $kT(k = 1, 2, \cdots, N)$ and its overhead is C_1.
(3) The respective forward and backward recoveries times are R_F and R_B.
(4) When errors occur at process k in $((k-1)T, kT]$, the process has the following two types of recoveries in Figure 2:
   (i) When M-L module works at some errors according to a failure distribution $F_L(t)$ with mean $1/\lambda_L$, the system is recovered by forward tool, and the process goes to $(k+1)$ process.
   (ii) When M-S module works at some errors according to a failure distribution $F_S(t)$ with mean $1/\lambda_S$, the system is recovered by backward tool, and the process repeats the same process k again.
   (iii) When M-L and M-S modules work at the same time, the system is recovered using forward recovery, and the process goes to $(k+1)$ process.

(5) The process N has the following two types of recoveries:
   (i) When there is no error, the process completes when process N succeeds, and its overhead is C_S.
   (ii) When M-L module works at some errors, the process is recovered by forward tool and its overhead is R_F.
   (iii) When M-S module works at some errors, the process repeats the same process N again using backward recovery and its overhead is R_B.
   (iv) When M-L and M-S modules work at the same time, the process is recovered and its overhead is R_F.

If the execution time S is divided into many parts, then checkpoints are generated at a short interval, their total overheads are large, and the system performance becomes low. But, if some errors are detected at a checkpoint, the process is re-executed from near the previous checkpoint. In this case, the re-execution time is small. On the other hand, if S is divided into a few parts, then checkpoints are generated at a long interval, their total overheads are small and the system performance becomes high. But, if some errors are detected at checkpoints, the process is re-executed from far the previous checkpoint. In this case, the re-execution time is large. Therefore, this is one kind of trade-off problem to generate a checkpoint interval, and is called a partition problem and is applied to practical partition methods (Nakagawa (2008)).

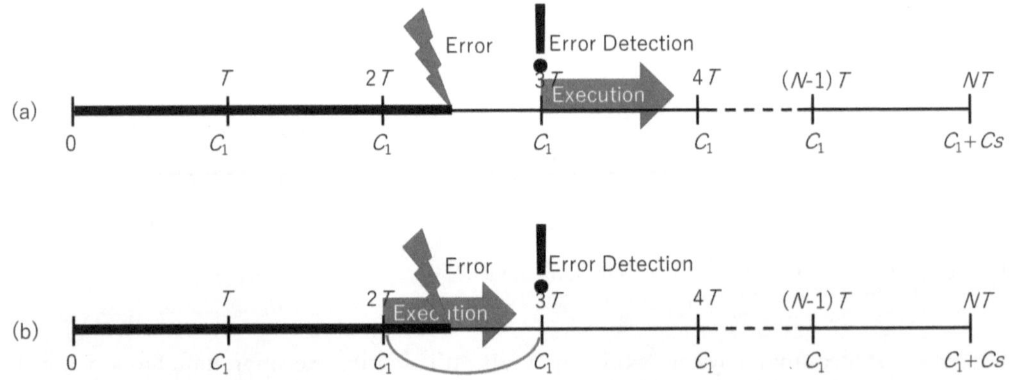

Figure 2. Recovery methods.

Using the assumptions (1)~(5), the mean execution time $\widetilde{L}_C(k)$ from process $k(k = 1, 2, \cdots, N)$ to complete the process is

$$\begin{aligned}\widetilde{L}_C(k) &= C_1 + T + \overline{F}_L(T)\overline{F}_S(T)\widetilde{L}_C(k+1) + F_L(T)\left[\widetilde{L}_C(k+1) + R_F\right]\\ &\quad + \overline{F}_L(T)F_S(T)\left[\widetilde{L}_C(k) + R_B\right] \qquad (k= 1, 2, \cdots, N-1),\\ \widetilde{L}_C(N) &= C_1 + T + \overline{F}_L(T)\overline{F}_S(T)C_S + F_L(T)R_F + \overline{F}_L(T)F_S(T)\left[\widetilde{L}_C(N) + R_B\right].\end{aligned} \tag{1}$$

Letting

$$A \equiv 1 - \overline{F}_L(T)F_S(T), \quad B \equiv C_1 + T + R_F F_L(T) + R_B\overline{F}_L(T)F_S(T),$$

(1) is rewritten as

$$\begin{aligned}A\widetilde{L}_C(k) &= B + A\widetilde{L}_C(k+1) \quad (k = 1, 2, \cdots, N-1),\\ A\widetilde{L}_C(N) &= B + \overline{F}_L(T)\overline{F}_S(T)C_S.\end{aligned} \tag{2}$$

In case of $N = 1$, from (2),

$$\widetilde{L}_C(1) = \frac{B + \overline{F}_L(T)\overline{F}_S(T)C_S}{A}.$$

In case of $N = 2$, from (2),

$$A\widetilde{L}_C(1) = B + A\widetilde{L}_C(2),$$

$$A\widetilde{L}_C(2) = B + \overline{F}_L(T)\overline{F}_S(T)C_S.$$

Thus,

$$\widetilde{L}_C(1) = \frac{2B + \overline{F}_L(T)\overline{F}_S(T)C_S}{A}.$$

In case of $N = 3$, from (2),

$$A\widetilde{L}_C(1) = B + A\widetilde{L}_C(2),$$

$$A\widetilde{L}_C(2) = B + A\widetilde{L}_C(3),$$

$$A\widetilde{L}_C(3) = B + \overline{F}_L(T)\overline{F}_S(T)C_S.$$

Thus,

$$\widetilde{L}_C(1) = \frac{3B + \overline{F}_L(T)\overline{F}_S(T)C_S}{A}.$$

Generally, we easily have

$$L_C(N) \equiv \widetilde{L}_C(1) = \frac{NB + \overline{F}_L(T)\overline{F}_S(T)C_S}{A} \quad (N = 1, 2, \cdots). \tag{3}$$

When $F_L(T) = 1 - e^{-\lambda_L T}$ and $F_S(T) = 1 - e^{-\lambda_S T}$, (3) is

$$L_C(N) = \frac{N\{C_1 + T + R_F(1 - e^{-\lambda_L T}) + R_B[e^{-\lambda_L T} - e^{-(\lambda_L+\lambda_S)T}]\} + C_S e^{-(\lambda_L+\lambda_S)T}}{1 - e^{-\lambda_L T} + e^{-(\lambda_L+\lambda_S)T}}$$

$$= \frac{N\{C_1 + S/N + R_F(1 - e^{-\lambda_L S/N}) + R_B[e^{-\lambda_L S/N} - e^{-(\lambda_L+\lambda_S)S/N}]\} + C_S e^{-(\lambda_L+\lambda_S)S/N}}{1 - e^{-\lambda_L S/N} + e^{-(\lambda_L+\lambda_S)S/N}} \tag{4}$$

$(N = 1, 2, \cdots)$.

Clearly, $\lim_{N\to\infty} L_C(N) = \infty$. Thus, there exists a finite N_C^* $(1 \le N_C^* < \infty)$ to minimize $L_C(N)$. Therefore, by computing $L_C(N)$ for $N = 1, 2, \cdots$, we can compute optimal N_C^* numerically.

2.2 *Dual-checking*

We adopt a dual-checking to see if the process is normal or not by comparing two states of processes which execute the same task.

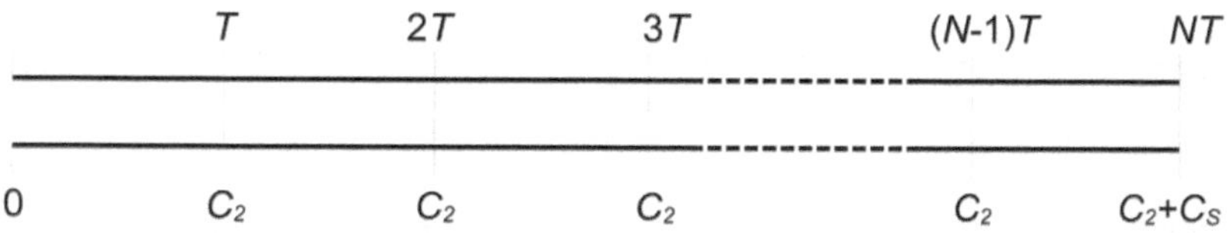

Figure 3. Checkpoint scheme for dual-checking.

We consider a computer system for a job with a native execution time S, and make the following assumptions which are the same as in Section 2.1 except the following ones:

(1) Checkpoints of dual-checking are placed at times $kT(k = 1, 2, \cdots, N)$ where $NT = S$, and its overhead is C_2.
(2) The respective failure distributions for M-L and M-S modules are $F_{2L}(t)$ with mean $1/(2\lambda_L)$ and $F_{2S}(t)$ with mean $1/(2\lambda_S)$.

Letting

$$A \equiv 1 - \overline{F}_{2L}(T)F_{2S}(T), \quad B \equiv C_2 + T + R_F F_{2L}(T) + R_B \overline{F}_{2L}(T)F_{2S}(T),$$

the mean execution time to complete the process is, from (3),

$$L_D(N) = \frac{NB + \overline{F}_{2L}(T)\overline{F}_{2S}(T)C_S}{A} \quad (N = 1, 2, \cdots). \tag{5}$$

When the probability that two processes have no error during $(0, t]$ are $\overline{F}_{2L}(t) = e^{-2\lambda_L t}$ and $\overline{F}_{2S}(t) = e^{-2\lambda_S t}$, respectively, (5) is

$$L_D(N) = \frac{N\left[C_2 + T + R_F\left(1 - e^{-2\lambda_L T}\right) + R_B\left(e^{-2\lambda_L T} - e^{-2(\lambda_L+\lambda_S)T}\right)\right] + e^{-2(\lambda_L+\lambda_S)T}C_S}{1 - e^{-2\lambda_S T} + e^{-2(\lambda_L+\lambda_S)T}}$$

$$= \frac{N\left\{C_2 + S/N + R_F\left(1 - e^{-2\lambda_L S/N}\right) + R_B\left[e^{-2\lambda_L S/N} - e^{-2(\lambda_L+\lambda_S)S/N}\right]\right\} + e^{-2(\lambda_L+\lambda_S)S/N}C_S}{1 - e^{-2\lambda_S S/N} + e^{-2(\lambda_L+\lambda_S)S/N}}$$

$$(N = 1, 2, \cdots). \tag{6}$$

Clearly, $\lim_{N\to\infty} L_D(N) = \infty$. Thus, there exists a finite N_D^* $(1 \le N_D^* < \infty)$ to minimize $L_D(N)$. Therefore, by computing $L_D(N)$ for $N = 1, 2, \cdots$, we can compute optimal N_D^* numerically.

2.3 *Majority-checking*

We adopt a majority-checking with $(2j + 1)$ modules to see if the process is normal or not, i.e., $(j + 1)$-out-of-$(2j + 1)$ system $(j = 1, 2, \cdots)$.

Assume that the error rates of modules are the same, and constant overhead for the comparison of states of three processes at each checkpoint $kT(k = 1, 2, \cdots, N)$ is C_3 (Naruse and Nakagawa 2022). Then, the probability that the process is correct during $(0, t]$ is

$$\overline{F}_{j+1}(t) = \sum_{k=j+1}^{2j+1} \binom{2j+1}{k} \left[\overline{F}(t)\right]^k [F(t)]^{2j+1-k} \quad (n = 1, 2, \cdots). \tag{7}$$

It is assumed that the error rates of modules are the same, and $F_L(t) = 1 - e^{-\lambda_L t}$ and $F_S(t) = 1 - e^{-\lambda_S t}$. When $j = 1$, from (7),

$$\overline{F}_{3L}(T) = 3e^{-2\lambda_L T} - 2e^{-3\lambda_L T}, \quad \overline{F}_{3S}(T) = 3e^{-2\lambda_S T} - 2e^{-3\lambda_S T}. \tag{8}$$

Letting

$$A \equiv 1 - \overline{F}_{3L}(T)F_{3S}(T), \quad B \equiv C_3 + T + R_F F_{3L}(T) + R_B \overline{F}_{3L}(T)F_{3S}(T),$$

the mean execution time $L_M(N)$ to complete the process is, from (3),

$$L_M(N) = \frac{NB + \overline{F}_{3L}(T)\overline{F}_{3S}(T)C_S}{A} \quad (N = 1, 2, \cdots). \tag{9}$$

Substituing (8) into (9) and setting $T = S/N$, we can obtain $L_M(N)$, and note that $\lim_{N\to\infty} L_M(N) = \infty$. Thus, there exists a finite N_M^* $(1 \le N_M^* < \infty)$ to minimize $L_M(N)$ in (9). Therefore, by computing $L_M(N)$ for $N = 1, 2, \cdots$, we can compute optimal N_M^* numerically.

2.4 *Numerical example*

We compute optimal N_C^*, N_D^* and N_M^* to minimize $L_C(N)$, $L_D(N)$ and $L_M(N)$, respectively, when $T = S/N$.

Table 1 presents optimal N_C^*, N_D^*, N_M^*, and the resulting execution times $L_C(N_C^*)$, $L_D(N_D^*)$ and $L_M(N_M^*)$ when $S = 10$, $C_1 = C_2 = C_3 = 0.1$, $C_S = 0.4$, $R_B = 0.5$ and $R_F = 0.6$ for λ_L and λ_H. For example, when $\lambda_L = 0.03$ and $\lambda_S = 0.01$, optimal N_C^* is 3 and $L_C(3)$ is 11.198. This indicates that N_C^*, N_D^* and N_M^* decrease with λ_L, and N_C^*, N_D^* and N_M^* increase with λ_S, and $L_C(N)$, $L_D(N)$ and $L_M(N)$ increase with λ_L and λ_S. Optimal N_C^*, N_D^* and N_M^* decrease with λ_L, but they increase with λ_S. Because when λ_S increase, backward recovery is increase, and $L_C(N)$, $L_D(N)$ and $L_M(N)$ increase.

The mean execution time of majority-checking model is the nearest original execution time S among three checkings. Although, a majority structure has the smallest overhead, but it requires more modules than the others.

It is assumed that overheads are getting large with the number of modules. We compute optimal number when $C_1 = 0.1$, $C_2 = 0.2$ and $C_3 = 0.3$. Table 2 presents optimal N_C^*, N_D^* and N_M^*, and the resulting execution times $L_C(N_C^*)$, $L_D(N_D^*)$ and $L_M(N_M^*)$ when $S = 10$, $C_1 = 0.1$, $C_2 = 0.2$, $C_3 = 0.3$, $C_S = 0.4$, $R_B = 0.5$ and $R_F = 0.6$ for λ_L and λ_S. This has the

Table 1. Optimal N_C^*, $L_C(N_C^*)$, N_D^*, $L_D(N_D^*)$, N_M^* and $L_M(N_M^*)$ when $S = 10.0$, $C_1 = C_2 = C_3 = 0.1$, $C_S = 0.4$, $R_B = 0.5$ and $R_F = 0.6$.

λ_L	λ_S	N_C^*	$L_C(N_C^*)$	N_D^*	$L_D(N_D^*)$	N_M^*	$L_M(N_M^*)$
0.01	0.01	3	11.134	4	11.506	2	10.683
0.02	0.01	3	11.166	4	11.572	2	10.696
0.03	0.01	3	11.198	3	11.626	2	10.716
0.04	0.01	2	11.217	1	11.560	2	10.741
0.05	0.01	2	11.226	1	11.393	1	10.750
0.10	0.01	1	11.016	1	10.944	1	10.723
0.01	0.01	3	11.134	4	11.506	2	10.683
0.01	0.02	4	11.471	6	12.017	3	10.847
0.01	0.03	6	11.741	8	12.439	4	10.991
0.01	0.04	6	11.977	9	12.817	5	11.125
0.01	0.05	7	12.193	10	13.168	5	11.242
0.01	0.10	11	13.126	15	14.699	8	11.789

Table 2. Optimal N_C^*, $L_C(N_C^*)$, N_D^*, $L_D(N_D^*)$, N_M^* and $L_M(N_M^*)$ when $S = 10.0$, $C_1 = 0.1$, $C_2 = 0.2$, $C_3 = 0.1$, $C_S = 0.4$, $R_B = 0.5$ and $R_F = 0.6$.

		$C_1 = 0.1$		$C_2 = 0.2$		$C_3 = 0.3$	
λ_L	λ_S	N_C^*	$L_C(N_C^*)$	N_D^*	$L_D(N_D^*)$	N_M^*	$L_M(N_M^*)$
0.01	0.01	3	11.134	3	11.873	1	10.980
0.02	0.01	3	11.166	3	11.909	1	10.975
0.03	0.01	3	11.198	1	11.882	1	10.968
0.04	0.01	2	11.217	1	11.669	1	10.961
0.05	0.01	2	11.226	1	11.500	1	10.953
0.10	0.01	1	11.016	1	11.046	1	10.925
0.01	0.01	3	11.134	3	11.873	1	10.980
0.01	0.02	4	11.471	4	12.581	2	11.306
0.01	0.03	6	11.741	6	13.155	3	11.627
0.01	0.04	6	11.977	7	13.667	3	11.858
0.01	0.05	7	12.193	7	14.137	4	12.127
0.01	0.10	11	13.126	11	16.187	6	13.205

same tendency to Table 1, and $L_C(N)$, $L_D(N)$ and $L_M(N)$ are large comparing with those of Table 1.

3 CONCLUSION

We have considered the execution times for computer systems built by two types of modules whose checkpoint structure are self-checking, dual and majority. We have obtained the total execution times, and discussed optimal execution times to minimize them in numerical examples.

Most computer systems are built by several modules and are required to work at high speed and quality. In practical systems, high performance cannot be achieved enough only by backward recovery. Also, depending on the computer system, some modules require forward recovery. Therefore, these schemes and techniques used in this paper would be helpful and useful for actual systems.

REFERENCES

Abd-El-Barr, M. (2007). *Reliable and Fault-Tolerant*. London: Imperial Colledge Press.

Gray, J., McJones P., Blasgen M., Lindsay B., Lorie R., Price T., Putzolu F., and Traiger I. (1981). The recovery manager of the system R database manager. *ACM Computing Surveys 13*(2), 223–242.

Jeon, S., Bang J., Byun K., and Lee S. (2012). A recovery method of deleted record for SQLite database. *Personal and Ubiquitous Computing 16*(6), 707–715.

Kim, H. and Shin K. G. (1996). Design and analysis of an optimal instruction-retry policy for TMR controller computers. *IEEE Transactions on Computers 45*(11), 1217–1225.

Lee, P. A. and Anderson T. (1990). *Fault Tolerance Principles and Practice. Dependable computing and fault-tolerant systems.* Wien: Springer.

Nakagawa, S., Fukumoto S., and Ishii N. (2003). Optimal checkpointing intervals of three error detection schemes by a double modular redundancy. *Mathematical and Computing Modeling* (38), 1357–1363.

Nakagawa, T. (2005). *Maintenance Theory of Reliability*. London: Springer.

Nakagawa, T. (2008). *Advanced Reliability Models and Maintenance Policies*. London: Springer.

Naruse, K. and Nakagawa T. (2020). Optimal checkpoint intervals, schemes and structures for computing modules. In *Reliability and Statistical Computing*, pp. 265–287. Springer.

Naruse, K. and Nakagawa T. (2022). Optimal design of checkpoint systems with general structures, tasks and schemes. In *Reliability and Maintainability Assessment of Industrial Systems*, pp. 73–91. Springer.

Naruse, K. and Nakagawa T. (2023). Optimal scheme models with imperfect checkpoint. *International Journal of Reliability, Quality and Safety Engineering 30*(04), 2350015.

Naruse, K. and Nakagawa T. (2024). Optimal scheme models with two kinds of errors. *International Journal of Reliability, Quality and Safety Engineering*, 2450010.

Ohara, M., Suzuki R., Arai M., Fukumoto S., and Iwasaki K. (2006, Sep). Analytical model on hybrid state saving with a limited number of checkpoints and bound rollbacks. *IEICE TRANSACTIONS on Fundamentals of Electronics, Communications and Computer Sciences 89*(9), 2386–2395.

Pradhan, D. and Vaidya N. (1992). Roll-forward checkpointing scheme: Concurrent retry with nondedicated spares. In *Proceedings of the IEEE Workshop on Fault-Tolerant Parallel and Distributed Systems*, pp. 166–174.

Pradhan, D. K. and Vaidya N. H. (1994). Roll-forward checkpointing scheme: A novel fault-tolerant architecture. *IEEE Transactions on Computers 43*(10), 1163–1174.

Siewiorek, D. P. and Swarz R. S. (1982). *The Theory and Practice of Reliable System Design*. Bedford, Massachusetts: Digital Press.

Zhang, Y. and Chakrabarty K. (2004). Dynamic Adaptation for Fault Tolerance and Power Management in Embedded Real-Time Systems. *ACM Transactions on Embedded Computing 3*(2), 336–360.

Ziv, A. and Bruck J. (1997). Performance optimization of checkpointing schemes with task duplication. *IEEE Transactions on Computers 46*, 1381–1386.

Ziv, A. and Bruck J. (1998). Analysis of checkpointing schemes with task duplication. *IEEE Transactions on Computers 47*, 222–227.

Recent Advances in Reliability and Maintenance Modeling – Okamura, Inoue & Xiao (Eds)
© 2025 the Author(s), ISBN: 978-1-032-79098-5

Joint optimization of maintenance and spare unit management with partial observations

N. Ogura, M. Kasuya & L. Jin
University of Electro-Communications, Tokyo, Japan

ABSTRACT: This research focuses on a series system consisting of non-identical units. A monitoring device provides partial observations related to system deterioration at equally spaced time periods. At the beginning of each period, the decision-maker determines whether to continue operation or perform maintenance based on deterioration and inventory information. When replacement is determined as the maintenance action, decisions regarding spare unit management—such as unit selection, inventory levels, and ordering timing—are made concurrently. This study proposes an optimal joint policy for maintenance and spare unit management to minimize expected total costs indefinitely. Additionally, the performance of the proposed policy is evaluated by comparing it with several benchmark policies.

1 INTRODUCTION

Condition-based maintenance is an effective and efficient way to improve system reliability and safety by performing maintenance only when necessary based on monitoring information. In condition-based maintenance, the action replacement replaces the deteriorated units using corresponding spare units. Spare unit management, including ordering, inventory control, and the selection of spare units brought to the maintenance site, mutually influences maintenance decisions. Therefore, to optimize the overall operation of the system, it is necessary to jointly optimize maintenance and spare unit management (Horenbeek *et al.* 2013; Hwang *et al.* 2019). Several researchers have shown interest in studies on condition-based maintenance that consider spare unit management. Xie *et al.* (2008) optimize maintenance policies under an (s, S) ordering policy, which indicates the order of $S(S > s)$ units when the inventory is reduced to s. Olde Keizer *et al.* (2017) formulated a maintenance optimization problem using Markov decision processes. They incorporated ordering actions into the model as one of the decisions and proposed a joint optimal policy of maintenance and spare unit ordering. They also presented the effectiveness of their proposed policy by comparing it with the (s, S) ordering policy. Some studies have optimized the choice of spare unit suppliers. Zheng *et al.* (2021) proposed a joint optimal policy for spare unit ordering and maintenance that considers spare unit suppliers with different ordering costs and lead times. Some numerical studies show the effect of system conditions on supplier selection and the effectiveness of the optimal policy. Zheng *et al.* (2023) optimize maintenance and spare unit management for k-out-of-n systems. They further presented the impact of information about spare unit inventory on maintenance costs by comparing policies that consider only system state or inventory. In most of the studies mentioned before, the decision maker makes decisions about maintenance and spare unit management based on the system's state. They also assumed that the system's state can always be completely observed. However, the monitoring device may output information that differs from the system's state due to device failures or noise from external sources. If the uncertainty in he monitoring information is not

DOI: 10.1201/9781003491309-21

taken into account, the risk cannot be correctly assessed, and appropriate maintenance and spare unit management may not be implemented. Therefore, it is important to consider the uncertainty of observation information.

Partially observable Markov decision process (POMDP) is a model generally used for sequential decision-making with partial observations. Jiang *et al.* (2015) considered optimal maintenance policy under partial observation for a multi-unit system consisting of non-identical units. Furthermore, Chen *et al.* (2018) clarified the structural properties of the optimal policy proposed by Jiang *et al.* (2015). Karabag *et al.* (2020) optimize a maintenance policy, including spare unit selection under partial observations. They also compare the proposed policy with two benchmark policies and show that the proposed policy can reduce the total expected cost.

This research proposes a joint policy for maintenance and spare unit management to minimize the expected total cost over an infinite time horizon under partial observation. Spare unit management considered in this research, including both spare unit selection and ordering which has not been discussed in other literature. Based on same simulation result, this research presents and discusses the impact of inventory on maintenance and spare unit selection. Moreover this research evaluated the effectiveness of the proposed policy by comparing it to several benchmark policies and obtained results indicating a reduction in that the total expected cost by about 21% to 24%.

2 PROBLEM DESCRIPTIONS

2.1 *Deterioration and spare units*

This research deals with a series system consisting of N non-identical units. Let s_i be the state of unit i, which takes a value from the set $S_i = \{0, 1, \ldots, \widetilde{s}_i\}$. The value of s_i reflects the deterioration of the unit, where $s_i = 0$ means that unit i is as-good-as-new, and $s_i = \widetilde{s}_i$ means that unit i is failed. Each unit $i : (\in \mathcal{I} = \{1, 2, \ldots, N\})$ has a unique defect threshold Δ_i: $(0 < \Delta_i < \widetilde{s}_i)$. If the unit's state is $s_i < \Delta_i$, it is considered normal; if $\Delta \leq s_i < \widetilde{s}_i$, it is considered defective. Let $p_i(s^{'}|s)$ denote the probability that unit i deteriorates from state s to s'. A matrix having $p_i(s|s')$ as the s-th row and s'-th column element is the state transition probability matrix for unit i, denoted by $\mathbf{P}_i$. The system's state is represented as a vector $\mathbf{s} = (s_1, s_2, \ldots, s_N)$ of which the elements indicate the states of N units. Let $q(\mathbf{s}'|\mathbf{s})$ denote the probability that the system state transitions from $\mathbf{s}$ to $\mathbf{s}'$. Also, the state transition probability matrix $\mathbf{Q}$ represents state transitions about the system. This research assumes that the state transitions of each unit are independent of each other. Then, the probability $q(\mathbf{s}'|\mathbf{s})$ that the system state transitions from $\mathbf{s}$ to $\mathbf{s}'$ is obtained by Equation 1.

$$q(\mathbf{s}'|\mathbf{s}) = \prod_{i \in \mathcal{I}} p_i(s_i'|s_i) \tag{1}$$

The deteriorated unit will be replaced by the corresponding spare unit if necessary. Let $I_i \in \{0, 1, \ldots, \widetilde{I}_i\}$ be the number of spare units in inventory for unit i and $\mathbf{I} = (I_1, I_2, \ldots, I_N)$ be the inventory of all types spare units. Here, $\widetilde{I}_i$ indicates the inventory limit. The state space $\mathcal{S}$ is represented by the state of the system and the inventory of all spare unit.

$$\mathcal{S} = \left\{(\mathbf{s}, \mathbf{I}) = (s_1, s_2, \ldots, s_N, I_1, I_2, \ldots, I_N)|s_i \in \{0, 1, \ldots, \widetilde{s}_i\}, I_i \in \{0, 1, \ldots, \widetilde{I}_i\}\right\} \tag{2}$$

2.2 *Partial observations*

To account for the uncertainties in observed information, this research considers belief π as a probability vector to describe the system's state. The decision maker obtains partial information about the system's state from the monitoring device at equally spaced time periods τ.

Let $\theta \in \Theta = \{0, 1, \ldots, \tilde{\theta}\}$ be the observation from the monitoring device, where a larger value indicates greater system deterioration. $\gamma(\theta|\mathbf{s})$ is the conditional probability that the monitoring device outputs the observation θ given the system's state $\mathbf{s}$. This represents the characteristics of the monitoring device. Let $\mathbb{S}$ be the set of all possible state vectors $\mathbf{s}$. The belief π consists of probabilities $\pi_{\mathbf{s}}$ for the system's state being $\mathbf{s} \in \mathbb{S}$. The set of beliefs is defined as follows,

$$\Pi = \left\{ \pi \in \mathbb{R}_{+}^{|\mathbb{S}|} \,|\, \sum_{\mathbf{s} \in \mathbb{S}} \pi_{\mathbf{s}} = 1 \right\}, \tag{3}$$

where $\mathbb{R}_{+}$ represents the set of positive real values. Given the belief vector $\boldsymbol{\pi}$, the probability of observing θ at the beginning of the next period can be obtained as follows,

$$\sigma(\theta|\pi) = \sum_{\mathbf{s}' \in \mathbb{S}} \sum_{\mathbf{s} \in \mathbb{S}} \pi_{\mathbf{s}} q(\mathbf{s}'|\mathbf{s}) \gamma(\theta|\mathbf{s}'). \tag{4}$$

When the θ is observed, the belief $\boldsymbol{\pi}$ is updated to $\boldsymbol{\pi}'$ based on Bayes' theorem, and the probability $\pi'_{\mathbf{s}'}$ of being in state $\mathbf{s}'$ in the updated belief is obtained as follows,

$$\pi'_{\mathbf{s}'} = T_{\mathbf{s}'}(\pi, \theta) = \frac{\sum_{\mathbf{s} \in \mathbb{S}} \pi_{\mathbf{s}} q(\mathbf{s}'|\mathbf{s}) \gamma(\theta|\mathbf{s}')}{\sigma(\theta|\pi)}. \tag{5}$$

2.3 *Actions and costs*

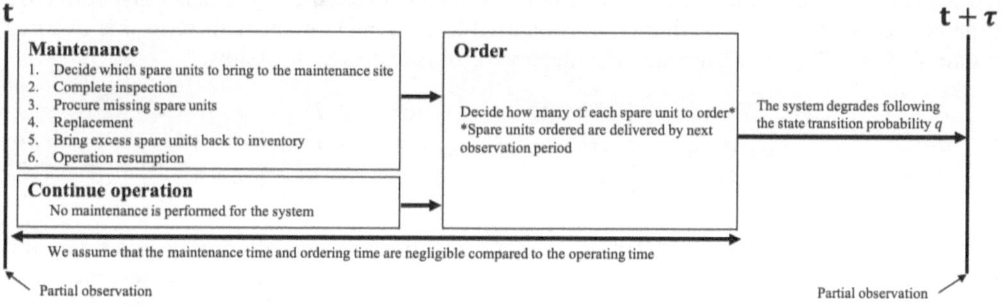

Figure 1. Flow of actions within one period.

At the beginning of each observation period, the decision-maker chooses whether to perform maintenance or continue operation, selects spare units to bring to the maintenance site from inventory, and orders spare units. As shown in Figure 1, during maintenance, a complete inspection of the system is conducted to obtain the states of all units. Subsequently, units with states equal to or greater than Δ_i are replaced with the spare units. In the case of continue operation, the system operates for one more period without any maintenance. The action space can be represented as $\mathcal{A} = \{0, 1\}$, where $a = 1$ indicates maintenance, and $a = 0$ indicates continue operation. Similar to Karabag *et al.* (2020), this research assumes that the maintenance time is negligible compared to the system's operating time and observation periods. Therefore, the system resumes operation immediately after maintenance.

The observation regarding the system's state obtained from the monitoring device is uncertain, and the decision-maker cannot directly ascertain the states of the units completely. Therefore, it is necessary to appropriately determine the spare units to bring to the site when performing maintenance. The selection of spare units is represented by a binary vector $\mathbf{g} = (g_1, g_2, \ldots, g_N)$, where 1 indicates bringing in spare units and 0 indicates not bringing in spare units. The set $\mathbf{G}$ of possible $\mathbf{g}$ configurations for spare unit selection can be expressed as

$$\mathbf{G} = \{(g_1, g_2, \ldots, g_N) | g_i \in \{0, 1\}, i \in \mathcal{I}\}. \tag{6}$$

The decision-maker decides the quantity $\boldsymbol{\omega} = (\omega_1, \omega_2, \ldots, \omega_N)$ of each spare unit and orders at the beginning of every observation period. However, the order quantity ω_i of unit i must not exceed the ordering limit $\widetilde{\omega}_i$ and the inventory limit $\widetilde{I}_i$. If maintenance is selected, the decision on the order quantity is made after the completion of maintenance. The ordered spare units arrive by the beginning of the next period and become available. Given the inventory $\mathbf{I}$, the set $\Omega_{(\mathbf{I})}$ of possible order quantities $\boldsymbol{\omega}$ for spare units can be expressed as

$$\Omega_{(\mathbf{I})} = \left\{(\omega_1, \omega_2, \ldots, \omega_N) | \omega_i \in \left\{0, 1, \ldots, \min(\widetilde{\omega}_i, \widetilde{I}_i - I_i)\right\}, i \in \mathcal{I}\right\}. \tag{7}$$

Similar to the maintenance time, this research assumes that the time for ordering is negligible. In this model, operation cost, maintenance cost, and spare unit management cost are considered. The operation cost, denoted as $C^{oc}(\mathbf{s})$, incurs per period based on the true state of the system. Maintenance costs consist of fixed maintenance cost $C^f(\mathbf{s})$ and variable maintenance cost $C^v(\mathbf{I}, \mathbf{s}, \mathbf{g})$. If maintenance is chosen, after selecting the spare units to bring in, a complete inspection is conducted. For all units with the states $s_i \geq \Delta_i$, a replacement is made with corresponding new spare units ($s_i = 0$). In this case, if there is at least one unit in a failure state within the system, it is considered corrective maintenance (CM) and incurs a fixed maintenance cost of c^{cm}. If there are no units in a failure state, it is considered preventive maintenance (PM) with a fixed maintenance cost of c^{pm}. Additionally, if only an inspection is performed without unit replacement, an inspection cost c^{im} is incurred as a fixed maintenance cost. Generally, corrective maintenance tends to have higher costs compared to preventive maintenance, so this research assumes $c^{im} \leq c^{pm} \leq c^{cm}$. The fixed maintenance cost is expressed as follows,

$$C^f(\mathbf{s}) = \begin{cases} c^{im} & \text{if} \quad s_i < \Delta_i \ \text{ for } \forall i \in \mathcal{I} \\ c^{cm} & \text{if} \quad s_i = \widetilde{s}_i \ \text{ for } \exists i \in \mathcal{I} \\ c^{pm} & \qquad \text{otherwise} \end{cases}. \tag{8}$$

Variable maintenance cost $C^v(\mathbf{I}, \mathbf{s}, \mathbf{g})$ is determined by the inventory, the system state, and the spare unit selection. If the necessary spare unit is not in inventory, an emergency procurement cost c_i^{eo} is incurred, and the spare units are transferred without lead time. After the spare units are brought in and replaced, a replacement cost c_i^r is incurred for each unit. In cases where the necessary spare units for replacement are not brought in, an emergency bring-in cost c^{eb} is incurred to bring in spare units from inventory. If the brought in spare units are not utilized, a bring-back cost c^b is incurred. The variable maintenance cost is expressed by Equation 9. Here, $\mathbf{1}_{\{.\}}$ is the indicator function, which is 1 when the condition is satisfied and 0 otherwise.

$$\begin{aligned} C^v(\mathbf{I}, \mathbf{s}, \mathbf{g}) &= \sum_{i \in \mathcal{I}} c_i^r \mathbf{1}_{\{s_i \geq \Delta_i\}} + c^b \sum_{i \in \mathcal{I}} g_i \mathbf{1}_{\{s_i < \Delta_i\}} + c^{eb} \sum_{i \in \mathcal{I}} (1 - g_i) \mathbf{1}_{\{s_i \geq \Delta_i\}} \\ &+ \left\{ \sum_{i \in \mathcal{I}} c_i^{eo} g_i \mathbf{1}_{\{I_i = 0\}} + \sum_{i \in \mathcal{I}} c_i^{eo} (1 - g_i) \mathbf{1}_{\{I_i = 0 \cap s_i \geq \Delta_i\}} \right\} \end{aligned} \tag{9}$$

Spare unit management cost $C^I(\mathbf{I}, \omega)$ consists of the procurement cost of spare units and the holding cost c^h. The procurement cost is composed of a fixed ordering cost c^l and the price c_i^p for each spare unit. The spare unit management cost is expressed as

$$C^I(\mathbf{I}, \omega) = c^h \sum_{i \in \mathcal{I}} I_i + c^l \mathbf{1}_{\{\omega_i > 0 \text{ for } \exists i\}} + \sum_{i \in \mathcal{I}} c_i^p \omega_i \ . \tag{10}$$

2.4 *Optimal joint policy of maintenance and spare unit management*

This research aims to minimize the total expected cost for maintenance and spare unit management measures, and a POMDP is used to optimize the decision-making problem. The total expected cost $V(\pi, \mathbf{I})$ over an infinite horizon given a belief π and a spare unit inventory $\mathbf{I}$ can be formulated as

$$V(\pi, \mathbf{I}) = \min \begin{cases} V^C(\pi, \mathbf{I}) & \text{if} \quad a = 0 \\ V^R(\pi, \mathbf{I}) & \text{if} \quad a = 1 \end{cases} , \tag{11}$$

where

$$V^C(\pi, \mathbf{I}) = \sum_{\mathbf{s} \in \mathbb{S}} \pi_{\mathbf{s}} C^{oc}(\mathbf{s}) + \min_{\omega \in \Omega} \left\{ C^I(\mathbf{I}, \omega) + \beta \sum_{\theta \in \Theta} \sigma(\theta | \pi) V(T(\pi, \theta), \mathbf{I} + \omega) \right\}, \tag{12}$$

$$V^R(\pi, \mathbf{I}) = \sum_{\mathbf{s} \in \mathbb{S}} \pi_{\mathbf{s}} C^f(\mathbf{s}) + \min_{\mathbf{g} \in \mathbf{G}} \left\{ \sum_{\mathbf{s} \in \mathbb{S}} \pi_{\mathbf{s}} C^v(\mathbf{I}, \mathbf{s}, \mathbf{g}) + \sum_{\mathbf{s} \in \mathbb{S}} \pi_{\mathbf{s}} V^C(M(\mathbf{s}), U(\mathbf{I}, \mathbf{s}, \mathbf{g})) \right\}. \tag{13}$$

Here, Equation 12 is the total expected cost of choosing to continue operation and β is the discount rate. Equation 13 is the total expected cost of choosing maintenance and $M(\mathbf{s})$ is the belief of the system after the replacement performed in state $\mathbf{s}$. Let $\pi_{\widehat{\mathbf{s}}}$ be an element of $M(\mathbf{s})$ as the probability of state $\widehat{\mathbf{s}}$. Since the replacement is performed for all units with state $s_i \geq \Delta_i$, $\pi_{\widehat{\mathbf{s}}}$ can be defined as follows,

$$\pi_{\widehat{\mathbf{s}}} = \begin{cases} 1 & \text{if} \quad \widehat{\mathbf{s}} = \mathbf{s} \circ \boldsymbol{\mu}_{\{s_i < \Delta_i\}} \\ 0 & \text{otherwise} \end{cases} (i \in \mathcal{I}). \tag{14}$$

Note that $\boldsymbol{\mu}_{\{.\}}$ is a column vector whose elements are 1 only if the condition is satisfied and 0 otherwise. The decision maker selects the spare unit $\mathbf{g}$ such that the expected cost is minimized.

The inventory update function is shown in Equation 15.

$$U(\mathbf{I}, \mathbf{s}, \mathbf{g}) = \mathbf{I} - \boldsymbol{\mu}_{\{I_i > 0 \cap s_i \geq \Delta_i\}} + \mathbf{g} \circ \boldsymbol{\mu}_{\{I_i = 0 \cap s_i < \Delta_i\}} \tag{15}$$

Here, the second term represents spare units taken out of inventory, and the third term represents spare units brought back to inventory. The optimal policy can be obtained by solving Equation 11 to determine the optimal action a^*, spare units order ω^*, and spare unit selection $\mathbf{g}^*$ given the states $\boldsymbol{\pi}$ and $\mathbf{I}$.

3 NUMERICAL STUDY

In the numerical study, we consider a series system consisting of two units. Each unit has three states, denoted as $S_1 = S_2 = \{0(\text{normal}), 1(\text{defective}), 2(\text{failed})\}$, with a defect threshold of $\Delta_i = 1\,(i = 1, 2)$. Additionally, it is assumed that each unit deteriorates independently following a Markov chain. Next, we assume that monitoring devices provide observations of three types ($\Theta = \{0(\text{normal}), 1(\text{defective}), 2(\text{failed})\}$). The transition probability matrices and the observation conditional probability $\gamma(\theta|\mathbf{s})$ are given by Equations 16 and Table 1, respectively.

Table 1. Observation conditional probability for each system's state **s**.

$\gamma(\theta\|\mathbf{s})$		**s**								
		(0, 0)	(0, 1)	(0, 2)	(1, 0)	(1, 1)	(1, 2)	(2, 0)	(2, 1)	(2, 2)
θ	0	0.90	0.05	0.05	0.05	0.05	0.05	0.05	0.05	0.05
	1	0.05	0.90	0.05	0.90	0.90	0.05	0.05	0.05	0.05
	2	0.05	0.05	0.90	0.05	0.05	0.90	0.90	0.90	0.90

$$\mathbf{P}_1 = \mathbf{P}_2 = \begin{bmatrix} 0.8 & 0.1 & 0.1 \\ 0 & 0.8 & 0.2 \\ 0 & 0 & 1 \end{bmatrix} \tag{16}$$

Refer to Karabag *et al.* (2020), Zheng *et al.* (2021), we set the costs as shown in Table 2. The emergency bring-in cost consists of additional transportation costs, while the emergency procurement cost includes the cost for express delivery.

The proposed policies for different spare unit inventories **I** are illustrated in Figure 2. The vertical axis represents π_1^2 (the probability of unit 2 being in a defective state), and the horizontal axis represents π_0^2 (the probability of unit 2 being in a normal state). As shown in Table 3, the symbols in the figure indicate maintenance actions, and the background pattern indicates orders.

Table 2. Maintenance and spare unit management costs ($i \in \mathcal{I}$).

	Operation cost			Fixed maintenance cost		
Parameter	c^n	c^d	c^f	c^{cm}	c^{pm}	c^{im}
Value	0	0	1000	300	100	100

	Variable maintenance cost				Spare unit management cost		
Parameter	c_i^r	c_i^{eo}	c^{eb}	c^b	c^l	c_i^p	c^h
Value	100	330	60	30	20	100	10

Table 3. Legend of the figures.

Symbol	Action	Selects spare units	Pattern	Orders spare units	Pattern	
●	Continue operation	-		Decided after replacement		$\boldsymbol{\omega} = (1,1)$
–	Replacement	$\mathbf{g} = (0,0)$		$\boldsymbol{\omega} = (0,0)$		$\boldsymbol{\omega} = (1,2)$
▶		$\mathbf{g} = (1,0)$		$\boldsymbol{\omega} = (0,1)$		$\boldsymbol{\omega} = (2,0)$
◀		$\mathbf{g} = (0,1)$		$\boldsymbol{\omega} = (0,2)$		$\boldsymbol{\omega} = (2,1)$
■		$\mathbf{g} = (1,1)$		$\boldsymbol{\omega} = (1,0)$		$\boldsymbol{\omega} = (2,2)$

Figures 2a-c represent the situations when there is no spare inventory for unit 2. In this scenario, when the probability of being in a defective state (π_1^2) is high, the region of continue operation becomes larger compared to the situation with available inventory for unit 2 as shown in Figures 2d-i. This indicates that delaying maintenance is preferable to avoid the

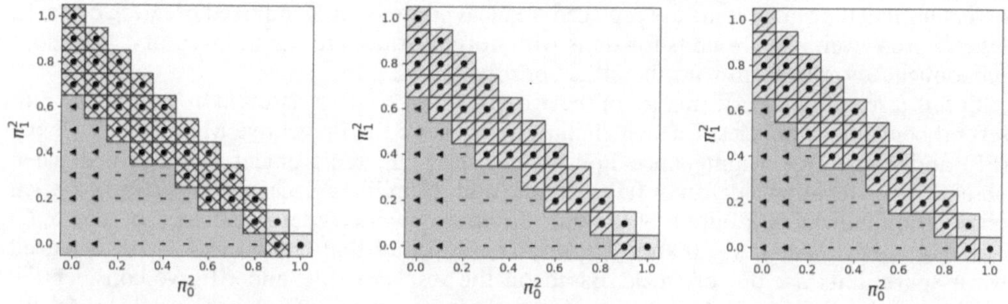

(a) $(I_1, I_2) = (0, 0)$ (b) $(I_1, I_2) = (1, 0)$ (c) $(I_1, I_2) = (2, 0)$

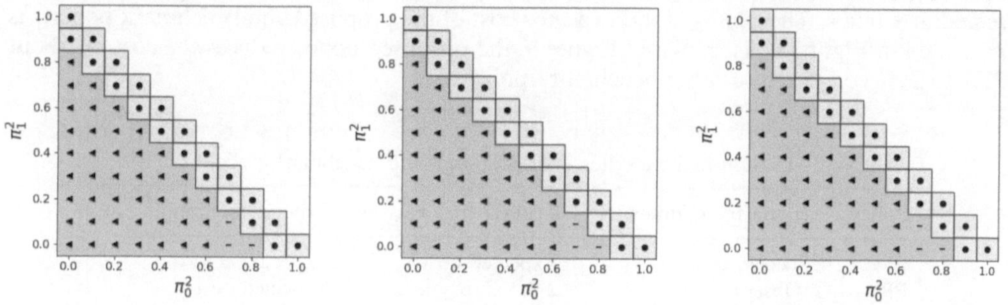

(d) $(I_1, I_2) = (0, 1)$ (e) $(I_1, I_2) = (1, 1)$ (f) $(I_1, I_2) = (2, 1)$

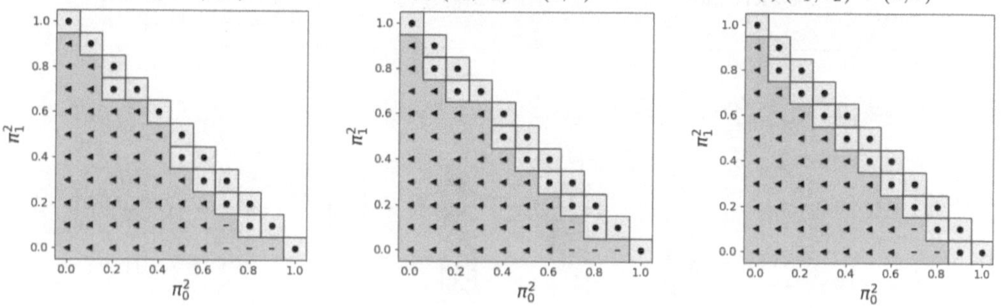

(g) $(I_1, I_2) = (0, 2)$ (h) $(I_1, I_2) = (1, 2)$ (i) $(I_1, I_2) = (2, 2)$

Figure 2. Proposed policy for each spare unit inventory ($\boldsymbol{\pi}^1 = (1, 0, 0)$).

occurrence of emergency procurement costs. In other words, when there is no spare unit inventory, the proposed policy tends to postpone maintenance, securing inventory before conducting replacements.

Next, we focus on the selection of spare units to bring to the maintenance site. When the probability of unit 2 being in a normal state (π_0^2) is high, the choice is made not to bring in spare units for unit 2 as shown in Figure 2. In cases where there is no spare unit inventory, the region for choosing not to bring in spare units expands. Conversely, in cases where there is sufficient spare unit inventory, that region becomes smaller. In such cases, spare units chosen for bringing to the maintenance site are emergency ordered. This reduces the advantage of bringing in spare units from the beginning. On the other hand, when there is sufficient inventory, spare units are actively utilized to reduce storage costs.

Finally, we focus on ordering. Orders are placed for spare units with no inventory. However, orders for spare units for unit 1 with no inventory are not placed in situations where the probability of unit 2 being normal is small as shown in Figure 2d, g. Postponing orders until deteriorated units are replaced, reduces storage costs and fixed ordering costs. In other words, even if spare units for units with normal states are not in inventory, it is more economical to order them with the orders of other units.

In this research, the performance of the proposed policy (IP) is assessed by comparing it to several benchmark policies. The benchmark policies are the Preventive Maintenance Policy (PP) and Corrective Maintenance Policy (CP). For PP, replacement is performed when observation indicating a defect or failure of the unit. For CP, replacement is performed when observation indicating a failure of the unit. In both policies, ordering is based on the (s, Q) ordering policy. We set $s = 0$ and $Q = 1$, which indicates that an order of 1 unit is placed when spare units are out of stock. Based on the respective PP and CP, we consider the policies to always bring in both types of spare units to the maintenance site policies as PP_f and CP_f. The decision-making criteria for the proposed and benchmark policies are presented in Table 4. The average total expected cost of the proposed and benchmark policies is presented in Figure 3. As shown in Figure 3, the proposed policy reduces the cost by about 21% to 24% compared to other benchmark policies.

Table 4. Decision-making criteria for proposed and benchmark policies.

Policy	Criteria for maintenance	Criteria for ordering	Criteria for bringing in
IP	Expected cost	Expected cost	Expected cost
PP	2*Observation	2*Inventory level	Expected cost
PP_f			Fixed: $\mathbf{g} = (1, 1)$
CP	2*Observation	2*Inventory level	Expected cost
CP_f			Fixed: $\mathbf{g} = (1, 1)$

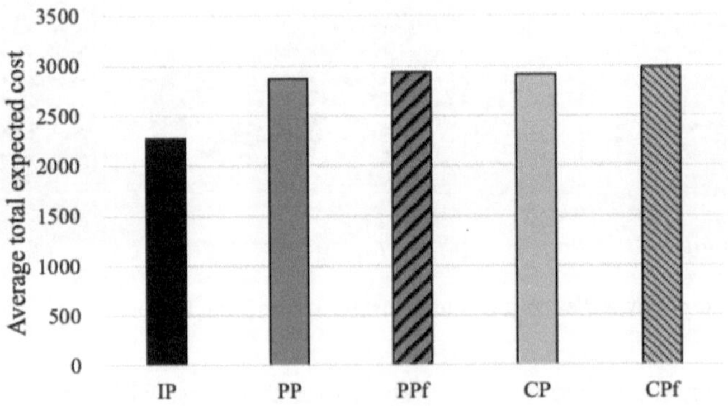

Figure 3. Average total expected cost for different policy.

4 CONCLUSION AND FUTURE RESEARCH

This research proposes an optimal joint policy for maintenance and spare unit management for a series system consisting of non-identical units under partial observation. The optimization problem to minimize the expected cost is formulated by a POMDP. In a numerical study, the impact of beliefs and inventories on the proposed policy is discussed in detail. Furthermore, this research compares the proposed policy with several benchmark policies and shows that the proposed policy can reduce the average total expected cost.

In this research, it was assumed that maintenance is completed immediately, and the lead time is constant. In real situations, maintenance and transportation time of spare units may take longer and may not always be constant. Future research includes considering maintenance time and random lead times. Additionally, the replacement threshold was prespecified in this research. Optimization of the replacement threshold may be considered as the next challenge.

REFERENCES

Chen, M., Jiang, Y. and Zhou, D. (2018). Decentralized maintenance for multistate systems With heterogeneous components. in *IEEE Transactions on Reliability* 67: 701–714.

Horenbeek, A., Buré, J., Cattrysse, D., Pintelon, L. and Vansteenwegen, P. (2013). Joint maintenance and inventory optimization systems: A review. *International Journal of Production Economics* 143: 499–508.

Hwang, J. and Samat, H. (2019). A review on joint optimization of maintenance with production planning and spare part inventory management. *IOP Conference Series: Materials Science and Engineering* 530(1): 012048.

Jiang, Y., Chen, M. and Zhou, D. (2015). A POMDP based decentralized maintenance for multi-state system with heterogeneous components. *Chinese Automation Congress (CAC)*: 2057–2062.

Karabag, O., Eruguz, A. and Basten, R. (2020). Integrated optimization of maintenance interventions and spare part selection for a partially observable multi-component system. *Reliability Engineering and System Safety* 200: 106955.

Olde Keizer, M., Teunter, R. and Veldman, J. (2017). Joint condition-based maintenance and inventory optimization for systems with multiple components. *European Journal of Operational Research* 257: 209–222.

Xie, J. and Wang, H. (2008). Joint optimization of condition-based preventive maintenance and spare ordering policy. *4th International Conference on Wireless Communications, Networking and Mobile Computing*: 1–5.

Zheng, M., Ye, H., Wang, D. and Pan, E. (2021). Joint optimization of condition-based maintenance and spare parts orders for multi-unit systems with dual sourcing. *Reliability Engineering & System Safety* 210: 107512.

Zheng, M., Lin, J., Xia, T., Liu, Y. and Pan, E. (2023). Joint condition-based maintenance and spare provisioning policy for a *k*-out-of-*n* system with failures during inspection intervals. *European Journal of Operational Research* 308: 1220–1232.

Recent Advances in Reliability and Maintenance Modeling – Okamura, Inoue & Xiao (Eds)
© 2025 the Author(s), ISBN: 978-1-032-79098-5

Deriving the confidence interval for the cumulative number of software faults using parametric simulation

O. Kento & X. Xiao
Tokyo Metropolitan University, Tokyo, Japan

ABSTRACT: Software reliability evaluation is important. Among various methods, interval estimation is easy to feed back to flexible decision making by providing upper and lower bound. Thus, we consider the interval estimation of software reliability in this paper. The limitation of existing studies is that creating resamples for constructing the confidence interval is time-consuming. We address this problem by introducing parametric simulation methods, and we verify the effectiveness of the proposed method by analyzing real data.

1 INTRODUCTION

Modern life, including the use of cars and home appliances, is closely related to software, the size of which increases as it becomes increasingly complex. Against this background, software reliability engineering was established to quantitatively measure and predict reliability in the software development process. Quantitative evaluation of reliability and quality assurance is highly important given the prevalence of self-driving car accidents and credit card payment service failures. Among software reliability models used in software reliability measurement and quantitative control of testing processes, the software reliability growth model based on non-homogeneous Poisson process (NHPP) has attracted considerable research interest because of its tractability and satisfactory goodness-of-fit. Point estimation and interval estimation are used for quantitative evaluation of software reliability to derive reliability measures. This paper focuses on interval estimation because point estimation cannot fully account for uncertainty.

2 EXISTING INTERVAL ESTIMATION METHOD

A series of studies conducted by Xiao and Dohi (2014, 2017, 2021) obtained empirical distribution from multiple estimates using the bootstrap method (Efron and Tibshirani (1993)). They considered the stochastic process behind software fault detection phenomena to be an NHPP. $N(t)$ is assumed to be a random variable representing the cumulative number of software faults detected at time t. If the stochastic process $\{N(t), t \geq 0\}$ forms an NHPP, then its probability function is expressed as follows:

$$Pr\{N(t) = n\} = \frac{\Lambda(t)^n}{n!} exp(-\Lambda(t)), \quad n = 0,\ 1, \ldots \tag{1}$$

where $\Lambda(t) = \int_0^t \lambda(s)ds$ is the mean value function of NHPP and is the expected value of the cumulative number of software faults detected at time t. $\lambda(t)$ is the intensity function of NHPP and is the software fault detection rate at time t.

Methods of Xiao and Dohi (2014, 2017, 2021) require generating NHPP arrivals from one software fault count data. More specifically, they use the following methods for simulation

 DOI: 10.1201/9781003491309-22

to generate the arrivals: a) thinning, b) order statistics (*OrdSta*), c) interval generation (*IntGen*), and d) time-scale transformation (*TimTra*). These methods are nonparametric simulation methods thereby resulting in a time-consuming sample generation. Therefore, this paper aims to develop a more efficient interval estimation method using the bootstrap method by applying parametric simulation methods.

3 PARAMETRIC SIMULATION-BASED INTERVAL ESTIMATION

Similar to Xiao and Dohi (2014, 2017, 2021), we use the bootstrap method (*Method 1*) proposed by Cowling *et al.* (1996) for stochastic processes.

Method 1. Generate the X_i^*'s as points of a Poisson process of intensity $\widehat{\lambda}$. Thus the resampled Poisson process on the interval $\mathcal{I} = (0,1)$ may be written as $X_1^*, \ldots, X_{N^*}^*$, where N^* has a Poisson distribution with intensity $\widehat{\Lambda}(1)$, and

$$\widehat{\Lambda}(x) = \int_0^x \widehat{\lambda}(y)dy. \tag{2}$$

To use this method, the key point is "determining the intensity $\widehat{\lambda}$" and "generating X_i^*s". To achieve these key points, the proposed method consists of the following steps. Each step is described in Section 3.1, Section 3.2, and Section 3.3. [left margin = 1 mm]

Step 1 A point estimate of the intensity function of NHPP is obtained.
Step 2 Resamples are created from arrivals generated using one of four parametric simulation methods.
Step 3 Confidence intervals are constructed from multiple resamples.

3.1 *Estimation of intensity function*

A point estimate of the continuous intensity function $\widehat{\lambda}(t)$ $(0 \leq t \leq t_e)$ must be computed from the software fault count data. t_e indicates the test end time. While many point estimation methods for NHPP are available, we use the wavelet shrinkage estimation (WSE) proposed by Xiao and Dohi (2013), which has high estimation accuracy. WSE is an estimation method that removes noise from the observed data to obtain a noise-free intensity function, and its procedure can be divided into three components: (1) obtain empirical wavelet coefficients using Haar wavelets, (2) remove noise from the empirical wavelet coefficients by using the thresholding method, and (3) estimate the intensity function using the wavelet coefficients with the noise removed. Xiao and Dohi (2014, 2017, 2021) applied linear spline interpolation given that WSE provides a point estimate of the intensity function in a discrete-time manner. In this paper, a nonlinear least-squares fitting method is applied.

3.2 *Simulation of NHPP*

The four parametric NHPP simulation methods used in this paper include the direct simulation (*COS*), simulation using gap statistics ($\alpha_1 < 0$) (*GS*−) and ($\alpha_1 > 0$) (*GS*+), and rejection and gap statistics (*RGS*). The first three methods were proposed by Lewis and Shedler (1976), and the last method was proposed by Lewis and Shedler (1979). Among them, the pseudo codes of *COS* and *RGS* are illustrated in Figures 1 and 2, respectively. Figure 3 shows the pseudo code of the rejection algorithm used in *RGS* (Figure 2). Lewis and Shedler (1976, 1979) discussed more details about this concept. Each simulation method is used to generate m sets of arrivals. These datasets are converted to software fault count data, and then m resamples are obtained. We construct empirical distributions for reliability measures and derive confidence intervals using m sets of software fault count data, originating from one of four simulation methods.

Algorithm 1 The pseudo code of *COS*

Require: $\alpha_1 \neq 0$
1: $C = e^{\alpha_1 t_e} - 1$, $\mu_0 = \frac{\lambda C}{\alpha_1}$
2: Generate a Poisson random variable n with μ_0
3: **if** n=0 **then**
4: exit; There are no events
5: **end if**
6: Generate n uniform variates and order them to get $U_{(0)} \leq \dots \leq U_{(0)}$
7: Get $T_1 = {\alpha_1}^{-1} log(1 + CU_{(1)}), T_2 = {\alpha_1}^{-1} log(1 + CU_{(2)}) \cdots$
8: Return $T_1, T_2, \dots$

Figure 1. The pseudo code of *COS*.

Algorithm 2 The pseudo code of *RGS*

Require: $\alpha_2 \neq 0$
1: **if** $\alpha_1 \geq 0$ *and* $\alpha_2 > 0$ **then**
2: $\gamma_0 = \alpha_0 - \alpha_2 t_e^2$ *and* $\gamma_1 = \alpha_1 + 2\alpha_2 t_e$
3: **end if**
4: **if** $\alpha_1 \leq 0$ *and* $\alpha_2 < 0$ **then**
5: $\gamma_0 = \alpha_0$ *and* $\gamma_1 = \alpha_1 + \alpha_2 t_e$
6: **end if**
7: **if** $\alpha_1 > 0$ *and* $\alpha_2 < 0$ **then**
8: **if** $\frac{-\alpha_1}{2\alpha_2} \geq t_e$ **then**
9: $\gamma_0 = \alpha_0$ *and* $\gamma_1 = \alpha_1 + \alpha_2 t_e$
10: **else**
11: $\gamma_0 = \alpha_0$ *and* $\gamma_1 = \frac{\alpha_1}{2}$ *and* $b = \frac{-\alpha_1}{2\alpha_2}$
12: **end if**
13: **end if**
14: **if** $\alpha_1 < 0$ *and* $\alpha_2 > 0$ **then**
15: **if** $\frac{-\alpha_1}{2\alpha_2} \geq t_e$ **then**
16: $\gamma_0 = \alpha_0$ *and* $\gamma_1 = \alpha_1$
17: **else**
18: $\gamma_0 = \alpha_0$ *and* $\gamma_1 = \alpha_1$ *and* $b = \frac{-\alpha_1}{2\alpha_2}$
19: **end if**
20: **end if**
21: Generate $m_1 \geq 0$ times-to-events using $GS-$ or $GS+$ with $exp(\gamma_0 + \gamma_1 t)$ on $(a, b]$
22: **if** $\alpha_1 \geq 0$ *and* $\alpha_2 > 0$ **then**
23: $c^* = exp(\alpha_0) - exp(\alpha_0 - \alpha_2 b^2)$
24: **end if**
25: **if** $\alpha_1 \leq 0$ *and* $\alpha_2 < 0$ **then**
26: $c^* = exp(\alpha_0) - exp(\alpha_0 + \alpha_1 b + \alpha_2 b^2)$
27: **end if**
28: **if** $\alpha_1 > 0$ *and* $\alpha_2 < 0$ **then**
29: **if** $\frac{-\alpha_1}{2\alpha_2} > t_e$ **then**
30: $c^* = exp(\alpha_0 + \alpha_1 b + \alpha_2 b^2) - exp(\alpha_0)$
31: **else**
32: $c^* = exp(\alpha_0 - \frac{\alpha_1^2}{4\alpha_2}) - exp(\alpha_0)$
33: **end if**
34: **end if**
35: **if** $\alpha_1 < 0$ *and* $\alpha_2 > 0$ **then**
36: **if** $\frac{-\alpha_1}{2\alpha_2} \geq t_e$ **then**
37: $c^* = exp(\alpha_0 + \alpha_1 b + \alpha_2 b^2) - exp(\alpha_0 + \alpha_1 b)$
38: **else**
39: $c^* = exp(\alpha_0 - \frac{\alpha_1^2}{4\alpha_2}) - exp(\alpha_0 - \frac{\alpha_1^2}{2\alpha_2})$
40: **end if**
41: **end if**
42: Generate $n_1 \geq 0$ random numbers using the algorithm of rejection
43: **if** $b = t_e$ **then**
44: set $m_2 = 0$ *and* $n_2 = 0$ and go to step 59
45: **else**
46: go to step 48
47: **end if**
48: $a = b$ *and* $b = t_e$
49: **if** $\alpha_1 > 0$ *and* $\alpha_2 < 0$ **then**
50: $\gamma_0 = \alpha_0 + \frac{\alpha_1 t_e}{2}$ *and* $\gamma_1 = \frac{\alpha_1}{2} + \alpha_2 t_e$
51: $c^* = exp(\alpha_0 - \frac{\alpha_1^2}{4\alpha_2}) - exp(\alpha_0 + \alpha_1 b + \alpha_2 b^2)$
52: **end if**
53: **if** $\alpha_1 < 0$ *and* $\alpha_2 > 0$ **then**
54: $\gamma_0 = \alpha_0 - \alpha_2 t_e^2$ *and* $\gamma_1 = \alpha_1 + 2\alpha_2 t_e$
55: $c^* = exp(\alpha_0 - \frac{\alpha_1^2}{4\alpha_2}) - exp(\alpha_0 - \frac{\alpha_1^2}{2\alpha_2} - \alpha_1 b - \alpha_2 b^2)$
56: **end if**
57: Generate m_2 times-to-events using $GS-$ or $GS+$ with $exp(\gamma_0 + \gamma_1 t)$ on $(a, b]$
58: Generate $n_2 \geq 0$ random numbers using the algorithm of rejection
59: **if** $m = m_1 + m_2 = 0$ *and* $n = n_1 + n_2 = 0$ **then**
60: exit; there are no events in $(0, t_e]$
61: **end if**
62: **if** $n = 0$ **then**
63: return times-to-events used $GS-$ or $GS+$ and exit
64: **end if**
65: **if** $m = 0$ **then**
66: return random numbers as required times-to-events used the algorithm of rejection and exit
67: **end if**
68: Merge times-to-events and random numbers, return as the required times-to-events and exit

Figure 2. The pseudo code of *RGS*.

Algorithm 3 The pseudo code of rejection algorithm

1: Generate a Poisson random variable n with $\Lambda^*(b) - \Lambda^*(a)$
2: **if** $n = 0$ **then**
3:    exit; There are no events
4: **else**
5:    set k=1
6: **end if**
7: Generate V uniformly distributed between 0 and 1 and set $T_k = a + (b - a)V$
8: Generate U independent of V and uniformly distributed between 0 and 1
9: **if** $U \leq \frac{f(T_k)}{g(T_k)} = \lambda^*(T_k)c^*$ **then**
10:    return T_k and go to step 14
11: **else**
12:    go to step 7
13: **end if**
14: **if** $k = n$ **then**
15:    exit
16: **else**
17:    set $k = k + 1$ and go to step 7
18: **end if**
19: Return $T_1, T_2, \ldots$

Figure 3. The pseudo code of rejection algorithm used in *RGS*.

3.3 *Construction of confidence intervals*

We achieve m sets of estimates $\widehat{\lambda}^*_{\mathbf{k}} = (\widehat{\lambda}^*_{k1}, \widehat{\lambda}^*_{k2}, \ldots\ldots, \widehat{\lambda}^*_{kn})$ after using WSE to m sets of resamples. The construction details are described in the study of Xiao and Dohi (2017).

The order statistics of the intensity function is defined by $0 = \widehat{\lambda}^*_{(0)i} \leq \widehat{\lambda}^*_{(1)i} \leq \widehat{\lambda}^*_{(2)i} \leq \cdots \leq \widehat{\lambda}^*_{(m)i}$ at time i and regarded as the complete sample from random variable λ^*_i, which represents the number of software faults detected at time i. If the cumulative distribution function (c.d.f.) of λ^*_i exists, then the empirical c.d.f. $G^i_{km}(\lambda)$ corresponding to the sample estimates $\widehat{\lambda}^*_{(k)i}$ $(k = 1, 2, \ldots, m)$ is expressed as follows:

$$G^i_{km}(\lambda) = \begin{cases} k/m & (\widehat{\lambda}^*_{(k)i} \leq \lambda \leq \widehat{\lambda}^*_{(k+1)i}) \\ 1 & (\widehat{\lambda}^*_{(m)i} \leq \lambda). \end{cases} \tag{3}$$

This empirical c.d.f. approaches the real (but unknown) c.d.f. of λ^*_i as $m \to \infty$ and $n \to \infty$, and is strongly consistent. From the empirical c.d.f. of λ^*_i, we can obtain not only the mean $E[\lambda^*_i]$ and its higher moments such as $E\left[(\lambda^*_i)^2\right]$, $E\left[(\lambda^*_i)^3\right]$, and $E\left[(\lambda^*_i)^4\right]$, but also the confidence interval with significant level p. The confidence interval of λ^*_i is defined as the quantile of Equation (3). Thus, the two-sided $100p\%$ confidence interval is expressed as follows: $(\widehat{\lambda}^*_{(k_L)i}, \widehat{\lambda}^*_{(k_U)i})$, where k_L and k_U are indexes corresponding to $100(1-p)\%$ and $100p\%$-quantiles, satisfying $G^i_{km}(\widehat{\lambda}^*_{k_L i}) = 100(1-p)$ and $G^i_{km}(\widehat{\lambda}^*_{k_U i}) = 100p$. A similar approach can be considered for the other reliability measures such as the mean value function.

4 EXPERIMENTS

The dataset used is obtained from the study of Lyu (1996). The total number of software faults is 133, and the end of the test is 62. Table 1 shows the time of sample generation using the eight simulation methods, four of which were employed in this paper; the four other approaches were employed in the study of Xiao and Dohi (2014, 2017, 2021). In the Table, "-" indicates that *IntGen* cannot be executed in the same manner as in the previous study. Among the four methods used in the proposal, only *RGS* has more digits in the elapsed time

Table 1. Time of sample generation.

	simulation method	time (seconds)	time (hours)
proposal	*COS*	6.15	0.0017
	GS−	7.72	0.0021
	GS+	7.96	0.0022
	RGS	11.91	0.0033
existing study	*TimTra*	103385.19	28.7181
	OrdSta	9823.20	2.7287
	Thinning	17362.77	4.8230
	IntGen	–	–

because it is executed by a combination of several algorithms, such as $GS-$ and $GS+$. The generation time of the proposed method is only approximately $1/800$ of *OrdSta*, which is the shortest one in the previous study. This finding is consistent with the advantage of parametric methods: low computational cost.

WSE (Xiao and Dohi (2013)) has over 1000 variants. This is because WSE consists of a combination of thresholding methods, thresholds, data transformations, and resolution levels. In this paper, we employ both hard (h) and soft (s) thresholding methods. The threshold used for non-data-transform-based WSE (H) is the level-dependent threshold (ldt) proposed by Kolaczyk (1997), while for data-transform-based WSE (A), the universal threshold (ut) proposed by Donoho and Johnstone (1994) is used. Here, A denotes the Anscombe transform (Makitalo (2011)). Additionally, possible resolution levels for our dataset are $J = 2, 3, 4$, and 5. Tables 2 and 3 show the interval estimation results obtained by the proposal and the existing study for the mean value function. Here, each % means the quartile of the test end time. More specifically, Table 2 represents the results, of which the point estimates of the intensity function were obtained using $H(h; ldt; J = 5)$. While, Table 3 shows the results, of which the point estimates of the intensity function were obtained using $A(h; ut; J = 3)$.

Table 2. Interval estimation results for the mean value function ($H(h; ldt; J = 5)$).

COS	mean	median	variance	skewness	kurtosis	Confidence Interval
25%	35.5677	35.4175	27.5495	0.1907	3.1200	(25.7200, 46.2930)
50%	71.2157	70.8697	68.2787	0.1519	3.0317	(55.5488, 88.0958)
75%	102.4530	102.2564	104.0662	0.1168	3.0550	(83.0417,123.1057)
100%	133.7531	133.5163	143.8805	0.0942	3.0601	(110.7990,158.2470)
GS−	mean	median	variance	skewness	kurtosis	Confidence Interval
25%	35.5961	35.4277	27.6678	0.2075	3.0841	(25.7652, 46.6362)
50%	71.2693	71.0803	66.4010	0.1421	3.0361	(55.8051, 87.8081)
75%	102.5357	102.4152	100.2842	0.1289	3.0244	(83.4839,122.6239)
100%	133.8113	133.6993	138.7267	0.0990	2.9658	(111.2991,157.3011)
GS+	mean	median	variance	skewness	kurtosis	Confidence Interval
25%	32.3757	32.2351	24.1877	0.1143	2.9458	(22.8767, 42.2277)
50%	66.8434	66.6488	61.1998	0.0784	2.9336	(51.8285, 82.4335)
75%	99.1209	98.9791	96.3035	0.0662	2.9342	(80.5399,118.5389)
100%	133.5924	133.4856	139.5679	0.0752	3.0351	(110.8941,157.0711)

(*continued*)

Table 2. Continued

OrdSta	mean	median	variance	skewness	kurtosis	Confidence Interval
25%	38.4633	38.2195	38.9334	0.2402	2.9413	(26.9318, 51.3518)
50%	70.0269	69.8322	68.6380	0.1462	3.1060	(54.1770, 86.8010)
75%	106.8390	106.6958	105.8869	0.1089	3.0693	(87.0216,127.7566)
100%	138.8630	138.6206	143.0119	0.0781	3.0479	(115.8582,163.1032)
Thinning	mean	median	variance	skewness	kurtosis	Confidence Interval
25%	38.4965	38.2854	38.7785	0.2162	3.0103	(27.0727, 51.1787)
50%	70.0588	69.9076	69.5362	0.1223	3.0391	(54.3208, 86.8868)
75%	106.9407	106.7598	107.9615	0.0849	3.0359	(86.9020,127.7320)
100%	138.8376	138.5737	146.4200	0.0838	2.9935	(115.8239,163.1429)

Table 3. Interval estimation results for the mean value function $(A(h; ut; J = 3))$.

COS	mean	median	variance	skewness	kurtosis	Confidence Interval
25%	23.5939	23.4210	24.3532	0.2399	3.0916	(14.4958, 33.8678)
50%	47.4143	47.1696	51.0219	0.1872	3.0330	(33.9289, 61.9949)
75%	68.4947	68.2660	74.9844	0.1588	3.0236	(52.0939, 85.8539)
100%	89.7993	89.6288	102.7508	0.1429	3.0570	(70.4856,110.1366)
GS−	mean	median	variance	skewness	kurtosis	Confidence Interval
25%	23.6457	23.4666	24.5689	0.2460	3.0922	(14.5200, 33.8970)
50%	47.4089	47.2637	50.0713	0.1823	3.0180	(34.1924, 61.9394)
75%	68.5085	68.3832	73.3790	0.1199	3.0114	(52.4995, 85.7825)
100%	89.8641	89.6361	99.9291	0.0965	2.9758	(70.7183,110.1033)
GS+	mean	median	variance	skewness	kurtosis	Confidence Interval
25%	21.7589	21.4911	22.7594	0.2691	3.0528	(12.9698, 31.6038)
50%	44.9358	44.7787	48.7181	0.1823	3.0527	(31.9062, 59.1712)
75%	66.6537	66.5145	72.0905	0.1447	3.0963	(50.3912, 83.7602)
100%	89.8051	89.6775	101.8250	0.1211	3.0489	(70.8622,110.0332)
OrdSta	mean	median	variance	skewness	kurtosis	Confidence Interval
25%	26.2377	26.0203	27.1080	0.2211	2.9646	(16.6797, 37.0507)
50%	45.9432	45.7730	47.9014	0.1871	3.0316	(33.1246, 59.9666)
75%	71.2733	70.9477	75.0088	0.1807	2.9652	(55.1212, 88.8182)
100%	91.5826	91.2457	100.4975	0.1802	2.9931	(72.8557,112.2397)
Thinning	mean	median	variance	skewness	kurtosis	Confidence Interval
25%	26.1710	25.9542	26.9623	0.2187	3.0229	(16.6435, 37.0385)
50%	45.8488	45.6242	48.1857	0.1848	3.0667	(32.9534, 60.0684)
75%	71.1908	70.9990	76.1102	0.1493	3.0283	(54.7596, 88.6796)
100%	91.4254	91.1587	102.6972	0.1277	3.0315	(72.2361,111.5671)

As can be seen in the variance column of Table 2, *GS*− shows smaller variances than the two existing methods, *OrdSta* and *Thinning*. Therefore, we can conclude that Table 2 suggests that *GS*− can generate arrivals with less variability. In addition, the skewness and kurtosis columns are close to 0 and 3 for all five methods, *COS*, *GS*−, *GS*+, *OrdSta*, and *Thinning*, thereby suggesting that the distribution approaches normality given the increase in the number of arrivals generated. Finally, three methods, *COS*, *GS*−, and *GS*+ have narrow confidence intervals at 25%. For *COS* and *GS*−, the confidence intervals at the 50%, 75%, and 100% are within ±5% of the existing methods *OrdSta* and *Thinning*, and within 10% of *GS*+. As indicated at the beginning of this paragraph, the low variance value for *GS*− generates results that are neither extremely narrow nor wide for the confidence intervals.

From Table 3, The skewness and kurtosis columns are close to 0 and 3 for all five methods, *COS*, *GS*−, *GS*+, *OrdSta*, and *Thinning*, indicating that a normal distribution is almost achieved owing to the increase in the number of arrivals generated. Looking at variance, *GS*− was the only method that obtained a smaller value at 3 quartile points than the existing methods, and the confidence intervals for all quartiles ranged ±5%. Therefore, the low variance value for *GS*− generates results that are neither extremely narrow nor wide for the confidence intervals.

Figures 4 and 5 show the confidence interval for the mean value function using $H(h; ldt; J = 5)$ and $A(h; ut; J = 3)$, respectively. It can be confirmed that the mean and the

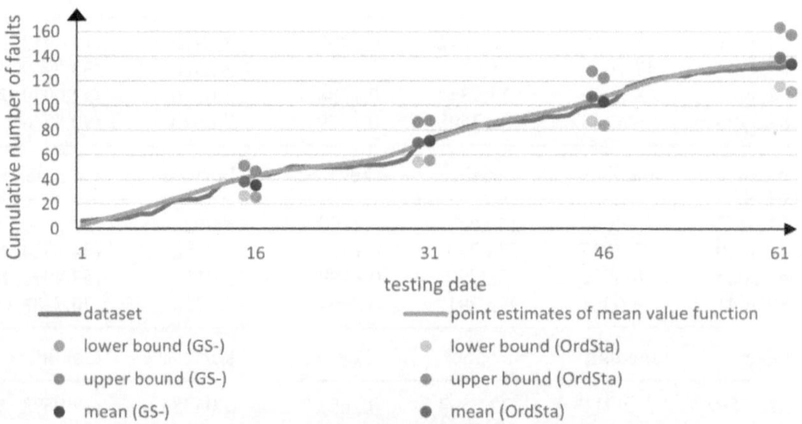

Figure 4. Confidence interval for the mean value function ($H(h; ldt; J = 5)$).

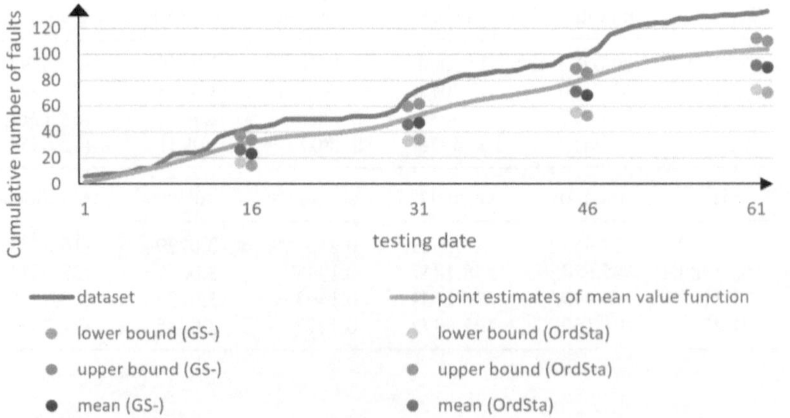

Figure 5. Confidence interval for the mean value function ($A(h; ut; J = 3)$).

lower and upper bounds of the proposed method (*GS*−) are very close to that of the existing method (*OrdSta*). In other words, Figures 4 and 5 suggest that the proposed method (*GS*−) can derive reliability measures that are not significantly different from that of the existing method (*OrdSta*). Note that as shown in Figure 5, there may be differences between observed data and point estimates, but this is due to the estimation accuracy of WSE and not due to the way the resamples are generated. Overall, the proposed method provides the same results as the existing methods in approximately $1/800$ of the time.

5 CONCLUSION

In this paper, we proposed a method for interval estimation of software reliability using the NHPP model. Resamples for calculating the confidence interval were created using parametric simulation methods. Data analysis confirmed that the proposed method can create resamples in approximately 1/800 of the time of the existing method, which maintains the same level of accuracy as the existing study. The common space of improvement for the WSE-based approaches is that they are easily influenced by the estimates of the intensity function. Future work includes investigating other simulation methods and validating the proposed method on a larger number of datasets. We also intend to use other estimation methods instead of WSE for obtaining point estimates.

REFERENCES

Cowling A., Hall P., and Phillips M. J., (1996). Bootstrap confidence regions for the intensity of a Poisson point process, *Journal of American Statistical Association*, vol. 91, no. 436, pp. 1516–1524.

Donoho D. L. and Johnstone I. M., (1994). Ideal spatial adaptation by wavelet shrinkage, *Biometrika*, vol. 81, no. 3, pp. 425–455.

Efron B. and Tibshirani R. J., (1993). *An Introduction to the Bootstrap*. New York, USA: Chapman & Hall.

Kolaczyk E. D., (1997). Nonparametric estimation of gamma-ray burst intensities using Haar wavelets, *The Astrophysical Journal*, vol. 483, no. 1, pp. 340–349.

Lewis P. A. W. and Shedler G. S., (1976). Simulation of nonhomogeneous Poisson processes with log linear rate function, *Biometrika*, vol. 63, no. 3, pp. 501–505.

Lewis P. A. W. and Shedler G. S., (1979). Simulation of nonhomogeneous poisson processes with degree-two exponential polynomial rate function, *Operations Research*, vol. 27, no. 5, pp. 1026–1040.

Lyu M. R. (ed.), (1996). *Handbook of Software Reliability Engineering*. McGraw-Hill, New York.

Makitalo M. and Foi A., (2011). Optimal inversion of the Anscombe transformation in low-count poisson image denoising, *IEEE Transactions on Image Processing*, vol. 20, no. 1, pp. 99–109.

Xiao X. and Dohi T., (2013). Estimating software intensity function based on translation-invariant Poisson smoothing approach, *IEEE Transactions on Reliability*, vol. 62, no. 4, pp. 930–945.

Xiao X. and Dohi T., (2013). Wavelet shrinkage estimation for non-homogenous Poisson process based software reliability models, *IEEE Transactions on Reliability*, vol. 62, no. 1, pp. 211–225.

Xiao X. and Dohi T., (2014). Interval estimation method for decision making in wavelet-based software reliability assessment, *IEICE Transactions on Information and Systems*, vol. E97-D, no. 5, pp. 1058–1068.

Xiao X., Zheng T. F., (2017). Simulation-based interval estimation of software reliability measures, *IEICE technical report*, vol. 117, no. 161, pp. 25–30.

Xiao X. and Dohi T., (2021). Interval estimation for non-parametric NHPP-based software reliability model via simulation-based bootstrap, Proceedings of the 8th International Conference on Dependable Systems and Their Applications (DSA2021), pp. 265–270.

Recent Advances in Reliability and Maintenance Modeling – Okamura, Inoue & Xiao (Eds)
© 2025 the Author(s), ISBN: 978-1-032-79098-5

Evaluation of software reliability models based on Hawkes processes

Nanxiang Qiu, Tadashi Dohi & Hiroyuki Okamura
Hiroshima University, Hiroshima, Japan

ABSTRACT: The Hawkes Process (HP) is widely used in seismology, finance, criminology, and other fields. This paper attempts to use HP in the field of software reliability, using 8 sets of software fault detection time data and a preliminary evaluation of HP-based Software Reliability Models (SRMs) with different excitation functions. These results are compared with the well-known Non-Homogeneous Poisson Process (NHPP)-based SRMs, and the results show that the goodness-of-fit of our proposed HP-based SRMs is able to be better than the NHPP-based SRMs in a fraction of cases.

Keywords: software reliability model, Hawkes process, software intensity function, excitation function, goodness-of-fit performance

1 INTRODUCTION

Software reliability modeling plays a crucial role in assessing the operational integrity of software by defining its reliability as the likelihood that it will function without failure for a designated period. This technique employs statistical methods and stochastic models to evaluate the reliability and fault occurrence of software during both the testing and operational stages. Furthermore, it provides developers and engineers with valuable insights into testing progress by tracking the accumulated faults detected. Reliable software performance is vital for the overall system's dependability and smooth operation, making the quantitative evaluation of software reliability a fundamental aspect of quality assurance, particularly in systems where failure is not an option. Over the past several decades, numerous software reliability models (SRMs) have been created and are extensively utilized to track and predict the number of faults uncovered through software testing (Lyu 1996; Musa & Okumoto 1987; Xie 1991).

The Non-Homogeneous Poisson Process (NHPP)-based SRMs (Abdel-Ghaly *et al.* 1986; Achcar *et al.* 1998; Erto *et al.* 2020; Goel 1985; Gokhale & Trivedi 1998; Musa & Okumoto 1984; Ohishi *et al.* 2009; Okamura & Dohi 2013; Zhao & Xie 1996) are widely used to describe the occurrence of software failures caused by software faults over the testing time, where the rate of event occurrence is allowed to vary. This flexibility makes NHPP particularly useful for modeling phenomena where event rates are not constant, such as the arrival of tasks in queueing systems or the occurrence of software failures over the lifecycle of development.

Building on the foundational concepts of NHPP, the Hawkes process(HP) (Hawkes 1971; Hawkes & Oakes 1974; Wang 2015; Rizoiu *et al.* 2022) introduces an additional layer of complexity by incorporating the self-exciting property, where each event can increase the likelihood of future events. This characteristic is particularly useful in fields where past events significantly influence future occurrences.

In the domain of software reliability, the application of the HP offers promising perspectives. Unlike traditional NHPP models which assume independence among events, the HP can effectively model the situation in software systems where faults or failures are not isolated incidents. The

DOI: 10.1201/9781003491309-23

occurrence of one fault can reveal or induce others, creating a chain reaction of defects. This makes the HP an ideal tool for predicting and managing the dynamics of software reliability, especially in complex systems where interactions among components can lead to cascading failures.

In this paper, we present a novel software reliability modeling framework based on the Hawkes process. To construct our model, we innovatively utilize the intensity functions from 11 models found within the Software Reliability Assessment Tool on Spreadsheet (SRATS) as the excitation functions of the proposed HP-based SRMs. This integration aims to capture the self-exciting nature of software faults more accurately. To evaluate the efficacy of our model, we conducted fitting tests using 8 sets of software fault detection time-domain data. Based on the numerical experiments conducted, the HP-based SRMs have shown to perform well in almost half of the data sets. These results are promising and provide us with hope for the potential of HP-based SRMs in predicting software failures in the future.

2 PRELIMINARY

2.1 *Overview of NHPP*

A Non-Homogeneous Poisson Process (NHPP) is widely utilized as a stochastic counting process that models the occurrence of discrete events, particularly when these events happen at varying rates over time. This process is especially relevant in fields like software reliability, where it models events such as software failures. In the domain of software reliability, an NHPP models the occurrences of software failures, which are detected during the testing phase. These failures are not uniformly distributed; rather, their occurrence is modeled to follow a Poisson distribution with a time-varying rate, which is a key distinction from the Homogeneous Poisson Process (HPP) where the rate is constant.

2.2 *Mathematical definition*

Let us consider a software testing scenario starting at $t = 0$. The cumulative number of detected software faults up to any time $t \geq 0$ is denoted by $\{N(t), t \geq 0\}$. The process $N(t)$ is characterized as an NHPP under the following conditions:

- $N(0) = 0$,
- $\{N(t), t \geq 0\}$ has independent increment,
- $\Pr\{N(t+t) - N(t) \geq 2\} = o(t)$,
- $\Pr\{N(t+t) - N(t) = 1\} = \lambda(t;\theta)t + o(t)$,

2.3 *Probabilistic and statistical properties*

The NHPP is governed by the following differential equations, which describe the evolution of the probability of observing exactly n faults by time t:

$$\frac{d}{dt}P_0(t) = -\lambda(t;\theta)P_0(t), \tag{1}$$

$$\frac{d}{dt}P_n(t) = \lambda(t;\theta)P_{n-1}(t) - \lambda(t;\theta)P_n(t), \quad n = 1, 2, \ldots \tag{2}$$

The initial conditions are $P_0(0) = 1$ and $P_n(0) = 0$ for $n \geq 1$. The resulting probability mass function for $N(t)$ is:

$$P_n(t) = \frac{[\Lambda(t;\theta)]^n}{n!}e^{-\Lambda(t;\mathbf{q})}, \quad n = 0, 1, 2, \ldots \tag{3}$$

where $\Lambda(t;\theta) = \int_0^t \lambda(s;\theta)\,ds$ serves as the mean value function, indicating the expected number of faults detected by time t.

2.4 *NHPP-based software reliability models*

Within the framework of Non-Homogeneous Poisson Process (NHPP) based software reliability models (SRMs), it is posited that each software fault manifests at random times, which are independent and identically distributed according to a continuous cumulative distribution function, denoted by $F(t;\pi)$, where π is a parameter vector.

Through basic algebraic derivations, the cumulative count of detected faults by time t, denoted by $N(t)$, can be characterized as an NHPP. The mean value function of this process, $\Lambda(t;\pi) = \omega F(t;\pi)$, incorporates both the initial fault expectation ω and the detection timing distribution $F(t;\pi)$, with $\theta = (\omega, \pi)$. The NHPP's mean value function satisfyingly converges to ω as t approaches infinity, given that $F(t;\pi)$ approaches 1 in the limit, suggesting a finite expected number of faults detectable over an extended period.

2.5 *Software reliability assessment tools*

In the realm of practical applications, Okamura and Dohi have contributed notably by developing a software reliability assessment tool named SRATS (Okamura & Dohi 2013), which is implemented on spreadsheet platforms. This tool incorporates eleven distinct distributions for time-to-failure in NHPP-based SRMs, enhancing its utility and adaptability in evaluating software reliability across diverse operational scenarios. We give the cumulative distribution functions and intensity functions of the 11 baseline models used in the SRATS in Table 1.

Table 1. Representative baseline intensity functions of NHPP-based SRMs.

Models	$(t;\pi),\ \pi = (b, c)$
Exponential dist. (exp)	$\Lambda(t;\pi) = \omega F(t;\pi)$ $F(t;\pi) = 1 - e^{-bt}$
Gamma dist. (gamma)	$\Lambda(t;\pi) = \omega F(t;\pi)$ $F(t;\pi) = \int_0^t \frac{c^b s^{b-1} e^{-cs}}{\Gamma(b)} ds$
Pareto dist. (pareto)	$\Lambda(t;\pi) = \omega F(t;\pi)$ $F(t;\pi) = 1 - \left(\frac{b}{b+t}\right)^c$
Truncated normal dist. (tnorm)	$\Lambda(t;\pi) = \omega \frac{F(t;\pi) - F(0;\pi))}{1 - F(0;\pi)}$ $F(t;\pi) = \frac{1}{\sqrt{2\pi b}} \int_{-\infty}^t e^{-\frac{(s-c)^2}{2b^2}} ds$
Log-normal dist. (lnorm)	$\Lambda(t;\pi) = \omega F(\ln t;\pi)$ $F(t;\pi) = \frac{1}{\sqrt{2\pi b}} \int_{-\infty}^t e^{-\frac{(s-c)^2}{2b^2}} ds$
Truncated logistic dist. (tlogist)	$\Lambda(t;\pi) = \omega \frac{F(t;\pi) - F(0;\pi))}{1 - F(0;\pi)}$ $F(t;\pi) = \frac{1}{1 + e^{-\frac{t-c}{b}}}$
Log logistic dist. (llogist)	$\Lambda(t;\pi) = \omega F(\ln t;\pi)$ $F(t;\pi) = \frac{1}{1 + e^{-\frac{t-c}{b}}}$
Truncated extreme-value max dist. (txvmax)	$\Lambda(t;\pi) = \omega \frac{F(t;\pi) - F(0;\pi))}{1 - F(0;\pi)}$ $F(t;\pi) = e^{-e^{-\frac{t-c}{b}}}$
Log-extreme-value max dist. (lxvmax)	$\Lambda(t;\pi) = \omega F(\ln t;\pi)$ $F(t;\pi) = e^{-e^{-\frac{t-c}{b}}}$
Truncated extreme-value min dist. (txvmin)	$\Lambda(t;\pi) = \omega \frac{F(0;\pi) - F(-t;\pi))}{F(0;\pi)}$ $F(t;\pi) = e^{-e^{-\frac{t-c}{b}}}$
Log-extreme-value min dist. (lxvmin)	$\Lambda(t;\pi) = \omega(1 - F(-\ln t;\pi))$ $F(t;\pi) = e^{-e^{-\frac{t-c}{b}}}$

*$\Gamma(\cdot)$: standard gamma function, erf $(\cdot)$: error function.

3 HAWKES PROCESS

Hawkes process is a type of self-exciting point process with clustering effects, often used in fields such as finance (Lallouache & Challet 2016), social media analysis (Lindstrom *et al.* 2020), and earthquake modeling. They extend the standard Poisson process by incorporating the history of the process into the intensity function, allowing past events to influence the likelihood of future events.

3.1 *Mathematical properties of Hawkes processes*

The defining characteristic of a Hawkes process is its conditional intensity function $\lambda(t)$, which depends not only on the current time t but also on the history of the process. Mathematically, it is given by:

$$\lambda(t) = \mu + \int_0^t \phi(t-s)dN(s), \tag{4}$$

where μ represents the baseline intensity, independent of past events, and $\phi(t-s)$ is the excitation function (or kernel) which describes how previous events at time s affect the current rate at time t. The function $dN(s)$ denotes increments to the counting process N, which tracks the number of events.

The integral form can be equivalently expressed in terms of the sum of impacts from all past events, reflecting the self-exciting nature of the process:

$$\lambda(t) = \mu + \sum_{i:T_i<t} \phi(t-T_i), \tag{5}$$

where T_i denotes the times of previous events before time t. This summation form clearly illustrates how each past event contributes to the intensity at time t through the kernel function $\phi(t-T_i)$, making the process inherently dependent on its own history.

3.2 *Excitation function*

The excitation function $\phi(t)$ typically takes a parametric form and is closely related to the density function of a particular distribution scaled by a parameter α. Common choices for $\phi(t)$ include (Lallouache & Challet 2016):

- **Exponential kernel:** $\phi(t) = \alpha e^{-\beta t}$, where $\alpha > 0$ increases the intensity and $\beta > 0$ controls the rate at which the effect of a past event decays over time.
- **Power-law kernel:** $\phi(t) = \frac{\alpha}{(1+\beta t)^{\gamma}}$, suitable for processes where influences decay polynomially over time.

3.3 *Role of parameter* α

The parameter α in the kernel function has significant implications for the behavior of the Hawkes process (Lesage *et al.* 2022). It scales the influence of past events on the current event rate, essentially controlling the degree of self-excitation or clustering in the process. Mathematically, the stability and characteristics of a Hawkes process are often analyzed through the spectral measure of α, and its influence can be summarized as follows:

- **defective case:** If $\alpha < 1$, the process is stable, and the intensity over time will eventually decay unless renewed by further events.
- **proper case:** When $\alpha = 1$, the intensity function is independent of α and is not considered in this paper.

- **excessive case:** For $\alpha > 1$, the intensity can grow indefinitely, leading to potentially explosive behavior where bursts of events can occur.

For example, we consider the case where the excitation function $\phi(t)$ follows an exponential decay, commonly modeled as $\phi(t) = \alpha e^{-\beta t}$ for $t \geq 0$. Here, $\beta > 0$ controls the decay rate, and $\alpha < 1$ ensures that the cumulative effect of past events does not lead to unbounded increases in intensity. Specifically, the impact of α can be understood through the renewal equation:

$$g(t) = \mu + \int_0^t \phi(t-s)g(s)ds, \tag{6}$$

where $\phi(t) = \alpha\beta e^{-\beta t}$ represents the rate at which past events influence the current rate at time t. The integral of $\phi(s)$ over $[0,\infty)$, given by:

$$\int_0^\infty \alpha\beta e^{-\beta s} ds = \alpha, \tag{7}$$

helps confirm that the total influence of past events remains bounded, contributing to a stable process as long as $0 < \alpha < 1$. In this case, the process exhibits stability where the intensity of events increases due to recent past events but does not explode over time. This is because the contribution of each past event diminishes exponentially, governed by the parameter β, and the scaling factor α ensures that these contributions do not accumulate to push the intensity beyond control.

For the purposes of this paper, we focus exclusively on the case where $0 < \alpha < 1$. This decision is rooted in our aim to model scenarios that are stable and realistic, avoiding the complexities and unbounded behaviors associated with supercritical regimes. This choice aligns with our analytical goals and the mathematical tractability of the models under consideration.

3.4 *Hawkes process-based software reliability models*

Building on the concepts of SRATS, which integrates 11 models of time-to-failure distributions within its framework, we can extend this to propose novel Software Reliability Models (SRMs) based on Hawkes processes. Each model within SRATS can be thought of

Table 2. Representative kernels of HP-based SRMs.

Kernels	$\phi(t;\pi),\ \pi = (b,c)$
Exponential dist. (exp)	be^{-bt}
Gamma dist. (gamma)	$\frac{c^b t^{b-1} e^{-ct}}{\Gamma(b)}$
Pareto dist. (pareto)	$\frac{bc\left(\frac{c}{c+t}\right)^{b-1}}{(c+t)^2}$
Truncated normal dist. (tnorm)	$\frac{e^{-\frac{(c-t)^2}{2b^2}}}{b\sqrt{2\pi}\left(1-\frac{1}{2}erf\left(\frac{c}{\sqrt{2}b}\right)\right)}$
Log-normal dist. (lnorm)	$\frac{e^{-\frac{(c-\log t)^2}{2b^2}}}{b\sqrt{2\pi}t}$
Truncated logistic dist. (tlogist)	$\frac{e^{-\frac{-c+t}{b}}}{b\left(1+e^{-\frac{-c+t}{b}}\right)^2\left(1-\frac{1}{1+e^{\frac{c}{b}}}\right)}$

(continued)

Table 2. Continued

Kernels	$\phi(t;\pi)$, $\pi=(b,c)$
Log logistic dist. (llogist)	$\frac{e^{-\frac{-c+\log t}{b}}}{b\left(1+e^{-\frac{-c+\log t}{b}}\right)^2 t}$
Truncated extreme-value max dist. (txvmax)	$\frac{e^{-e^{-\frac{-c+t}{b}}-\frac{-c+t}{b}}}{b\left(1-e^{-e^{\frac{c}{b}}}\right)}$
Log-extreme-value max dist. (lxvmax)	$\frac{e^{-e^{\frac{c-\log t}{b}}+\frac{c-\log t}{b}}}{bt}$
Truncated extreme-value min dist. (txvmin)	$\frac{e^{e^{\frac{c}{b}}-e^{-\frac{-c-t}{b}}-\frac{-c-t}{b}}}{b}$
Log-extreme-value min dist. (lxvmin)	$\frac{e^{-e^{-\frac{-c-\log t}{b}}-\frac{-c-\log t}{b}}}{bt}$

* $\Gamma(\cdot)$: standard gamma function, erf$(\cdot)$: error function.

as providing a distinct kernel function for the Hawkes process, where the density functions from these models are adapted to serve as the excitation functions controlled by parameter α. We show 11 different SRMs and corresponding kernels in Table 2. This structured approach allows us to progressively build on the concept of Hawkes processes, adapting it to the domain of software reliability through the integration of software reliability models.

4 EXPERIMENTS

4.1 *Parameter estimation*

To optimize the NHPP-based SRMs and Hawkes Process (HP)-based SRMs, we employ Maximum Likelihood Estimation techniques to estimate parameters. These techniques are pivotal for accurately estimating the parameters of our models based on observed data. For the NHPP and HP, the parameter vectors are defined as $\theta_{\text{NHPP}}=(\omega,b,c)$ and $\theta_{\text{HP}}=(\mu,\alpha,b,c)$.

Assuming that $t_0=0$, we register the time of occurrence for each software fault, recording up to n time points in each data set.The general form of the log-likelihood function for these processes, given observed events over a time interval $[0,T]$, is:

$$\ln L(\theta)=\sum_{i=1}^{n}\log\lambda(t_i;\theta)-\int_0^T\Lambda(s;\theta)\,ds, \tag{8}$$

where $\lambda(t;\theta)$ represents the intensity function dependent on the parameter vector θ. And the cumulative intensity function is the integral of the intensity function defined by $\Lambda(T;\theta)=\int_0^T\lambda(s;\theta)\,ds$.

4.2 *Data sets*

We examine eight time-domain datasets of software fault detection (TDS1~TDS8) as documented in Table 3, derived from actual software development projects. Notably, datasets TDS1, TDS2, and TDS6 were collected from real-time command and control systems. Dataset TDS3, which comes from a military application, is critical given that any failures could lead to significant consequences. The data for TDS4, TDS7, and TDS8 were gathered from faults occurring in single-user workstations.

Table 3. Software fault-detection time data sets.

Data	No. faults	Maximum testing length	Source
TDS1	54	108708	SYS2[Musa 1979]
TDS2	24	1095.88	SYS4[Musa 1979]
TDS3	41	4312598	SYS27[Musa 1979]
TDS4	129	89040	CSR2[Lyu 1996]
TDS5	197	50236822	SYS4[Musa 1979]
TDS6	136	88682	SYS1[Musa 1979]
TDS7	104	15369.5	SRC3[Lyu 1996]
TDS8	397	108890	SRC1[Lyu 1996]

4.3 *Goodness-of-fit performances*

Evaluating the goodness-of-fit of an SRM is crucial for understanding how closely a theoretical probability model aligns with the actual observed data. An effective model fit suggests that the model successfully captures the essential patterns in the data, whereas a suboptimal fit might indicate that the model fails to encapsulate key aspects of the observed phenomena. To quantify the fit of our proposed models, we utilize the Akaike Information Criterion (AIC), defined as:

$$\text{AIC} = -2 \ln L(\widehat{\theta}) + 2 \times (\text{number of parameters}) \tag{9}$$

Here, $\ln L(\widehat{\theta})$ represents the log-likelihood function maximized over the parameter space. A smaller AIC value is preferable, indicating a model with a better goodness-of-fit performance.

In the experiment, we divided each dataset into three distinct lengths (20%, 50%, and 80%) to simulate the SRM's performance during different stages of software testing. In Table 4, we compare the minimum AIC with 8 time-domain data sets, where the bold numbers mark the best SRM with the smallest AIC in each data set.

Table 4. Goodness-of-fit performance based on AIC.

	20% Test Length			
	NHPP		HP	
	Best Baseline Model	AIC	Best Kernel Model	AIC
TDS1	lxvmax	**143.239**	exp	143.609
TDS2	lxvmax	35.7883	tlogist	**28.156**
TDS3	exp	**187.583**	exp	189.583
TDS4	llogist	**291.964**	txvmax	293.335
TDS5	exp	**1046.92**	exp	1048.920
TDS6	tlogist	313.746	txvmax	**296.130**
TDS7	exp	**191.227**	exp	192.424
TDS8	txvmax	**752.937**	tnorm	756.527
	50% Test Length			
	NHPP		HP	
	Best Baseline Model	AIC	Best Kernel Model	AIC
TDS1	exp	403.368	tnorm	**386.353**
TDS2	exp	123.477	tnorm	**114.971**

(continued)

Table 4. Continued

50% Test Length				
	NHPP		HP	
	Best Baseline Model	AIC	Best Kernel Model	AIC
TDS3	tnorm	**445.247**	gamma	448.960
TDS4	lxvmax	764.771	exp	**742.417**
TDS5	exp	2602.910	pareto	**2547.040**
TDS6	gamma	861.946	exp	**821.137**
TDS7	exp	527.080	tnorm	**526.903**
TDS8	lnorm	1901.030	exp	**1881.350**
80% Test Length				
	NHPP		HP	
	Best Baseline Model	AIC	Best Kernel Model	AIC
TDS1	lxvmax	**691.677**	lnorm	703.095
TDS2	tlogist	176.266	exp	**147.727**
TDS3	exp	769.836	txvmax	771.130
TDS4	lxvmax	1372.520	pareto	**1327.630**
TDS5	exp	**4203.050**	exp	4205.050
TDS6	llogist	1478.500	exp	**1453.410**
TDS7	lxvmax	**921.784**	exp	922.061
TDS8	llogist	3465.000	pareto	**3426.85**

After analyzing the results, it was found that out of the 24 numerical experiments conducted, the HP-based SRMs performed the best in terms of goodness-of-fit in 13 cases. This suggests that the HP-based SRMs offer better performance in more than half of the cases. Assuming that 20%, 50%, and 80% of the test lengths were for the pre-testing, mid-testing, and post-testing phases, respectively, it was observed that the HP-based SRMs did not perform as well as the NHPP-based SRMs in the pre-testing phase. However, in the mid-testing phase, the HP-based SRMs outperformed the NHPP-based SRMs in 7 out of 8 data sets, which is a significant majority. Finally, in the later stages of testing, both HP-based SRMs and NHPP-based SRMs exhibited optimal results in 4 of the 8 data sets, indicating that their performance was closely matched.

5 CONCLUSION

In this paper, we have proposed novel SRMs based on the Hawkes process, applying it to the domain of software fault detection. This novel approach leverages the self-exciting nature of the Hawkes process to model the temporal dynamics of software failures more accurately.

Our study involved optimizing 11 different Hawkes process-based SRMs and comparing their goodness-of-fit performances with traditional Non-Homogeneous Poisson Process (NHPP)-based SRMs. The results indicate that in the majority of cases, the Hawkes process-based SRMs exhibited superior goodness-of-fit performance compared to the NHPP-based SRMs. This finding underscores the potential of the Hawkes process as a robust tool in the field of software reliability, suggesting that it can offer significant improvements over conventional methods.

However one of the challenges with Hawkes process-based SRMs lies in their mathematical complexity. Specifically, the inherent properties of the Hawkes process complicate the derivation of explicit expressions for the mean value function (Cui *et al.* 2020), making fault prediction more challenging. Future research will need to address these mathematical hurdles to enhance the predictive capabilities of these models. Developing analytical or numerical methods to approximate or directly compute the mean value function could significantly advance the practical application of Hawkes process-based SRMs in predicting software reliability.

REFERENCES

Abdel-Ghaly A. A., Chan P. Y., and Littlewood B. (1986). Evaluation of competing software reliability predictions, *IEEE Transactions on Software Engineering*, vol. SE-12, no. 9, pp. 950–967.

Achcar J. A., Dey D. K., and Niverthi M. (1998). A Bayesian approach using non-homogeneous Poisson processes for software reliability models, in *Frontiers in Reliability*, A. P. Basu, K. S. Basu, and S. Mukhopadhyay (eds.), pp. 1–18, World Scientific, Singapore.

Cui, L., Hawkes, A., and Yi, H. (2020). An elementary derivation of moments of Hawkes processes. *Advances in Applied Probability*, 52(1), 102–137.

Erto P., Giorgio M., and Lepore A. (2020). The generalized inflection S shaped software reliability growth model, *IEEE Transactions on Reliability*, vol. 69, no. 1, pp. 228–244.

Goel A. L. (1985). Software reliability models: assumptions, limitations and applicability, *IEEE Transactions on Software Engineering*, vol. SE-11, no. 12, pp. 1411–1423.

Gokhale S. S., and Trivedi K. S. (1998). Log-logistic software reliability growth model, in *Proceedings of the 3rd IEEE International Symposium on High-Assurance Systems Engineering (HASE-1998)*, pp. 34–41, IEEE CPS.

Hawkes, A. G., 1971. Spectra of some self-exciting and mutually exciting point processes. *Biometrika*, 58(1), 83–90.

Hawkes, A. G., and Oakes, D. (1974). A cluster process representation of a self-exciting process. *Journal of Applied Probability*, 11(3), 493–503.

Lallouache, M., and Challet, D. (2016). The limits of statistical significance of Hawkes processes fitted to financial data. *Quantitative Finance*, 16(1), 1–11.

Lesage, L., Deaconu, M., Lejay, A., Meira, J. A., Nichil, G., and State, R. (2022). Hawkes processes framework with a Gamma density as excitation function: application to natural disasters for insurance. *Methodology and Computing in Applied Probability*, 24(4), 2509–2537.

Lindstrom, A., Lindgren, S., and Sainudiin, R. (2020). Hawkes processes on social and mass media: a causal study of the blacklivesmatter movement in the summer of 2020. *In Proceedings of the 12th International Conference on Data Science, Technology and Applications-DATA* (pp. 77–88).

Lyu M. R. (ed.). (1996). *Handbook of Software Reliability Engineering*, McGraw-Hill, New York.

Musa J. D. (1979). Software Reliability Data, Technical Report in Rome Air Development Center.

Musa J. D., and Okumoto K. (1984). A logarithmic Poisson execution time model for software reliability measurement, in *Proceedings of the 7th International Conference on Software Engineering (ICSE-1984)*, pp. 230–238, ACM.

Musa J. D., Iannino A., and Okumoto K. (1987). *Software Reliability, Measurement, Prediction, Application*, McGraw-Hill, New York.

Ohishi K., Okamura H., and Dohi T. (2009). Gompertz software reliability model: estimation algorithm and empirical validation, *Journal of Systems and Software*, vol. 82, no. 3, pp. 535–543.

Okamura H. and Dohi T. (2013). SRATS: Software reliability assessment tool on spreadsheet, in *Proceedings of the 2013 IEEE 24th International Symposium on Software Reliability Engineering (ISSRE-2013)*, pp. 100–107, IEEE CPS.

Rizoiu, M. A., Soen, A., Li, S., Calderon, P., Dong, L. J., Menon, A. K., and Xie, L. (2022). Interval-censored Hawkes processes. *Journal of Machine Learning Research*, 23(338), 1–84.

Wang, Q. (2015). Applications of Hawkes' process in finance (Doctoral dissertation, Tilburg University).

Xie M. (1991). *Software Reliability Modeling*, World Scientific, Singapore.

Zhao M., and Xie M. (1996). On maximum likelihood estimation for a general non-homogeneous Poisson process, *Scandinavian Journal of Statistics*, (1996), vol. 23, no. 4, pp. 597–607.

Recent Advances in Reliability and Maintenance Modeling – Okamura, Inoue & Xiao (Eds)
 ISBN: 978-1-032-79098-5

Paper solution algorithm for the portfolio optimization problems of real estate investment trust using genetic algorithm

Hiroto Saito & Tomoaki Akiba
Chiba Institute of Technology, Narashino, Japan

ABSTRACT: This study considers the portfolio optimization problem of diversified investment in companies that treat Real Estate Investment Trusts (REIT) as financial instruments. Portfolio optimization involves finding an optimal portfolio for investors. By considering the trade-off between the two indicators of risk and return, I find a way to minimize risk under a constant return. In this study which uses a genetic algorithm, the portfolio optimization problem is solved based on stock selection. The Markowitz model is used to determine the optimal investment ratio. In this study, I solved a portfolio based on stock selection used a genetic algorithm and it has challenged CVaR to determine the optimal investment ratio.

1 INTRODUCTION

In recent years, the monetary policy known as the "low interest monetary policy" has kept the interest rates of bank deposits and loans very low. This has led to a significant interest in risk assets such as stocks, particularly from a limited number of wealthy people and people who are concerned about the future. It is easier to increase assets by investing compared savings; however, investing also entails higher degree of risk [1].

Therefore, we have decided to focus on the disadvantages, such as, "A high possibility of losing money" and "Not knowing what to buy," which were given as reasons [2]. We thought that if We could provide resolutions to these disadvantages, more people would invest, and economic activity would become more invigorated.

This study addresses the problem of portfolio optimization for diversified investments in companies that handle real estate investment trusts (REIT) as financial instruments. Moreover, we use genetic algorithms to find the optimal portfolio solution that minimizes risk and obtain returns among companies that handle REIT.

In this study, we solved a portfolio based on stock selection used a genetic algorithm and it has challenged CVaR to determine the optimal investment ratio.

2 ASSET MANAGEMENT

2.1 *Risk and return of financial instrument*

The return of a financial instrument refers to "the income obtained from asset management." On the other hand, the risk of a financial instrument refers to the "uncertainty (unpredictability) of the return." [3]

2.2 *REIT*

An REIT is a financial instrument that invests in real estate with funds collected from investors and distributes dividends to investors using rental income and profits from real estate sales. REIT also have the following characteristics: There are two types of REIT: single purpose and multipurpose [4].

DOI: 10.1201/9781003491309-24

Single-purpose REIT operate in one type of real estate, such as office buildings, residential properties, commercial facilities, logistics facilities, hotels, etc. Multi-use REIT, operate with multiple types of real estate, and are of two types: composite-purpose REIT and comprehensive-purpose REIT [5].

3 PREVIOUS RESEARCH

3.1 *Formulation of the diversified investment problem [6]*

This section provides the definition of a portfolio.

Here, we define the following symbols.

n: Number of assets, i,j: Assets, x_i: Investment amount, ξ_i: Rate of return
μ_i: Average (return), σ_i^2: Variance (risk), $\boldsymbol{C}$: Covariance matrix, σ_i: Standard deviation
ρ_{ij}: Correlation coefficient, γ: Lower limit of return, $r(\boldsymbol{x},\boldsymbol{\xi})$: Portfolio return
$\bullet^T$: Bold letters represent vectors and matrices

We create a portfolio for one period using a predetermined set of n assets. Let $x_i \in \Re,\ \ i = 1,\ldots,n$ be the amount invested in asset i using own funds as a unit. When using only own funds for investment, the total investment amount $x_i \in \Re$ is

$$x_1 + x_2 + \cdots + x_n = 1 \tag{1}$$

In addition, to prohibit short selling, we set $0 \leq x_i$.

A portfolio is expressed as a vector $\boldsymbol{\xi} = (\xi_1,\ldots,\xi_n) \in \Re^n$ with investment amount $x_i \in \Re$. Also, based on the Modern Portfolio Theory, the rate of return $\xi_i \in \Re$, of asset i is distributed with mean μ_i Let σ_i^2 be a random variable that follows a normal distribution.

$$\xi_i \sim \mathcal{N}(\mu_i,\ \sigma_i^2) \tag{2}$$

The vector $\boldsymbol{\xi} = (\xi_1,\ldots,\xi_n) \in \Re^n$ of the rate of return $\xi_i \in \Re$ in Equation (2) is the average of $\boldsymbol{\mu} = (\mu_1,\ldots,\mu_n) \in \Re^n$ and follows a multivariate normal distribution with the covariance matrix $\boldsymbol{C} \in \Re^{n\times n}$.

$$\xi \sim \mathcal{N}(\boldsymbol{\mu},\boldsymbol{C}) \tag{3}$$

The covariance matrix C in Equation (3) can be derived from the standard deviation σ_i of the rate of return and the correlation coefficient ρ_{ij}, as shown in Equations (4), (5), and (6).

$$\boldsymbol{D} = \begin{bmatrix} \sigma_1 & 0 & \cdots & 0 \\ 0 & \sigma_2 & \cdots & 0 \\ \vdots & \vdots & \ddots & \vdots \\ 0 & 0 & \cdots & \sigma_n \end{bmatrix} \tag{4}$$

$$\boldsymbol{R} = \begin{bmatrix} 1 & 0 & \cdots & \rho_{1n} \\ \rho_{21} & 1 & \cdots & \rho_{2n} \\ \vdots & \vdots & \ddots & \vdots \\ \rho_{n1} & \rho_{n2} & \cdots & 1 \end{bmatrix} \tag{5}$$

$$\boldsymbol{C} = \boldsymbol{DRD} \tag{6}$$

It is assumed that the rate of return $\boldsymbol{\xi} \in \Re^n$ is independent and follows the normal distribution in Equation (3). Actual data on the rate of return on stocks are known to be normal. The efficient market hypothesis in financial economics, also argues that current prices reflect all past prices and information, and that the rate of return is independent.

The return due to portfolio $\boldsymbol{x} \in \Re^n$ is,

$$r(\boldsymbol{x},\boldsymbol{\xi}) = \sum_{i=1}^{n} \xi_i x_i = \boldsymbol{\xi x}^T \tag{7}$$

Since the rate of return $\boldsymbol{\xi} \in \Re^n$ in Equation (3) is a random variable, the return $\boldsymbol{r}(\boldsymbol{x}, \boldsymbol{\xi}) \in \Re$ in Equation (7) is also a random variable. Furthermore, from the linearity of the normal distribution, $\boldsymbol{r}(\boldsymbol{x}, \boldsymbol{\xi}) \in \Re$ follows a normal distribution.

$$\boldsymbol{r}(\boldsymbol{x}, \boldsymbol{\xi}) \sim \mathcal{N}(\mu_r(\boldsymbol{x}), \sigma^2(\boldsymbol{x})) \tag{8}$$

The average (expected value) of the normal distribution in Equation (8) is,

$$\mu_r(\boldsymbol{x}) = \sum_{i=1}^{n} \mu_i x_i = \boldsymbol{\mu} \boldsymbol{x}^T \tag{9}$$

Also, the variance of the normal distribution in Equation (8) is,

$$\sigma^2(\boldsymbol{x}) = \boldsymbol{x} \boldsymbol{C} \boldsymbol{x}^T \tag{10}$$

3.2 *CVaR model equation [7]*

I now explain a model that uses the CVaR as a risk measure in a portfolio selection problem.

CVaR is the average loss when the portfolio return loss exceeds a certain probability level β. Let β-VaR be the minimum α when the portfolio return loss does not exceed α with probability β, then β-CVaR represents the expected value of loss when the loss of portfolio return exceeds α. Because VaR and CVaR represent the magnitude of loss, VaR and CVaR are treated as negative values when considering portfolio returns.

$-\phi_\beta$ is calculated as the value obtained by subtracting the conditional expected value of the shortfall below $-\alpha_\beta$ from $-\alpha_\beta$. β-VaR is not larger than β-CVaR. Therefore, a small CVaR portfolio will have a small VaR value.

When the portfolio return rate $R(x)$ follows a continuous distribution with a probability density function, β-VaR can be defined as follows.

$$\beta\text{-VaR}(R)(= \alpha_\beta(R)) \equiv min\left\{ \alpha \middle| \int_{-\alpha}^{\infty} f(R)dR \geq \beta \right\} \tag{11}$$

From Equation (11)

$$\int_{-\infty}^{-\alpha_\beta(R)} f(R)dR = 1 - \beta \tag{12}$$

Using $\alpha_\beta(R)$, β-CVaR can be defined as follows.

$$\beta\text{-CVaR}(R)(= \phi_\beta(R)) \equiv \frac{-\int_{-\infty}^{-\alpha_\beta(R)} Rf(R)dR}{\int_{-\infty}^{-\alpha_\beta(R)} f(R)dR} = -\frac{1}{1-\beta}\int_{-\infty}^{-\alpha_\beta(R)} Rf(R)dR \tag{13}$$

To show how to formulate this using discrete data such as historical and scenario data, I expand the formula as follows.

$$\begin{aligned}
\phi_\beta(R) &= \frac{\alpha}{1-\beta}\int_{-\infty}^{-\alpha_\beta(R)} f(R)dR - \frac{1}{1-\beta}\int_{-\infty}^{-\alpha_\beta(R)} (R+\alpha_\beta)f(R)dR \\
&= \alpha_\beta + \frac{1}{1-\beta}\int_{-\infty}^{-\alpha_\beta(R)} |R+\alpha_\beta|_{\text{-}} f(R)dR = \alpha_\beta + \frac{1}{1-\beta}\int_{-\infty}^{\infty} |R+\alpha_\beta|_{\text{-}} f(R)dR \\
&= \alpha\beta + \frac{1}{1-\beta}E[|R+\alpha_\beta|_{\text{-}}]
\end{aligned} \tag{14}$$

Using the portfolio return rate r_t at event t,

$$|r_t + \alpha_\beta|_{\text{-}} \ = \ \min\{u_t | r_t + \alpha_\beta + u_t \geq 0, \ u_t \geq 0\} \quad (t = 1, \ldots, S) \tag{15}$$

Therefore, β-CVaR can be calculated as follows.

$$\phi_\beta(R) = \alpha_\beta + \frac{1}{1-\beta} \cdot min\left\{\frac{1}{S}\sum_{t=1}^{S} u_t | r_t + \alpha_\beta + u_t \geq 0,\ \ u_t \geq 0;\ \ t = 1, \ldots, S\right\} \tag{16}$$

Therefore, the CVaR model is defined as follows.

$$\begin{bmatrix} min \quad \alpha_\beta + \dfrac{1}{(1-\beta)S}\displaystyle\sum_{t=1}^{S} u_t \\ s.t \displaystyle\sum_{j=1}^{n} r_{jt} x_i + \alpha_\beta + u_t \geq 0 \ (t = 1, \ldots, S) \\ u_t \geq 0 \ (t = 1, \ldots, S) \\ \bar{r}_p \geq r_E \\ \displaystyle\sum_{j=1}^{n} x_j = 1 \\ x_j \geq 0 \ (j = 1, \ldots, n) \\ \alpha_\beta : \text{ No sign restriction} \end{bmatrix} \tag{17}$$

3.3 *Diversified investment problem using REIT*

Since REIT are financial instruments that can be traded like stocks, I believe that the symbols and Formulas (1) to (10) in Section 3.1 can be applied. The symbols used in this section are as follows.

$\boldsymbol{g}(\boldsymbol{x},\boldsymbol{\xi})$: REIT portfolio return

Let $x_i \in \Re,\ i = 1, \ldots, n$ be the amount invested in asset *i* using own funds as a unit. When only one's own funds are used for investment, the total investment amount $x_i \in \Re$ is as shown in Equation (1). Also, $0 \leq x_i$ to prohibit short selling.

From Equation (7), the return on a portfolio investment in an REIT is,

$$g(\boldsymbol{x}, \boldsymbol{\xi}) = \sum_{i=1}^{n} \xi_i x_i = \boldsymbol{\xi x}^T \tag{18}$$

Since the rate of return $\boldsymbol{\xi} \in \Re^n$ in equation (3) is a random variable, the return $\boldsymbol{g}(\boldsymbol{x}, \boldsymbol{\xi}) \in \Re$ in Equation (14) is also a random variable. Furthermore, from the linearity of the normal distribution, $\boldsymbol{g}(\boldsymbol{x}, \boldsymbol{\xi}) \in \Re$ follows the following normal distribution.

$$\boldsymbol{g}(\boldsymbol{x}, \boldsymbol{\xi}) \sim \mathcal{N}(\mu_g(\boldsymbol{x}), \sigma^2(\boldsymbol{x})) \tag{19}$$

The average (expected value) of the normal distribution in Equation (14) is,

$$\mu_g(\boldsymbol{x}) = \sum_{i=1}^{n} \mu_i x_i = \boldsymbol{\mu x}^T \tag{20}$$

Also, the variance $\sigma^2(\boldsymbol{x})$ is given by Equation (10).

Using the Markowitz model of Equation (11), the conflicting equation for the diversified investment problem is expressed as follows:

$$\begin{bmatrix} min\ \sigma^2(\boldsymbol{x}) = \boldsymbol{xCx}^T \\ s.t \quad \mu_g(\boldsymbol{x}) = \boldsymbol{\mu x}^T \geq \gamma, \\ \quad x_1 + x_2 + \cdots + x_n = 1, \\ \quad 0 \leq x_i,\ i = 1, \ldots, n \end{bmatrix} \tag{21}$$

4 GENETIC ALGORITHM

The portfolio optimization problem in this study is one of the combinatorial optimization problems and is characterized by an exponential increase in the number of combinations as the number of stocks handled increases. For the experiments, we implemented the problem using the python language and compiled the executable program using Visual Studio Code. The computing environment was a PC with an Intel Core i7-12700k 3.60GHz CPU, 32GByte RAM, and Windows11 Education OS, and it was found to take approximately 17billion hours to solve. Since it is practically impossible to solve this problem, this study uses a genetic algorithm to solve the problem in realistic time.

A genetic algorithm (GA) is an algorithm based on the evolution of life, in which individuals that can adapt to the environment survive and those that cannot adapt to the environment perish. Multiple "individuals" are prepared, each with a genetic representation of a candidate solution, and individuals with high fitness are preferentially selected to search for a solution by repeating operations such as a crossover and sudden mutation [8]. Figure 1 shows the overall flow of the experiment, and the genetic algorithm is used. The GA are shown in orange in Figure 1.

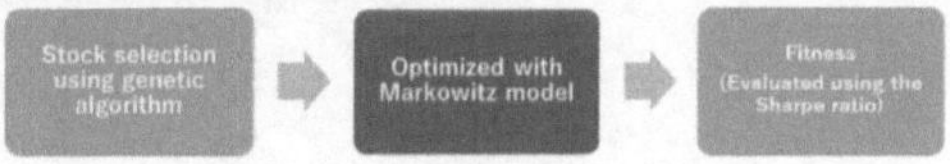

Figure 1. Experimental flow and genetic algorithm.

5 RESULTS AND DISCUSSION

5.1 *This research*

This study, we use GA to select stocks and solve portfolio optimization problems. We also conduct preliminary experiments to examine the accuracy of the GA compared to the enumeration. The calculations are performed by implementing algorithms using the language, Python.

5.2 *Data used [9]*

The data used in this paper were provided by Prop Tech Plus REITDB Service Co. Ltd.

5.3 *Experimental method*

5.3.1 *Preliminary experiment*

First, we perform an enumeration of the five stocks that can be obtained and find the optimal combination and investment ratio among them. Next, we calculate the GA using the stocks used in the enumeration, find the optimal stock among them, and determine the combination and investment ratio. We use 8987, 8986, 8985, 8984, and 8979, as the stocks.

5.3.2 *Main experiment*

Below, we find the optimal portfolio from 52 stocks.

5.3.2.1 *Stock selection using genetic algorithm* First, we define the conditions for GA. The Sharpe ratio is used to evaluate the adaptation. Roulette selection, and uniform crossover are used for selection and crossover, respectively. The probabilities of crossovers and mutations are 95% and 5%, respectively.

The Sharpe ratio is an indicator that allows a comparison of the extent to which the return obtained from safe assets (assets assumed to have zero risk) is higher than that obtained from risk-taking operations. Generally, the annualized rate of return, standard deviation, and risk-free asset rate of the portfolio are used. The larger the number, the more efficient the return against risk.

$$\text{Sharpe ratio} = \frac{\text{Portfolio return} - \text{risk-free assets}}{\text{standard deviation of portfolio return}} \tag{22}$$

5.3.2.2 *Solving portfolio optimization problems* Based on the results of stock selection using a GA, we solve a distributed investment problem, which is a quadratic programming problem, using the Python Library CVXOPT.

5.4 *Results*

The results of the preliminary experiment are shown in the Figure 2.

Figure 2 compares the portfolio in the case of enumeration and that using the GA.

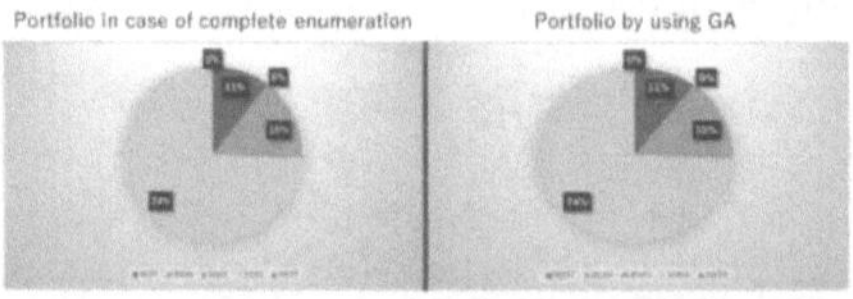

Figure 2. Portfolio the enumeration case and by using GA comparison.

Let us now compare the enumeration and GA scenarios. Because the GA does not restrict stock selection, all are selected. However, if you look at the investment ratio, it is almost the same as the enumeration scenario, Therefore, it can be said that a deserves GA good solution.

In the diversified investment problem using REIT in Equation (21), we use the daily data described in Section 5.2 [9]. For all days from January 1, 2017 December 31, 2021, we use bar data. The number of assets for which we obtain five years of daily bar data is 52 stocks; therefore, we conduct the experiment with n = 52.

The GA parameter settings are 1,000 for the number of individuals, 1,000 for the number of generations, 200 for the number of elite individuals, and 700 for the maximum number of stagnations. As this study does not impose any restrictions on selection, all stocks are selected. In descending order of ratio, 8984 (Daiwa House REIT Investment Corporation) is approximately 38%, 8961 (Mori Trust General REIT Corporation) is approximately 27%, 3463 (Ichigo Hotel REIT Corporation) is approximately 14%, and 3283 (Nippon Prologis REIT Investment Corporation) is approximately 38%. (Corporate) is approximately 12%, and 3269 (Advanced Residence Investment Corporation) is approximately 9%. The portfolio variance is 0.49%. We created a portfolio based on these five stock selections and ratios. Figure 3 illustrates the created portfolio.

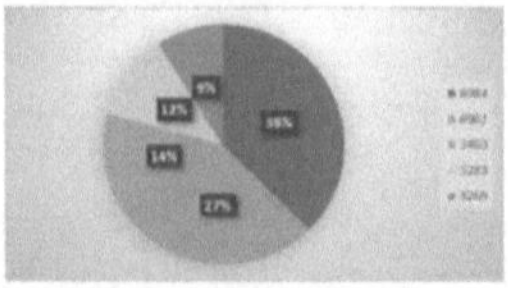

Figure 3. Portfolio results.

5.5 *Consideration of the findings*

From the results of the experiment, we obtain a portfolio that is composed of "comprehensive purpose," "hotel specialized type," "logistics facility specialized type," and "residential specialized type." Next, we consider the characteristics. Comprehensive purpose has securities code

8984 (Daiwa House REIT Investment Corporation) and 8961 (Mori Trust Sogo REIT Corporation), which increase the effect of diversification and reduces risk. The ratio of investments in the portfolio is high. Securities code 8984 is a company that mainly handles logistics facilities and housing. Since logistics facilities and housing have low risk and stable profits, we believe that they are selected with the highest investment ratio. Securities code 8961 is a company that mainly handles office buildings. Office buildings are easily affected by economic conditions. Therefore, high yields can be expected when the economy is good, but there is a risk of loss when the economy is bad. We believe it is selected with the second highest value because of this. We believe that the hotel specialization type is selected with the third highest value under securities code 3463 (Ichigo Hotel REIT Corporation) because it is dependent on management and has high risk, but also a high investment return. The logistics facility specialization type is selected with security code 3283 (Nippon Prologis REIT Investment Corporation) and has low risk and stable returns. Therefore, you can expect a certain amount of dividends, which is why we think it came in fourth. Residential-specialized stocks have a stock code of 3269 (Advance Residence Investment Corporation) and are characterized by being less affected by economic conditions and having stable returns, and the second and third stocks are selected with higher risks. We believe that this residential type is chosen because of its stable earnings.

6 CONCLUSION

This study, we address the problem of portfolio optimization for diversified investments in companies that handle REIT as financial instruments and find the optimal portfolio that can minimize risk and obtain returns among companies that handle REIT. We found a solution using a genetic algorithm and was able to create a REIT portfolio from the results.

In response to the reasons given in Chapter 1 for not investing, the high likelihood of losing money and not knowing what to buy-the results obtained seem to clarify how to make purchases that minimize risk and yield returns.

Future issues will be "stock selection" and "combination of stocks and REIT." In stock selection, although We have introduced restrictions on selection, the accuracy decreases when there are restrictions. Although We focus on one type of product, portfolios can also be combined with other types of financial instruments; possible challenges that my arise include combining stocks and REIT.

ACKNOWLEDGEMENTS

This work was supported by JSPS KAKENHI Grant Numbers JP22K04603.

REFERENCES

[1] Japan Securities Dealers Association (in Japanese) https://www.jsda.or.jp/shiryoshitsu/toukei/kojn_isiki.html
[2] Nomura Asset Management (in Japanese) https://www.nomura-am.co.jp/corporate/surveys/
[3] Japan Securities Dealers Association (in Japanese) https://www.jsda.or.jp/jikan/
[4] Sumitomo Mitsui Trust Asset Management (in Japanese) https://www.smtam.jp/pickup/reit/outline/
[5] Wealth Engine https://www.chiba-tv.com/media/archives/219
[6] Yukiko Orito: Application of differential evolution algorithm to portfolio optimization problem using loans (2021) (in Japanese)
[7] Norio Masaraki: Portfolio optimization and mathematical programming (2005) (in Japanese)
[8] Seita Saito: Derivation of optimal layout of k-window system using genetic algorithm (2013) (in Japanese)
[9] Prop Tech plus Co., Ltd. REITDB Service (in Japanese)
[10] SMBC Nikko Securities (in Japanese) https://www.smbcnikko.co.jp/terms/japan/si/J0267.html

Recent Advances in Reliability and Maintenance Modeling – Okamura, Inoue & Xiao (Eds)
© 2025 the Author(s), ISBN: 978-1-032-79098-5

Assessing creation methods of word embedding models for analyzing and repairing classical Japanese literature

S. Kusabiraki & Y. Shimmura
Aichi University Graduate School, Aichi, Japan

K. Yamamoto
Sky Co., Ltd., Japan

K. Iwata
Aichi University, Aichi, Japan

ABSTRACT: This study not only proposes a valuable model for predicting missing words in classical Japanese literature but also suggests the potential of this model to be instrumental in repairing the literature. It could significantly advance the field of Natural Language Processing research in the context of historical literature. In recent years, Natural Language Processing has been applied to artificial intelligence programs such as ChatGPT and literary works. However, Natural Language Processing research in Japan has mainly focused on modern Japanese, and research in Japanese classical literature has yet to progress enough. Our research takes a novel approach by attempting to forecast missing words in classical Japanese literature. It creates several language models based on three pieces of classical literature using the Skip-gram of fastText. We employ LOOCV (leave-one-out cross-validation) to validate each model's accuracy. The results highlight significant differences between the modern language model and our proposed models, which we attribute to the historical context. Next, the experiment demonstrates the efficiency of our model creation method in predicting a missing word. The results of the experiments show that our proposed method can predict words similar to a missing word.

Keywords: Big Data, Data Mining & Analytics, Prediction

1 INTRODUCTION

In recent years, remarkable progress has been made in the evolution of artificial intelligence programs, such as ChatGPT (OpenAI 2024), driven by Natural Language Processing (NLP) applications. In particular, ChatGPT is attracting attention as an AI model that can interact as naturally as humans. It can answer questions, summarize texts, and generate simple programming code. Soon, companies and schools are expected to use NLP in chat-bot services to measure operational efficiency and support learning.

The applications of NLP extend beyond artificial intelligence programs to literary works such as stylistic analysis and classification of works. Stylistic analysis involves understanding the structure of sentences and the writer's style. The classification of works mainly involves a broad classification of categories. In addition, it determines the authenticity of the work and the period in which it was made. In this way, the applications of NLP are being pursued from various perspectives. However, research in "Japanese classical literature" has not advanced much. Research in natural language processing for modern Japanese has been advancing. Murai (Murai, 2016) stated that it is possible to analyze frequently occurring words in texts,

 DOI: 10.1201/9781003491309-25

extract keywords, and group and compare words. Some websites create word clouds that correspond to Japanese (User Local, Inc. 2024). Additionally, there are many examples of analyzing Japanese using word2vec. On the other hand, very little research has been conducted on Japanese classical literature. The reason is that there is no useful natural language processing model for classical analysis.

This study not only proposes a valuable model for predicting missing words in classical Japanese literature but also suggests the potential of this model to be instrumental in repairing the literature. The models undergo a rigorous testing process using keywords derived from classical Japanese literature, ensuring whether the proposed model can extract more appropriate words than the pre-trained modern model trained on Wikipedia. In addition, the experiment in the following steps demonstrates the efficacy of predicting a missing word in our model. (1) preparing an intentional word-missing text, i.e., removing a word. (2) creating a model using the word-missing text. (3) finding similar words to the missing word with the model. (4) investigating whether the extracted words are appropriate by considering the relationship between them and the missing word. The results of the experiments show that our proposed method can predict words similar to a missing word.

2 RELATED WORK

2.1 *Japanese writing style*

Japanese is a language written without spaces (delimiters) between words. Figure 1 shows the same sentence in English and Japanese. In stark contrast, English sentences have spaces meticulously placed between each word (Figure 1 line1). On the other hand, in Japanese sentences with the same meaning, such visual separations do not exist (Figure 1 line2). The sentence must be segmented into the most minor meaningful units called morphemes to insert delimiters among words. This process, known as "tokenization," is illustrated by the slashes in line 3 of Figure 1.

In English, there are 26 alphabetic letters, along with other symbols. In contrast, Japanese is broadly categorized into three types: "hiragana," "katakana," and "kanji." Particularly, kanji comprises over ten thousand characters. Additionally, distinctions such as "voiced sounds," "nasal sounds," and "semi-voiced sounds" exist within hiragana and katakana. Consequently, Japanese is considered more challenging to analyze compared to languages such as English.

English : My name is Kento.

Japanese : 私の名前はケントです。

Tokenization : 私 / の / 名前 / は / ケント / です / 。

Figure 1. Example sentences and "Tokenization".

2.2 *Word embeddings*

In natural language processing (NLP), a one-hot vector is employed. This vector is a matrix (or vector) of size $1 \times N$, where N represents the total number of words in the text. This vector distinguishes each word in the text from all others. Specifically, the vector contains 0s in all its cells except for one cell, which contains only a single value. This unique configuration of 1s and 0s uniquely identifies the corresponding word. One-hot vectors can easily convert words into vectors, but these vectors do not have meaning. The vectors are called local representations and are unsuitable for word-to-word comparisons. This approach is unwieldy because of the enormous number of dimensions.

This paper adopts word embeddings to consider words' meanings. Word embeddings are also a way of representing words as numerical vectors. These vectors capture the meaning of the words and their relationships to other words in a text. There are two main approaches to creating word embeddings: CBOW and Skip-gram models. In the CBOW model, the target word is presumed by surrounding words; on the other hand, the Skip-gram model predicts what words will appear around the word. In other words, CBOW is trained to predict a single word from context words, whereas Skip-gram tries to predict several context words from a single input word. The difference is shown in Figure 2. This study uses the Skip-gram model because it shows higher accuracy (but slower) than the CBOW model, and the previous study (Yamamoto and Iwata 2024) used it.

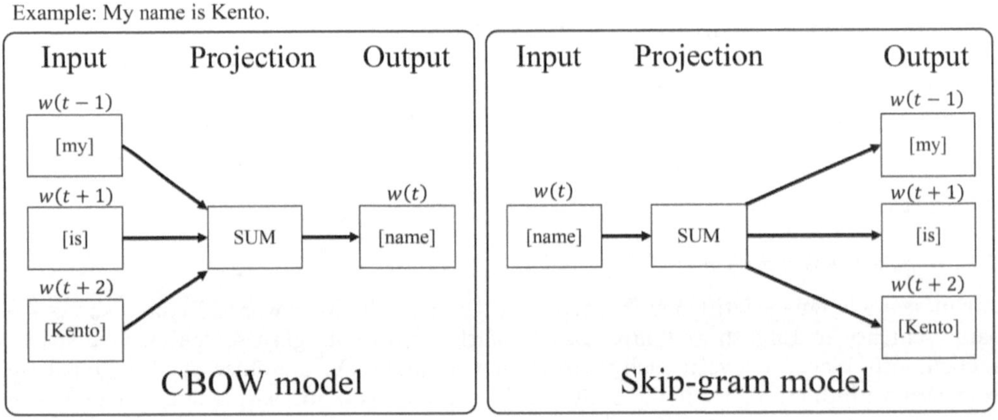

Figure 2. Difference between CBOW model and skip-gram model.

2.3 *Previous research*

In the previous study (Yamamoto and Iwata 2024), words appearing in "The Kojiki" were analyzed ("The Kojiki" is the oldest extant history book in Japan.) "The Kojiki" was morphologically analyzed, and important words were extracted. Subsequently, similar words to these critical terms were identified using a model derived from "The Kojiki" and a contemporary language model. It is a testament to our research that for words such as "神" (gods)[1] and "イザナギ (one of the Japanese god names)," the unique characteristics of these words in the Kojiki were successfully captured, marking a significant milestone in our study. For example, these words resulted in related words such as "appear" and "omniscient." These words can be seen as words close to gods.

Consequently, employing a similar approach in Japanese classical literature could create more useful classical language models. Therefore, this study aims to increase the number of texts examined and create a valuable classical language model.

3 CREATION MODELS FOR CLASSICAL JAPANESE LITERATURE

The processes of creation of a language model for classical Japanese literature and verification of it are as follows (these details are described in subsection 3.2):

(1) Preparing classical text data. This phase removes unnecessary parts for analysis.

[1]English words in parentheses after Japanese words have the almost the same meaning as Japanese words.

(2) Tokenizing the text data. As mentioned in subsection subsection 2.1, Japanese text must be segmented.
(3) Counting up the nouns in the segmented text. These carefully selected nouns serve as crucial keywords.
(4) Creating a language model based on the text data. A language model is a machine learning model trained to consider a probability distribution over words. The model treats each word as a vector and can calculate the similarity of words.
(5) Providing the exact keywords to the model created from the data and the pre-trained modern language model and outputting the top 10 similar words or phrases for each keyword.
(6) Rigorously comparing these top 10 results to verify the model's accuracy, ensuring its reliability.

3.1 *Target text*

While the previous study dealt only with the "Kojiki", this study deals with three kinds of literature. The details are shown in Table 1. These three pieces of classical literature have appeared in Japanese higher education textbooks and are essential for studying Japanese classical literature (Bible Be Available (Revival), unknown a, unknown b, unnown c; TOKYO SHOSEKI CO., LTD 2024).

Table 1. The information of target literature.

Title	Author	Publication year	Category
The Taketori Monogatari (Tale of the Bamboo Cutter)	Unknown	Around 10th century	Fable
The Makura no Soshi (Pillow Book)	Sei Shonagon	About 1001	Essay
The Tosa Nikki (Tosa Diary)	Ki no Tsurayuki	About 934	Diary

3.2 *Details of creating and verifying classical models*

3.2.1 *Text preprocessing*

Before handling text data, removing irrelevant parts for analysis was necessary. The processes are summarized below:

- Deleting titles and subtitles.
- Deleting of furigana and ruby.
- Deleting of blank lines.
- Deleting of various symbols.

3.2.2 *Tokenization*

When dealing with Japanese literature and conducting natural language processing, it was necessary to perform tokenization, dividing the text into individual words (in Japanese, "wakachigaki"). To achieve this, the morphological analysis engine MeCab was utilized. MeCab is an open-source morphological analysis engine developed through a joint research unit project between the Graduate School of Informatics at Kyoto University and the Nippon Telegraph and Telephone Corporation Communication Science Laboratories (Graduate School of Informatics, Kyoto University 2013]. This engine can arrange Japanese text files for analysis.

3.2.3 *Keywords extraction*

Only nouns are extracted from classic literature, and the nouns that appear frequently are used as keywords. Ambiguous words such as "this" and "that" (relative pronouns) were excluded from the keyword extraction process due to their lack of specificity.

3.2.4 *Model creation*

For model creation, the Skip-gram of fastText was utilized. FastText is one of the natural language processing libraries developed by Facebook AI Research (Bojanowski *et al.* 2017). It enables the representation of words as vectors, allowing the measurement of similarity between words based on their distances. The number of dimensions of a model is how many different vectors represent a term, and it depends on the text file size. In this study, the vectors have 100 dimensions and 100 epochs, which is the number of times the target literature is learned. The model creation uses only nouns occurring more than or equal to four times because these nouns occupy more than 85% of the text.

The modern language model used in this study is the pre-trained model on Wikipedia, with 300 dimensions, provided by fastText (Facebook Inc. 2016).

3.2.5 *Detection of similar words*

The model outputs the top 10 similar terms for the keywords. The similarity is determined by "cosine similarity," a measure between two non-zero vectors defined by an inner product space. In other words, the cosine similarity indicates the cosine of the angle between the vectors and does not depend on the magnitudes of the vectors. The cosine similarity always has the interval [−1, 1]. The larger the cosine similarity, the smaller the angle between two vectors, which means the two vectors are more similar.

3.2.6 *Comparison of the results*

This study conducted LOOCV (leave-one-out cross-validation) to assess the accuracy of each model created. Figure 3 presents an illustrative diagram of this process. The literature marked in orange was used to create the model, and keywords were given from non-used literature. A classical $\alpha\beta$ model, which was a fusion of texts from classical α and classical β, was meticulously crafted. Its accuracy was then thoroughly evaluated using keywords from classical γ. Similarly, the $\beta\gamma$ model was evaluated with keywords from classical α, and the $\gamma\alpha$ model with keywords from classical β. The contemporary language model was also provided with respective keywords for comparison purposes. Although simplified in Figure 3, α is the Taketori Monogatari, β is the Makura no Soshi, and γ is the Tosa Nikki.

Literature			Models	Keywords
α	β	γ	$\alpha\beta$ model	γ
α	β	γ	$\beta\gamma$ model	α
α	β	γ	$\gamma\alpha$ model	β

Figure 3. Simplified example of the LOOCV.

Table 2 presents a partial result of cross-validation, which analyzed words in the Taketori Monogatari using the $\beta\gamma$ model (the $\beta\gamma$ model was made from the Makura no Soshi and the Tosa Nikki). The "Cosine similarity" indicates the proximity between keywords and the outputted words, with values closer to 1 suggesting more remarkable similarity to the keywords. The first word, "荒れ" (dilapidated), denotes desolation, fitting the depiction of dilapidated structures in the classical text, hence the result. The second word, "あるじ" (master), though written in hiragana, is interpreted as "主" (master), suggesting the head of a household or owner, aligning with the interpretation as the master of the house. Additionally, the tenth word, "住ま" (live in), likely derives from the verb "住ま" (live in).

Table 2. Similar words and cosine similarity for “家” (home).

Rank	Similar words	Cosine Similarity
1	荒れ(dilapidated)	0.44707
2	あるじ (master)	0.43827
3	所 (spot)	0.41954
4	除目 (bureaucrat)	0.40139
5	焼け (burn)	0.39991
6	なり (determine)	0.38468
7	べき (should)	0.37684
8	詣で (visit to a shrine)	0.37638
9	清 (pure)	0.37133
10	住ま (live in)	0.36963

These results demonstrate the ability to capture the meaning of phrases even when utilizing keywords from different kinds of classical literature. Therefore, classical models generated using this method are deemed valuable and functional.

4 PREDICTING MISSING WORDS IN CLASSICAL JAPANESE LITERATURE

The following steps demonstrate the model proposed in section 3, which efficiently predicts missing words in classical Japanese literature.

(1) Preparing an intentional word-missing text. In each piece of literature, one word is intentionally deleted. In the Taketori Monogatari, “かぐや姫,” in the Tosa Nikki, “海賊” and in the Makura no Soshi, “蔵人” are deleted.
(2) Creating a model using the word-missing text.
(3) Finding similar words to the missing word with the model. The list of similar words is restricted to noun-only words from the list of all similar words.
(4) Investigating whether the extracted words are appropriate by considering the relationship between them and the missing word.

5 RESULTS AND CONSIDERATION

This section deals with the most salient results using the classical and modern language models to determine whether they can find similar words for some classical terms. In addition, the section shows the results of predicting missing words in classical Japanese literature and discusses them.

5.1 *Model accuracy*

Table 3 presents the analysis results for the term “姫” (princess). “姫” appeared 81 times in the literature (the total word count was 13,202). There were differences in the similar words in the classical and modern language models. The classical model extracts the words “君” (load) and “大臣” (minister), titles for high-society people in a past era. In fact, “姫君” (a polite expression for princess) is used in the classical literature. The term “綱” (rope) is also related to proper nouns and not a noun like “rope.” During the period of the classical kinds of literature examined in this study, the kanji “綱” was contained in the names of relatively influential persons. The names influence the results. These phrases are unique to the classical pieces of literature, and the results show the importance of the proposed models for analyzing archaic words. On the other hand, the modern model elects words related to women.

Table 3. Analysis of "姫" (princess) using the $\beta\gamma$ model.

Rank	Results ($\beta\gamma$ model)	Results (modern model)
1	匣 (box)	天使 (angel)
2	綱 (rope)	女王 (queen)
3	みえ (see, watch)	ウサ (rabbit)
4	君 (you)	乙女 (maiden)
5	大臣 (minister)	茶々 (tease)
6	うらやましけれ (envious)	竜 (dragon)
7	奉ら (present)	妖精 (fairy)
8	清涼 (fresh)	ひめ (princess)
9	まだ (still)	女神 (goddess)
10	おほき (goodly)	華 (flower)

Table 4 presents the analysis results for the term "車" (car). "車" appeared 92 times in the literature (the total word count was 79,446). The modern model lists words closely resembling modern vehicles. However, there is no such word in the classical model. Considering the historical context when these classical kinds of literature were created, "車" (car) would likely refer to "牛車" (ox cart), a type of ox-drawn carriage. During that period, ox-carts were only accessible to a few people, such as royalty. As a result, "持" (have, use), "皇子" (prince), and "天人" (People living in the capital) are likely listed as similar terms due to this influence. These are characteristics of the classical model.

Table 4. Analysis of "車" (car) using the $\gamma\alpha$ model.

Rank	Results ($\gamma\alpha$ model)	Results (modern model)
1	持 (have)	車両 (vehicle)
2	皇子 (prince)	乗用車 (passenger vehicle)
3	しり (back)	車種 (type of car)
4	石 (stone)	自動車 (car)
5	門 (gate)	自家用車 (my car)
6	天人 (people living in the capital)	列車 (tarin car)
7	ながら (without change)	車輌 (vehicle)
8	あるじ (master)	新車 (new car)
9	道 (road)	外車 (foreign car)
10	山崎 (one of the place names)	クルマ (car)

Table 5. Analysis of "歌" (song) using the $\alpha\beta$ model.

Rank	Results ($\alpha\beta$ model)	Results (modern model)
1	よみ (recite)	曲 (songs)
2	詠む (recite)	唄 (songs)
3	よむ (recite)	歌詞 (lyrics)
4	題 (theme)	歌声 (singing voice)
5	よま (recite)	合唱 (chorus)
6	と (and)	歌曲 (lied)
7	か (or)	詩 (verse)
8	を (particle)	歌唱 (sing)
9	は (particle)	詞 (lyrics)
10	よろしう (beseem)	校歌 (school song)

Table 6. Results of predicting of missing words (only noun).

Rank	The Taketori Monogatari	The Makura no Soshi	The Tosa Nikki
Missing word	かぐや姫 (the princess's name)	蔵人 (chamberlain)	海賊 (pirates)
1	かぐや姫	滝 (falls)	和泉 (one of the place names)
2	つる (auxiliary verb)	筋 (case)	やみ (dark)
3		寄せ (trust)	ぬれ (auxiliary verb)
4		臂 (arm)	なみ (wave)
5		綾 (textile)	神 (gods)
6		御曹司 (son of a noble)	海賊
7			出 (set out)

Table 5 presents the analysis results for the term "歌" (song). "歌" appeared 53 times in the literature (the total word count was 7,788). The modern model chooses many words closely related to music as a form of singing. On the contrary, the classical model elects "詠む" (recite), related to composing poetry, instead of music song. The result is also due to the historical context. During that period, these poems were called "和歌" (ancient style poem) and were widely popular among nobles and the general public. Therefore, the related word is associated with poems rather than music. These results show some intriguing differences between the modern language model and the classical literature model. For instance, in the term "歌" (song) analysis, the modern model listed many words closely related to music as a form of singing. However, the classical model listed "詠む" (recite), a term more associated with composing poetry than music. This discrepancy is likely due to the historical context, where these poems were called "和歌" (ancient style poem) and were widely popular among nobles and the general public. Therefore, the associated word is more related to poetry than music. These findings underscore the high accuracy of our method in creating the classical model.

5.2 *Predicting missing words*

The results of the predicting missing words are shown in Table 6. In these results, only words that MeCab decided as nouns were picked from the words predicted with the context's vectors. In the Taketori Monogatari and the Tosa Nikki, the correct words that were "かぐや姫" (the princess's name) and "海賊" (pirates) were successfully predicted, respectively. However, in the Makura no Soushi, "蔵人" (chamberlain) was not picked up. The failure to predict a missing word in the Makura no Soushi could be caused by the number of characters. In the Makura no Soshi, with its 120,000 characters, some words have various meanings compered with words in the Taketori Monogatari (15,000 characters) and the Tosa Nikki (10,000 characters).

In addition, "つる" picked from the Taketori Monogatari can be understood as the animal "鶴" (crane), but the crane is never used in the story. Most of "つる" is used as auxiliary verbs in the literature. The misunderstanding is due to the tokenization "wakachigaki" using MeCab, which cannot correctly recognize auxiliary verbs in classical Japanese literature.

Overall, the results indicate that the model provides candidates for missing words and reduces the human effort required to find them without prior information.

6 CONCLUSION AND FUTURE WORK

This study proposed a valuable model for predicting missing words in classical Japanese literature and evaluated the potential of this model to be instrumental in repairing the literature. Our research involved creating several models by fastText based on three pieces of classical literature. The models allow users to learn classical literature representations and classifiers. A model of modern language was also used to identify differences in the meaning of words and phrases in modern and ancient times. This model was provided by fastTxet and created based on Wikipedia. The accuracy of each model was not just assessed, but meticulously evaluated through LOOCV (leave-one-out cross-validation). We provided the exact keywords for both the model created from classical data and the modern language model. We then output the top 10 similar words or phrases for each keyword, comparing these top 10 results. This rigorous evaluation method ensures the reliability and robustness of our method.

The results unveiled some fascinating disparities between the modern language and classical literature models. These disparities serve as a testament to the exceptional precision of our method in crafting the classical model. The experiments in predicting missing words showed that our proposed method could find candidates similar to a missing word. The method can reduce the effort required to find missing words, even without prior information.

However, we are confronted with two intricate issues. The first issue revolves around the tokenization "wakachigaki". The MeCab occasionally needs to improve in accurately identifying archaic words. For instance, the adjective "をかし" (tasteful) in archaic language, meaning "charming" or "interesting", is often misinterpreted in modern language as "を/かし," with "を" as a case particle and "かし" as a verb such as "lend." To tackle this, we need an extensive dictionary that includes numerous words not commonly used in modern languages, corresponding to archaic words.

The second issue is that the model cannot sometimes analyze specific terms (words) or predict appropriate missing words. The cause of this issue is the difference in the length of classical literature. The texts analyzed in this study varied significantly in length, with the Taketori Monogatari around 15,000 characters, the Makura no Soshi around 120,000 characters, and the Tosa Nikki around 10,000 characters. Addressing this issue requires an investigation into determining an optimal length for literature. This approach will enable us to deal with more literature and provide a more comprehensive analysis. However, in doing so, we should note the distribution of words' appearance frequency and develop strategies to address it.

ACKNOWLEDGMENTS

This research was partially supported by the joint research project "Modeling and Analyzing Information Security Measures for Server Systems" of the Institute of Managerial Research, Aichi University.

REFERENCES

Bible Be Available (Revival) unknown a. *The Makura no Soshi (in Japanese)*. https://classicstudies.jimdofree.com/ (Last accessed: May 13, 2024).

Bible Be Available (Revival) Unknown b. *The Taketori Monogatari (in Japanese)*. https://classicstudies.jimdofree.com/ (Last accessed: May 13, 2024).

Bible Be Available (Revival) Unknown c. *The Tosa Nikki (in Japanese)*. https://classicstudies.jimdofree.com/ (Last accessed: May 13, 2024).

Bojanowski, P., E. Grave, Joulin, A. and Mikolov, T. (2017). *Enriching Word Vectors with Subword Information*. https://arxiv.org/abs/1607.04606. (Last accessed: May 13, 2024).

Facebook Inc. (2016). Wiki word vectors. https://fasttext.cc/docs/en/pretrained-vectors.html. (Last accessed: May 13, 2024).

Graduate School of Informatics, Kyoto University (2013). *MeCab: Yet Another Part-of-Speech and Morphological Analyzer*. https://taku910.github.io/mecab/. (Last accessed: July 10, 2023).

Murai, H. (2016). Recent trends in computational narratology and its application for the future (in Japanese). *Information Processing Society of Japan, SGI Technical Report 2016-CH-112*(5), 1–5.

OpenAI (2024). *Chat Generative Pre-trained Transformer*. https://chat.openai.com. (Last accessed: May 13, 2024).

TOKYO SHOSEKI CO., LTD (2024). New Edition Advanced Classics (in Japanese). https://ten.tokyo-shoseki.co.jp/text/hs/kokugo/16589/. (Last accessed: May 13, 2024).

User Local, Inc. (2024). *User Local AI Text Mining (in Japanese)*. https://textmining.userlocal.jp. (Last accessed: May 13, 2024).

Yamamoto, K. and Iwata, K. (2024). *Fluctuation Analysis of the Meaning of Words in the Kojiki Using Distributed Representations of Words (in Japanese)*. https://saturn.aichi-u.ac.jp/center/com.html. *Journal of Aichi University Media Center* 33(1), 1–30.

Recent Advances in Reliability and Maintenance Modeling – Okamura, Inoue & Xiao (Eds)

Performability assessment measure based on deep learning for open source software

Y. Tamura, H. Takeda, S. Heima, S. Miyamoto & L. Zhou
Yamaguchi University, Ube, Yamaguchi, Japan

A. Anand
University of Delhi, Delhi, India

P.K. Kapur
Amity University, Noida, UP, India

S. Yamada
Tottori University, Tottori, Tottori, Japan

ABSTRACT: We focus on the edge computing. Several Open Source Software (OSS) are included in the software environment of edge computing. Then, it is important to consider the characteristics of OSS in terms of the edge computing. We focus on the reliability assessment measures based on the deep learning for the edge computing. In particluar, we propose the performability as reliability assessment measure from the proposed deep learning model. Also, this paper shows several numerical examples in terms of the performability by using the actual fault big data sets.

1 INTRODUCTION

The open source software (OSS) is used by many companies. Also, various commercial software are developed by using several OSS components. Especially, the OSS components are embedded in the cloud computing (Ibrahim *et al.* 2018). Moreover, several OSS components are embedded as the edge components (Ahmad *et al.* 2019; Caprolu *et al.* 2019; Ngoko *et al.* 2017; Ozcan *et al.* 2019). It is important to consider the performability in the network based OSS. Then, we need to consider various external factors from the standpoint of edge computing. Our research group has been proposed the method of OSS reliability assessment in the past (Yamada, S. & Tamura Y. 2016). In particular, several methods of OSS reliability assessment based on the deep learning have been proposed (Tamura, Y. *et al.* 2023, 2024; Yamada S. and Tamura Y. 2022). This paper proposes the method of OSS reliability based on the deep learning considering the performability assuming the edge OSS operation. Then, we consider that the edge OSS operation depends on various factors of edge computing. We apply the Wiener process to the data preprocessing. Furthermore, we propose the performability considering the noisy cases based on the proposed deep learning model. This paper shows several numerical examples of performability based on the proposed deep learning model.

2 OSS PERFORMABILITY BASED ON DEEP LEARNING MODELING FOR THE EDGE COMPUTING

We focus on the performability of edge computing based on OSS. In the past, our research group has proposed the performability assessment measures for open source solutions

DOI: 10.1201/9781003491309-26

(Tamura, Y. and Yamada S. 2013). Based on the approach of performability, we consider the performability for edge computing. We consider the discrete model based on deep learning. Then, we consider the following objective functions as the performability.

$$Per(i) = \frac{i - MTTR(i)C(i)}{i}, \tag{1}$$

where $MTTR(i)$ the downtime of edge computing at i-th fault. Also, $C(i)$ is the cumulative number of detected faults. Generally, the Mean Time to Repair (MTTR) means the mean time to correction. In this paper, we define that the type of density function for correction time is the exponential distribution. Then, the expected value of exponential distribution with the rate parameter λ is $\frac{1}{\lambda}$.

Therefore, we can obtain the performability for edge computing is given as follows:

$$Per(i) = \frac{\lambda i - C(i)}{\lambda i}, \tag{2}$$

Then, we consider Figure 1 as the network structure based on our deep learning model.

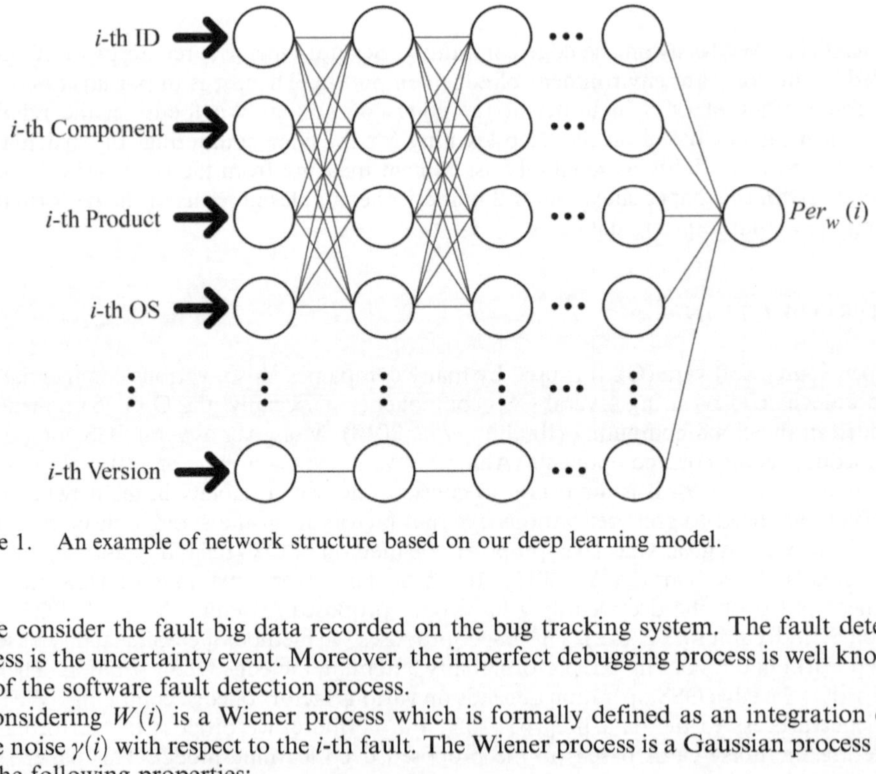

Figure 1. An example of network structure based on our deep learning model.

We consider the fault big data recorded on the bug tracking system. The fault detection process is the uncertainty event. Moreover, the imperfect debugging process is well known as one of the software fault detection process.

Considering $W(i)$ is a Wiener process which is formally defined as an integration of the white noise $\gamma(i)$ with respect to the i-th fault. The Wiener process is a Gaussian process and it has the following properties:

$$Pr[W(0) = 0] = 1, \tag{3}$$

$$E[W(i)] = 0, \tag{4}$$

$$E[W(i)W(i')] = Min[i, i'], \tag{5}$$

where Pr[·] and $E[\cdot]$ represent the probability and expectation, respectively. We assume that the performability is approximately based on Wiener process. We consider the following equation from Equation (6) as the objective variable.

$$Per_w(i) = \frac{\lambda i - C(i)}{\lambda i} + W(i), \tag{6}$$

where $W(i)$ is the Wiener process at i-th fault. Equation (6) is applied to the data pre-processing and the output values of the training data and testing one.

3 NUMERICAL EXAMPLES

This paper focuses on *OpenStack* Project (The OpenStack project) with several edge components. This paper shows several numerical examples by using the data sets collected from the bug tracking system assuming the edge OSS computing.

We show the estimated performability in case of testing data 10%, 30%, and 50% in Equation (6) and $\lambda = 1$, respectively. Moreover, we show the estimated performability in case of testing data 10%, 30%, and 50% in Equation (6) and $\lambda = 2$, respectively. Furthermore, we show the estimated performability in case of testing data 10%, 30%, and 50% in Equation (6) and $\lambda = 3$, respectively.

We have found that the proposed performability based on deep learning model considering the Wiener process can estimate more pessimistic than the testing data. Moreover, we found that the performability fits to the testing data, when the estimation period becomes long.

In particular, the fluctuation of performability becomes large by using the deep learning model. The OSS managers can understand as the stability of edge computing from the fluctuation of performabiliy.

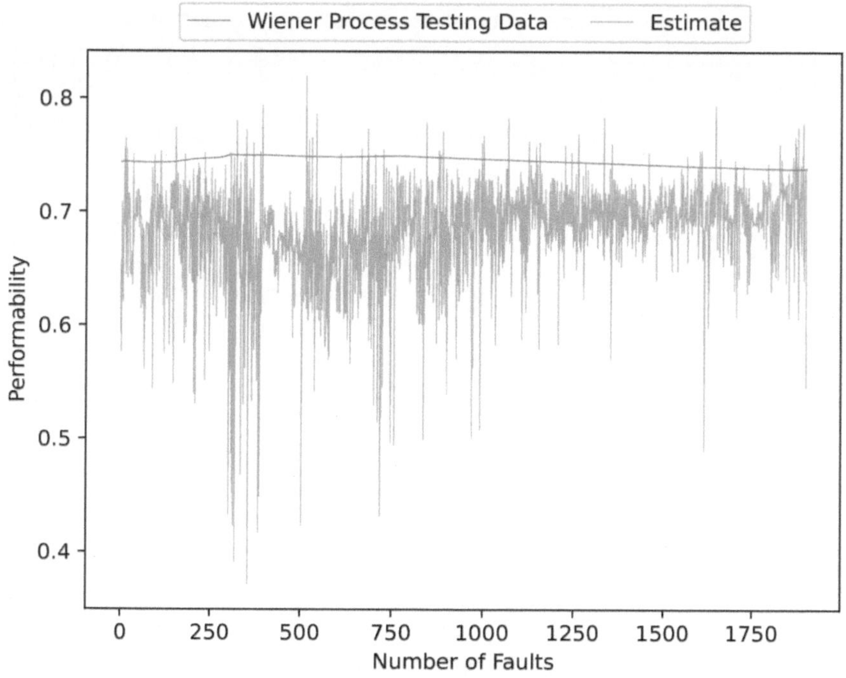

Figure 2. Estimated performability in case of $\lambda = 1$ and 10% testing data.

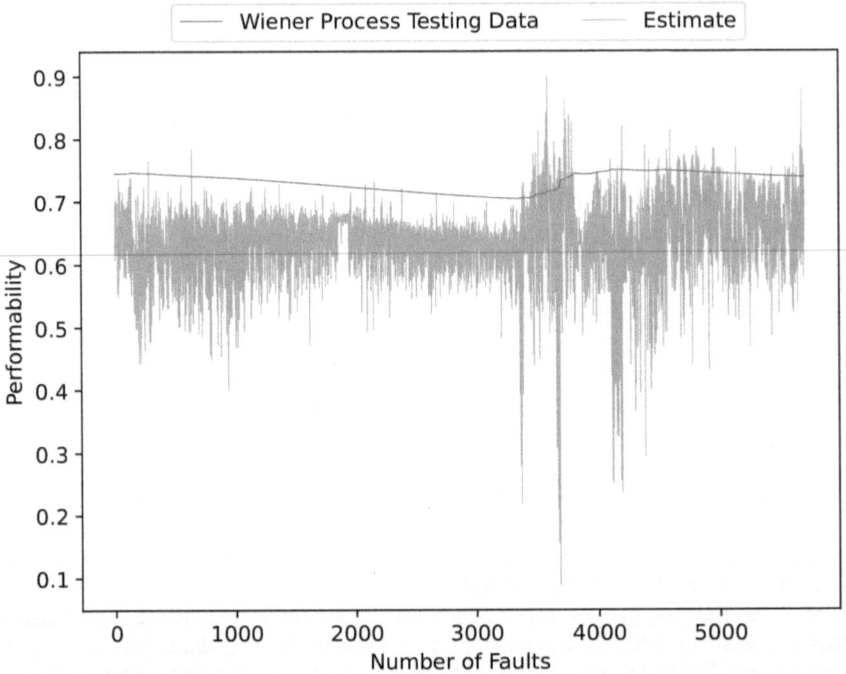

Figure 3. Estimated performability in case of $\lambda = 1$ and 30% testing data.

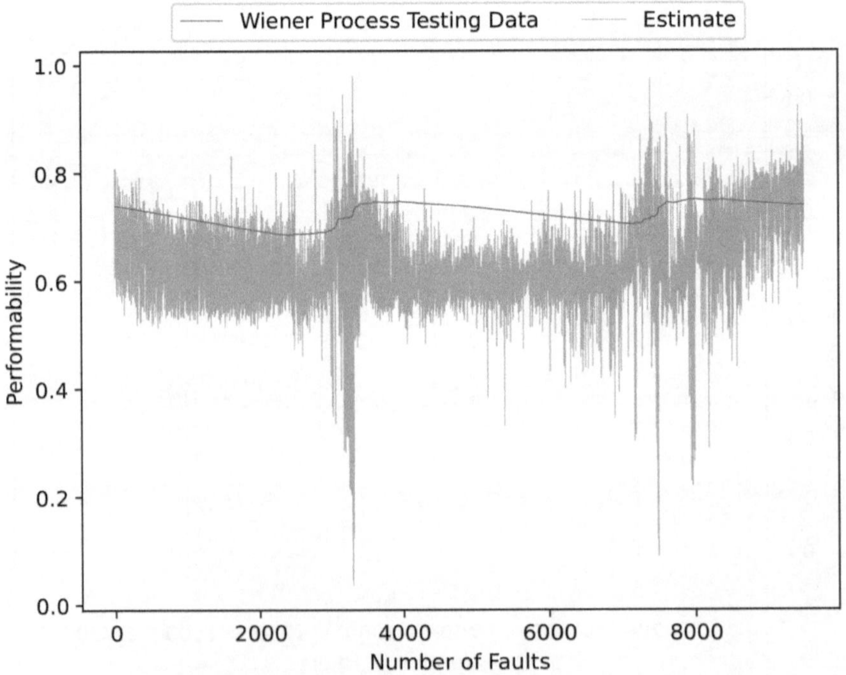

Figure 4. Estimated performability in case of $\lambda = 1$ and 50% testing data.

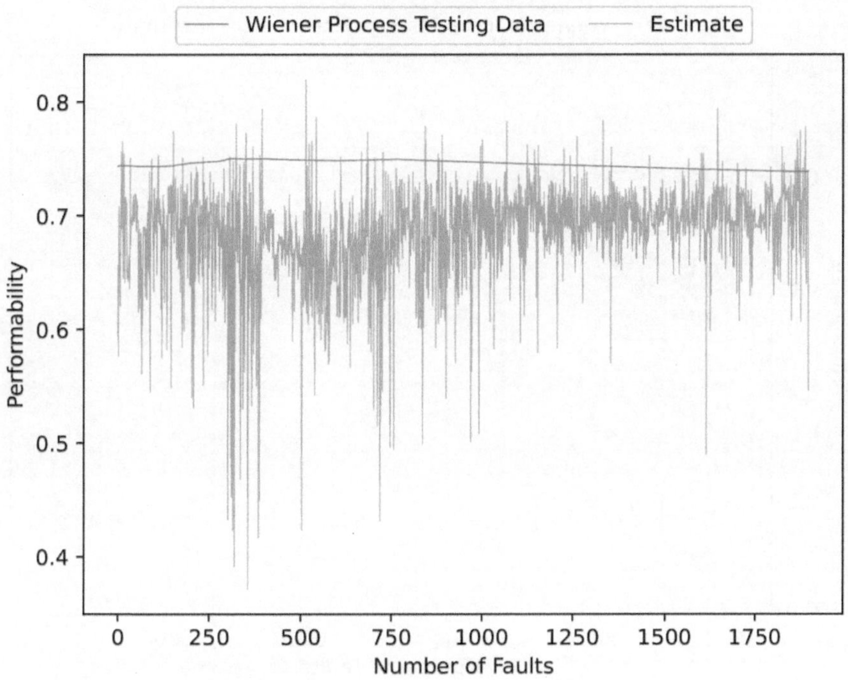

Figure 5. Estimated performability in case of $\lambda = 2$ and 10% testing data.

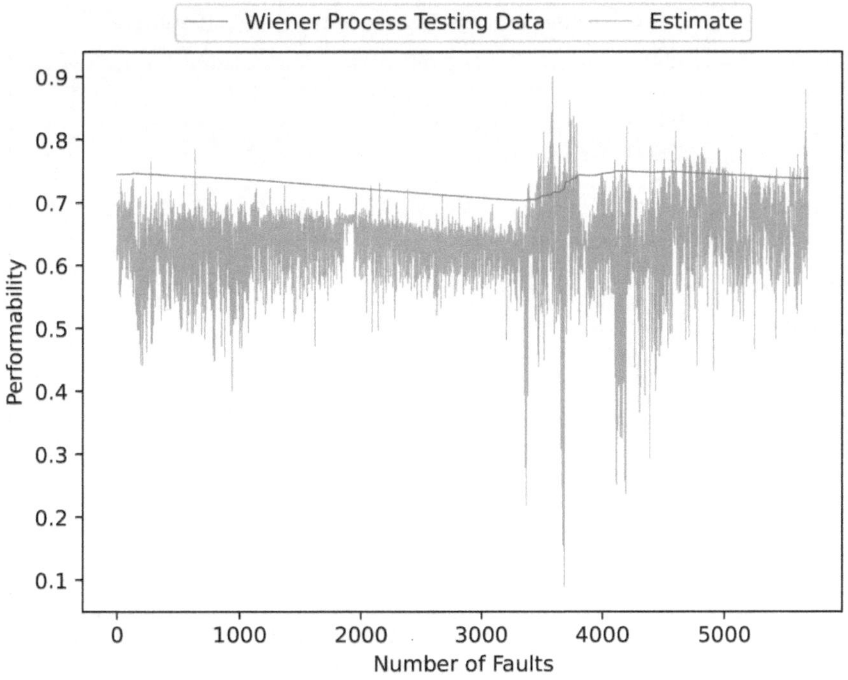

Figure 6. Estimated performability in case of $\lambda = 2$ and 30% testing data.

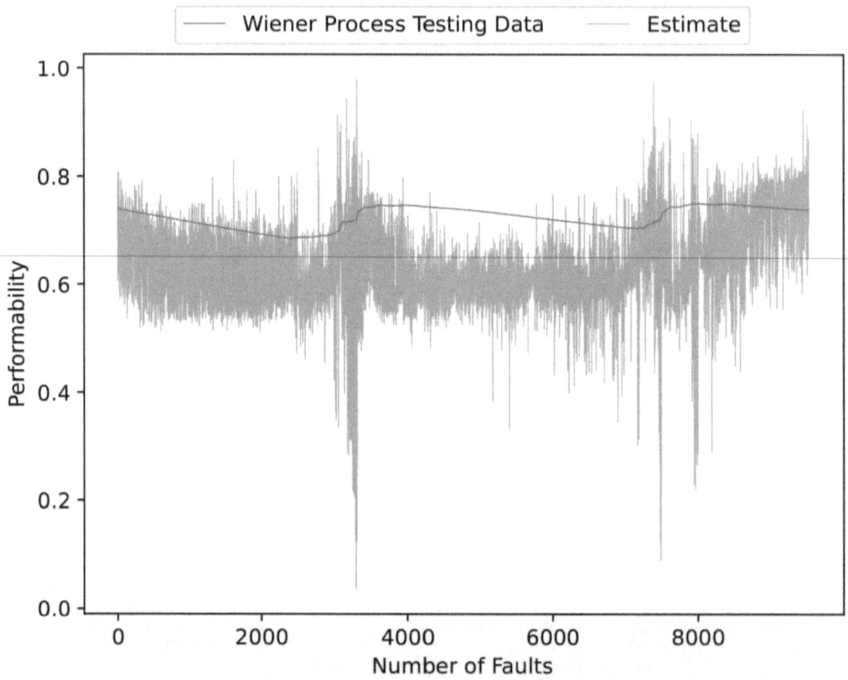

Figure 7. Estimated performability in case of $\lambda = 2$ and 50% testing data.

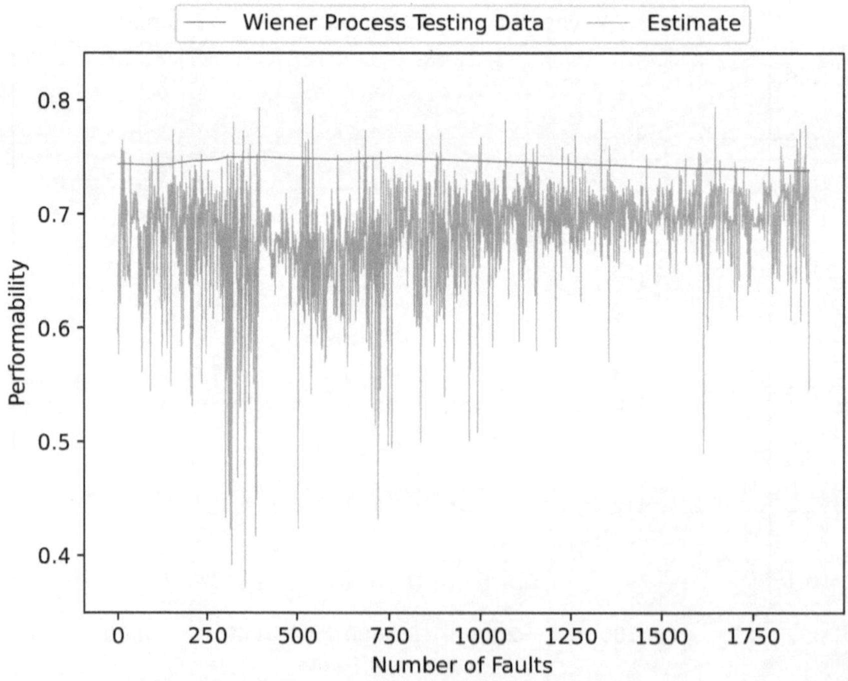

Figure 8. Estimated performability in case of $\lambda = 3$ and 10% testing data.

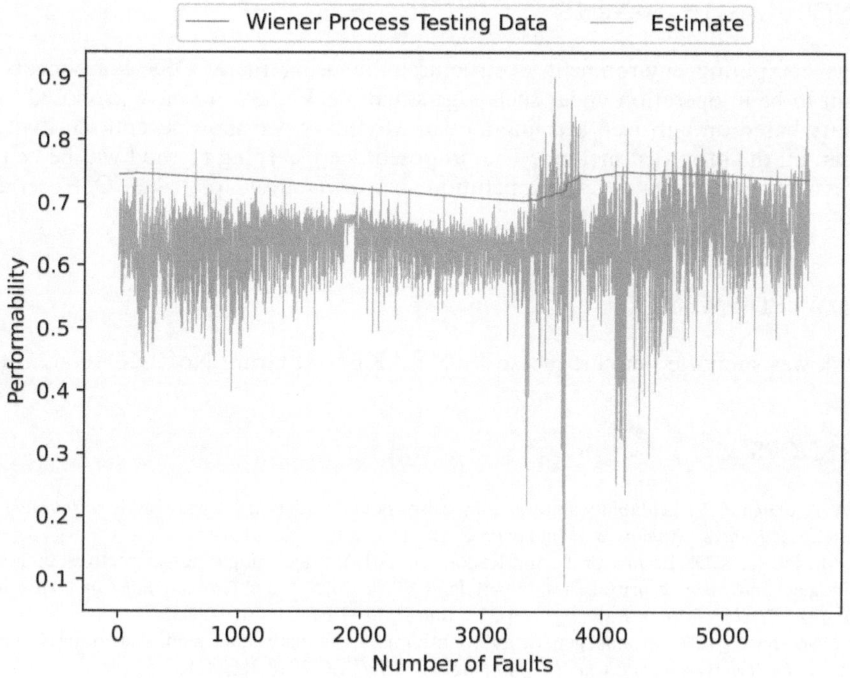

Figure 9. Estimated performability in case of $\lambda = 3$ and 30% testing data.

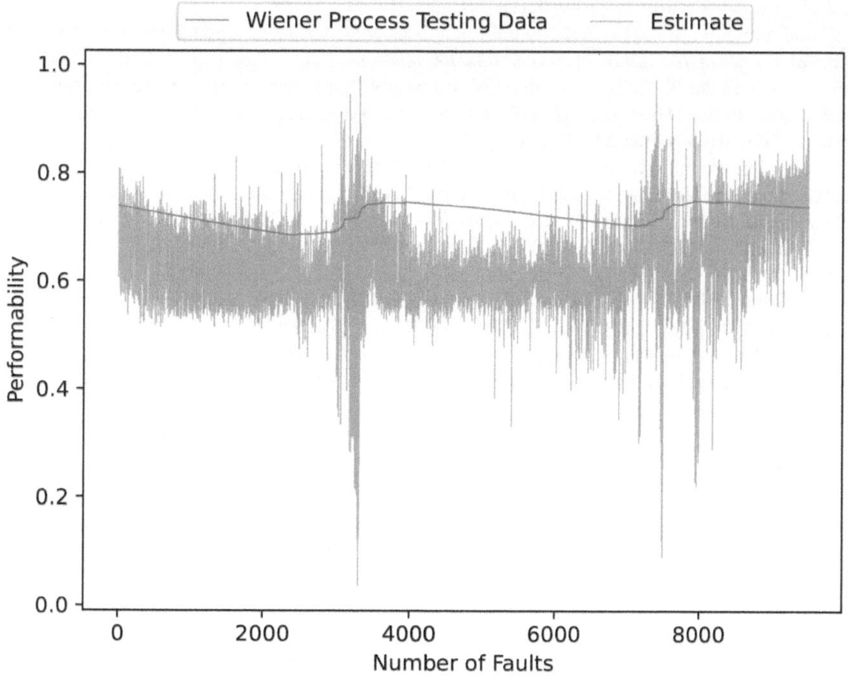

Figure 10. Estimated performability in case of $\lambda = 3$ and 50% testing data.

4 CONCLUDING REMARKS

The edge computing environment is structured by using many OSS components. It is important to be in operation under such edge situation. Finally, we have proposed the performability based on our deep learning model. Moreover, we show several sensitivity analysis based on the proposed method. The proposed deep learning method will be helpful as the assessment measures of the operation performability for edge OSS service in operation phase.

ACKNOWLEDGMENTS

This work was supported in part by the JSPS KAKENHI Grant No. 23K11066 in Japan.

REFERENCES

Ahmad, A. A. *et al.* (2019). Scalability analysis comparisons of cloud-based software services, *Journal of Cloud Computing: Advances, Systems and Applications*, 10.1186/s13677-019-0134-y

Caprolu, M., Pietro, R.D., Lombardi, F. and Raponi, S. (2019). Edge computing perspectives: architectures, technologies, and open security issues, *Proceedings of the 2019 IEEE International Conference on Edge Computing (EDGE)*, Milan, Italy, pp. 116–123, doi: 10.1109/EDGE.2019.00035.

Ibrahim, I. M. *et al.* (2018). A robust generic multi-authority attributes management system for cloud storage services, *IEEE Transactions on Cloud Computing*, 10.1109/TCC.2018.2867871.

Ngoko, Y. and Cérin, C. (2017). An edge computing platform for the detection of acoustic events, *Proceedings of the 2017 IEEE International Conference on Edge Computing (EDGE)*, Honolulu, HI, USA, pp. 240–243, doi: 10.1109/IEEE.EDGE.2017.44.

Ozcan, M. O., Odaci, F., and Ari, I. (2019). Remote debugging for containerized applications in edge computing environments, *Proceedings of the 2019 IEEE International Conference on Edge Computing (EDGE)*, Milan, Italy, pp. 30–32, doi: 10.1109/EDGE.2019.00021.

Tamura, Y. and Yamada, S. (2013) Performability analysis considering debugging behaviors for open source solution, *International Journal of Performability Engineering*, Vol. 9, No. 1, pp. 13–21.

Tamura, Y. and Yamada, S. (2022). Prototype of 3D reliability assessment tool based on deep learning for edge OSS computing, *Mathematics*, Vol. 10, No. 9, Multidisciplinary Digital Publishing Institute, Switzerland, DOI: 10.3390/math10091572, pp. 1–12.

Tamura, Y. Ueki, R. Anand, A., and Yamada, S. (2023). Estimation and comparison of mean time between failures based on deep learning for OSS fault big data, *International Journal of System Assurance Engineering and Management*, Springer, Latest Articles, https://doi.org/10.1007/s13198-023-01907-2, pp. 1–16.

Tamura, Y., Miyamoto, S., Zhou, L., Anand, A., Kapur, P. K., and Yamada, S. (2024) OSS reliability assessment method based on deep learning and independent Wiener data preprocessing, *Journal International Journal of System Assurance Engineering and Management*, Vol. ahead-of-print No. ahead-of-print, Springer, https://doi.org/10.1007/s13198-024-02288-w, pp. 1–9.

The OpenStack project, *Build the future of Open Infrastructure*, https://www.openstack.org/

Yamada, S. and Tamura Y. (2016). *OSS Reliability Measurement and Assessment*, Springer International Publishing, Switzerland, 2016.

Recent Advances in Reliability and Maintenance Modeling – Okamura, Inoue & Xiao (Eds)

On planning accelerated life tests for model selection

T. Wakimoto
University of Electro-Communications, Tokyo, Japan

W. Yamamoto
Keio University, Kanagawa, Japan

L. Jin
University of Electro-Communications, Tokyo, Japan

ABSTRACT: Accelerated life testing aims to predict the lifetime of items under normal conditions using an accelerated lifetime model based on chemical theory. The application of accelerated life testing for reliability evaluation requires that the model is known. However, it is not always possible to select a model with confidence in advance, such as in the development of new materials. When information about the relationship between stress and lifetime distribution is ambiguous, it becomes necessary to identify the model family. Wakimoto *et al.* (2023) proposed using the EIC as a criterion for identification. In their study, they focused on examining the identification performance of the model family without considering experimental costs. Therefore, this study considers the design of experiments for model family identification to evaluate the identification performance of EIC under more realistic settings. Furthermore, we compare the identification performance of EIC under designs with equal allocations and experimental plans.

1 INTRODUCTION

Lifetime testing provides both theoretical and practical frameworks for quantifying the lifetime of product systems. In modern times, highly reliable product systems are designed to operate for several years to decades. Conducting lifetime tests under normal usage conditions for these products is extremely costly and time-consuming. Accelerated life testing addresses this issue by intentionally accelerating the degradation of products to shorten the test time. Accelerated life testing involves conducting lifetime tests under more severe conditions than usual to accelerate degradation. Factors that affect the estimation of lifetime are called stresses, including temperature, humidity, and voltage (Escobar and Meeker 2006; Kececioglu 2002). Accelerated life testing uses the relationship where a change in stress S leads to a scale transformation of the lifetime $t = t_0/k(S)$ to predict the lifetime of items under normal conditions. The model of the acceleration factor $k(S)$ is called an accelerated lifetime model. Accelerated lifetime models can be expressed as log-linear models by taking the logarithm and representing them as linear combinations. Several accelerated lifetime models are based on chemical theory. Models that consider a single stress include the Arrhenius model and the Eyring model, while models that consider several stresses include the generalized Eyring model (Bagdonaviciius and Mikulin 2001; Escobar and Meeker 2006; Kececioglu 2002; Nelson 2009). Thus, there are several model families of accelerated lifetime models.

Accelerated life testing estimates the parameters of the accelerated lifetime model and predicts the lifetime distribution under normal conditions using the estimated parameters. For prediction, the application of accelerated life testing for reliability evaluation assumes that the

DOI: 10.1201/9781003491309-27

accelerated lifetime model is known. However, it is not always possible to assume a model with confidence in advance, such as in the development of new materials. When information about the relationship between stress and lifetime distribution is ambiguous, it becomes necessary to identify the model family in addition to estimating the model parameters.

For data from lifetime tests with censoring, it is necessary to use information criteria rather than graphical methods for selecting distributions and models (Bagdonaviciius and Levuliene 2019; Park and Padgett 2005; Vonta and Alex 2007). Wakimoto *et al.* (2023) proposed using the Extended Information Criterion (EIC) for identifying different model families, showing that EIC performs better than AIC and BIC in numerical examples. These examples used Kullback-Leibler divergence (K-L divergence) to measure the distance between distributions (Wakimoto *et al.* 2023). However, these numerical examples required different experimental settings for each item, resulting in high costs, which is not practical.

Accelerated life testing involves significant resource expenditure for stress application. Additionally, a single environmental chamber can only experiment at one stress level at a time. Under such constraints, design of experiments (DoE) using specific optimality criteria are necessary to efficiently obtain lifetime data. Escobar and Meeker (1995) proposed DoE for multi-stress accelerated life testing (MALT) under constant stress. Park and Yum (1996) discussed optimal settings for accelerated life testing with two stresses and modeled the interaction between the two stresses using the generalized Eyring model. This study considers DoE for the problem of model family identification.

This paper is structured as follows. Section 2 describes the identification of model families and DoE for model selection. Section 3 explains the EIC used as the identification criterion and the setting for model family identification. Section 4 evaluates the DoE for model selection and their performance under the identified plans. Section 5 discusses the conclusions, considerations, and future challenges of this study.

2 IDENTIFICATION OF MODEL FAMILIES

Suppose existing theory is represented by a model M_0. For example, $X \sim Weibull(m, \eta)$, where the shape parameter η varies depending on several covariates $x_1, \ldots, x_p$. This model is denoted as M_0 when it is represented as a log-linear model under appropriate variable transformations for each covariate:

$$\log \eta = \beta_0 + \beta_1 \tilde{x}_1 + \cdots + \beta_p \tilde{x}_p \quad (1)$$

Consider a model family $\mathscr{M}$. The initial model is denoted as $\mathscr{M}_0 = M_0$. The family of alternative models is denoted as $\mathscr{M}_k = \{M_{k0}, M_{k1}, \ldots, M_{kl_k}\} \backslash \mathscr{M}_0, k = 1, \ldots, m$. In this paper, M_{kj} represents the entire set of functions expressed as linear combinations of j k-th degree spline functions $(x - x_j)_+^k$, and $\mathscr{M}_k$ includes all functions in M_{kj} for $0 \leq j \leq l_k$ excluding those in M_0.

We are particularly interested in the cases of $k = 1$ and $k = 3$. As l_1 and l_3 approach infinity, $\mathscr{M}_1$ and $\mathscr{M}_3$ both provide good approximations of nonlinear functions. However, physical laws sometimes exhibit different slopes for relatively few partitions of the covariate space, and what was previously considered linear may be revealed as nonlinear. To address such issues, we assume l_1 and l_3 are small. In this situation, $\mathscr{M}_1$ and $\mathscr{M}_3$ provide distinct model families.

In this study, we refer to distinguishing such problems as model family identification, as opposed to model selection. Model selection involves choosing explanatory variables to minimize AIC or BIC for a given model. In contrast, model family identification involves choosing a model family to minimize EIC, not the selection of explanatory variables.

Wakimoto *et al.* (2023) proposed using EIC for model family identification in the context of accelerated lifetime models. Numerical examples showed that EIC performed better than AIC and BIC in distinguishing between model families $\mathscr{M}_0$ and $\mathscr{M}_1$, as well as $\mathscr{M}_0$ and $\mathscr{M}_3$. It was also shown that $\mathscr{M}_3$ was more likely to be selected than $\mathscr{M}_1$ among the three model families $\mathscr{M}_0$, $\mathscr{M}_1$, and $\mathscr{M}_3$.

For $X \sim Weibull(m, \eta)$, where the shape parameter η varies depending on covariates x_1, x_2, the true model assuming the traditional linear relationship is

$$\log \eta_L(x_1, x_2) = \beta_0 + \beta_1 \tilde{x}_1 + \beta_2 \tilde{x}_2 \tag{2}$$

expressed as a log-linear model belonging to $\mathcal{M}_0$. A true model with few thresholds for covariates and different slopes for each partition belongs to $\mathcal{M}_1$, expressed as a piecewise linear model:

$$\log \eta_S(x_1, x_2) = \begin{cases} \beta_{10} + \beta_{11}x_1 + \beta_{12}x_2 & x_1 < \alpha, x_2 < \beta \\ \beta_{20} + \beta_{21}x_1 + \beta_{22}x_2 & x_1 \geq \alpha, x_2 \geq \beta \end{cases} \tag{3}$$

where α and β are the thresholds for each covariate. A true model with a non-linear relationship that was previously considered linear belongs to $\mathcal{M}_3$, expressed as an additive model allowing nonlinearity for each covariate effect:

$$\log_A \eta(x_1, x_2) = g(x_1) + g(x_2) \tag{4}$$

For each assumed true model, estimation is performed using seven models belonging to $\mathcal{M}_0$: M_{00}, M_{10}; $\mathcal{M}_1$: M_{11}, M_{12}; and $\mathcal{M}_3$: M_{30}, M_{31}, M_{32}.

Model family identification uses the Bootstrap Information Criterion (EIC). The Bootstrap method allows EIC to be applied to various models without complex analytical procedures under very relaxed assumptions (Efron 1993; Konishi and Kitagawa 2008). EIC is expressed as

$$\text{EIC} = -2 \log L(\hat{\theta}, \hat{\beta}) + b_G(\theta) \tag{5}$$

where $b_G(\theta)$ is estimated using bootstrap repetitions as

$$b_G(\theta) = E_{\mathbf{X}^*}\left\{\log L(\mathbf{X}^* \middle| \hat{\theta}^*, \hat{\beta}^*) - \log L(\mathbf{X} \middle| \hat{\theta}^*, \hat{\beta}^*)\right\} \tag{6}$$

referred to as the bias penalty (Ishiguro *et al.* 1997). The model family with the minimum EIC is selected. Therefore, even if the model family identification result matches the true model family, the true model and the estimated model may not match. All models used in this estimation can be represented as Generalized Additive Models for Location, Scale, and Shape (GAMLSS) (Simon 2017). The problem of identifying three model families with few basis functions l reduces to the problem of selecting basis functions in GAMLSS.

Wakimoto *et al.* (2023) conducted simulations assuming different settings for each item to identify model families without considering experimental design effects. This setting incurs high costs and is not practical, necessitating realistic DoE for model family identification. The next section considers such realistic settings.

3 EXPERIMENTAL DESIGN FOR MODEL FAMILY IDENTIFICATION

Consider the experimental design that maximizes the estimation accuracy of EIC. In experimental design, the target of optimization is the estimation accuracy of unknown parameters of the accelerated lifetime model and the prediction accuracy of reliability indicators. Since EIC uses the log-likelihood of the model for estimation, higher estimation accuracy of model parameters results in higher estimation accuracy of EIC. An experimental design for model family identification maximizes the estimation accuracy of model parameters. One criterion for optimizing parameter estimation accuracy is the D-optimality criterion. D-optimality minimizes the determinant of the asymptotic variance-covariance matrix of the parameters.

Reducing the number of experimental points through experimental design leads to cost savings and time reduction. In practice, test time is limited, and budgets are fixed, so DoE must meet these constraints. Reducing experimental points reduces the number of test runs and condition settings, leading to shorter test times and cost savings. However, reducing experimental points makes the model representation monotonic. Accelerated life testing assumes the accelerated lifetime model is known, and DoE must be optimized with the minimum number of experimental points required to estimate the model. Data obtained under DoE estimate all candidate model families for identification. Estimating the most flexible model family $\mathcal{M}_3$ requires at least five points for each of the absolute temperature and relative humidity. Therefore, this study takes 25 experimental points in a grid by taking five levels at equal intervals within each range for absolute temperature and relative humidity.

Consider an experimental design that minimizes the D-optimality criterion by allocating items to 25 experimental points. This design is an integer programming problem with the D-optimality criterion as the objective function. However, this integer programming problem is complex and difficult to solve analytically. Heuristic methods like genetic algorithms (GA) are used to search for suboptimal solutions.

4 EXPERIMENTAL DESIGN FOR MODEL SELECTION

Using a genetic algorithm, we consider an experimental design that minimizes the D-optimality criterion by allocating items to 25 experimental points. For each true model family $\mathcal{M}_0$, $\mathcal{M}_1$, and $\mathcal{M}_3$, we compare the estimated D-optimality criterion for designs with equal allocations and DoE, and report the results. Next, we examine the identification performance of EIC under D-optimal designs through simulations. For each true model family, we identify the model family using EIC under designs with equal allocations and D-optimal designs and report the results.

4.1 *Experimental design*

The experimental settings are as follows. The ranges of absolute temperature T and relative humidity RH are $T = 80°\text{C} \sim 135°\text{C}$ and $RH = 80\% \sim 100\%$. The number of items is $n = 100$, and censoring is not considered. The number of bootstrap repetitions for EIC bias estimation is $B = 200$. Twenty-five experimental points are arranged in a grid (Figure 1).

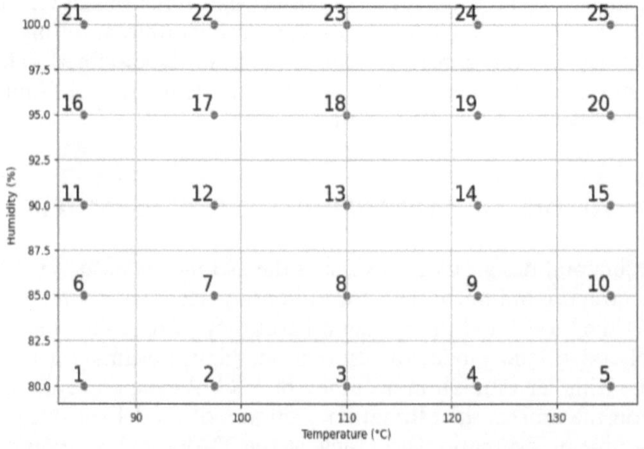

Figure 1. Experimental points arranged in the temperature and humidity space.

Using a genetic algorithm, we search for an experimental design that minimizes the D-optimality criterion by allocating 100 items to 25 experimental points. The design with the minimum D-optimality criterion is selected as the suboptimal design after 50 iterations of estimation with the genetic algorithm. The parameters used in the genetic algorithm are as follows:

- Crossover rate: 0.9
- Mutation rate: 0.1
- Population size: 50
- Number of iterations: 100

The resulting experimental design is shown in Figure 2. The Temperature axis represents temperature, the Humidity axis represents humidity, and the Count axis represents the number of items allocated to each experimental point.

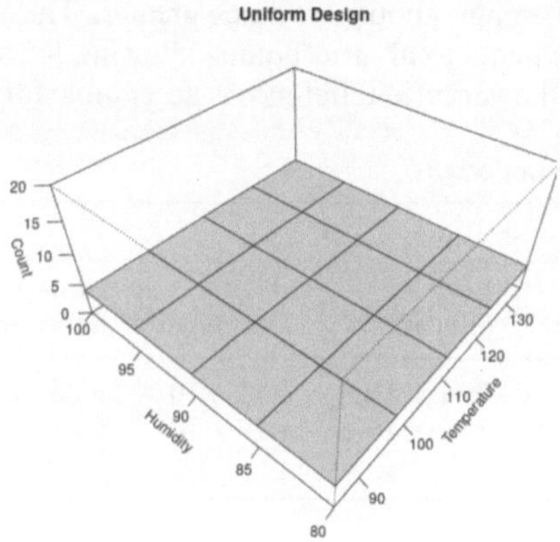

(a) Design with equal allocations

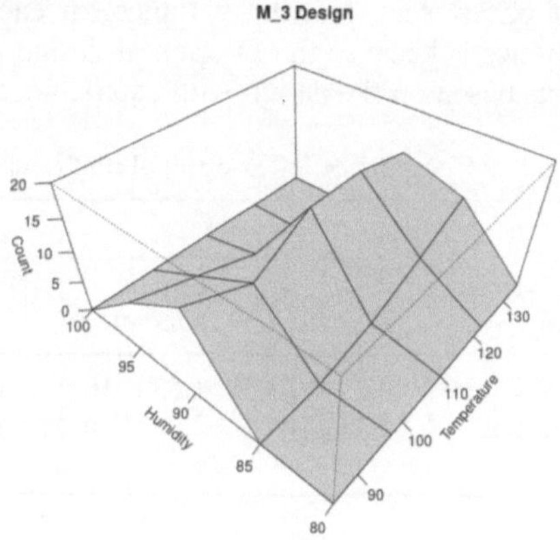

(b) $\mathcal{M}_0$ is the true

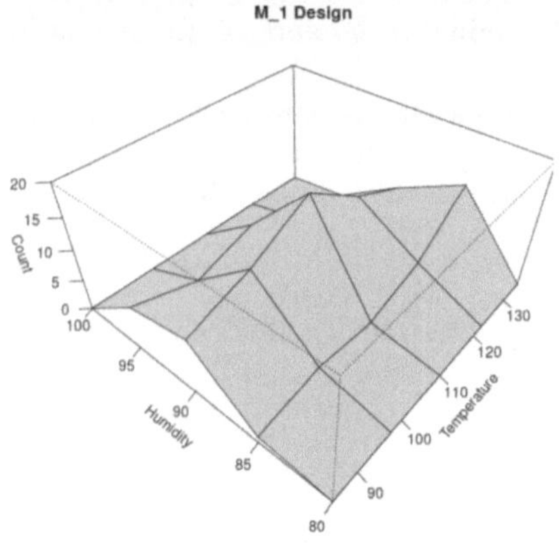

(c) $\mathcal{M}_3$ is the true

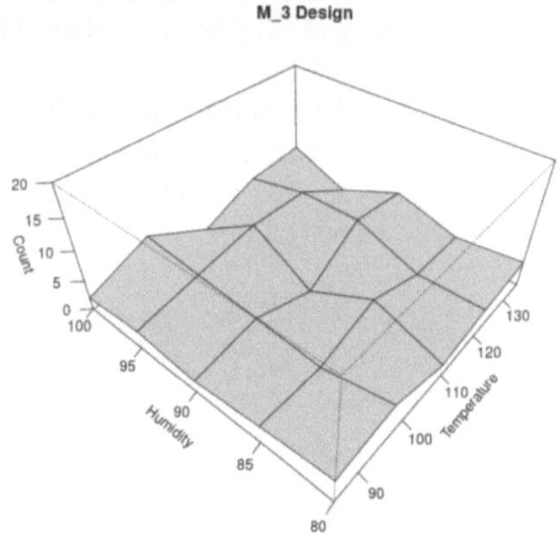

(d) $\mathcal{M}_1$ is the true

Figure 2. Design of experiments.

Figures 2b-2c show the sample points for each true model. For $\mathcal{M}_0$, samples are widely taken across the temperature range for 90% humidity. For $\mathcal{M}_1$, samples are scattered across both temperature and humidity, with many points concentrated at the low temperature and low humidity edges. For $\mathcal{M}_3$, samples are scattered across the temperature range for 90% humidity, with most points concentrated around $(T, RH) = (110, 90)$, close to the threshold $(T, RH) = (105, 90)$.

We compare the D-optimal designs and designs with equal allocations and examine the performance of the samples. The design with equal allocations follows the Poisson process-based design by Wakimoto *et al.* (2023) but coarser (Wakimoto *et al.* 2023). The mean and minimum values of the D-optimality criterion for each true model family are shown in Table 1. For $\mathcal{M}_0$, the difference between the D-optimal design and the design with equal allocations is a factor of 10^5. Since the true model has four parameters, the variance for each parameter is about 18 times greater. For $\mathcal{M}_1$, the mean value of the uniform distribution is 3.26×10^3, a factor of 10^5 difference compared to the D-optimal design. This difference is because the models estimated with the design with equal allocations did not converge before the estimation was cut off. Therefore, a design that places items near the threshold is necessary for $\mathcal{M}_1$. For $\mathcal{M}_3$, the difference is a factor of 10^3, the smallest among the three model families. The true model has eight parameters, and the variance for each parameter is about 2.5 times greater. The small difference is because the D-optimal design allocates items to all grid points, allowing for simulations based on the design with equal allocations if the variance difference is acceptable for $\mathcal{M}_3$.

Table 1. Comparison of D-optimal and equal allocation designs.

	$\mathcal{M}_0$		$\mathcal{M}_3$		$\mathcal{M}_1$	
True Model Family	D-optimal	Equal allocations	D-optimal	Equal allocations	D-optimal	Equal allocations
Mean Value	9.25×10^{-15}	1.63×10^{-10}	1.30×10^{-12}	7.41×10^{-9}	6.95×10^{-15}	3.26×10^3
Minimum Value	1.31×10^{-19}	2.92×10^{-11}	3.84×10^{-13}	7.43×10^{-12}	6.12×10^{-17}	0.41

4.2 *Identification performance of EIC*

We perform simulations using the D-optimal designs and designs with equal allocations obtained in Section 4.1 for each true model family. The assumed true models are known in terms of function and number of knots. Models belonging to $\mathcal{M}_1$ and $\mathcal{M}_3$ for estimation use models up to $j \leq 2$.

Figures 3 to 5 show the identification results of model families using D-optimal designs when each model family is the true model. The most significant change in the identification results was when $\mathcal{M}_1$ was the true model. When $\mathcal{M}_1$ was the true model, the selection ratio of $\mathcal{M}_3$ using EIC exceeded 60.

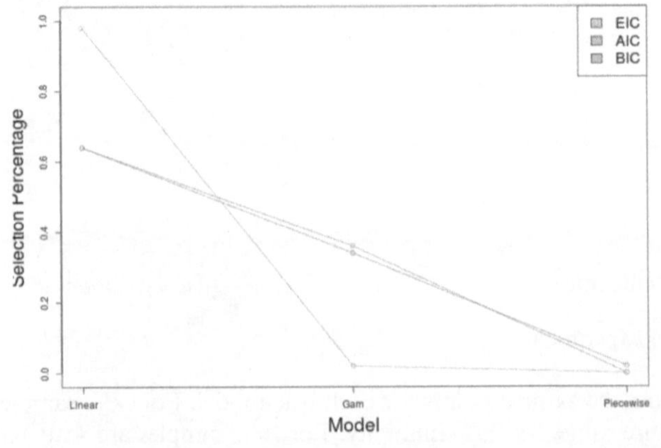

Figure 3. Model family selection ratios when $\mathcal{M}_0$ is the true model using the experimental design.

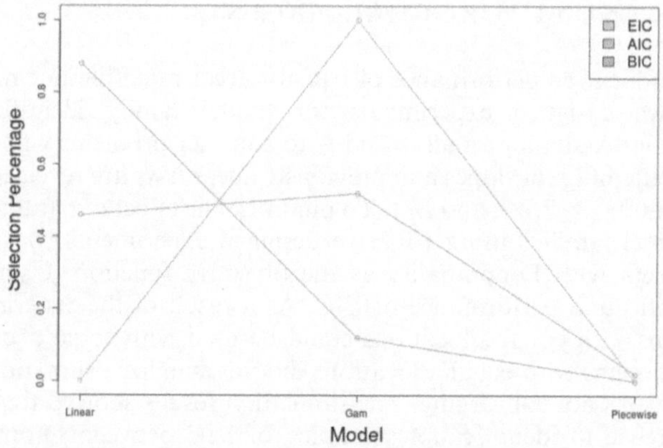

Figure 4. Model family selection ratios when $\mathscr{M}_3$ is the true model using the experimental design.

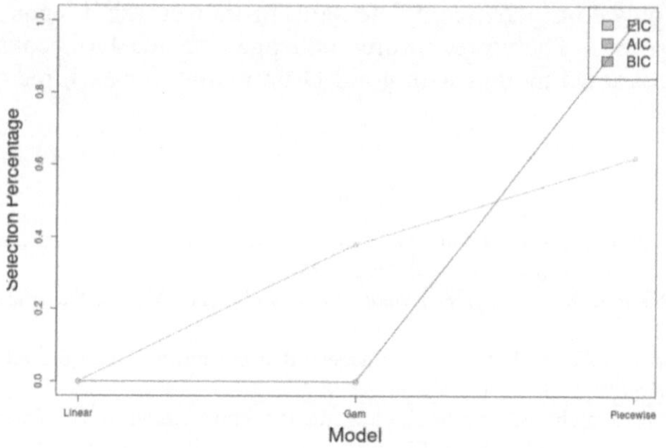

Figure 5. Model family selection ratios when $\mathscr{M}_1$ is the true model using the experimental design.

4.3 *Summary*

We compared D-optimal designs and designs with equal allocations for each model family. As a result, for $\mathscr{M}_0$ and $\mathscr{M}_3$, parameter estimation accuracy slightly decreased, but designs with equal allocations could be expected to provide some estimation accuracy. However, for $\mathscr{M}_1$, designs with equal allocations could not be used. In designs with equal allocations, models containing knots such as M_{11} did not converge. This is because a design that places many items near the knots is necessary for estimating the knots. Therefore, if a true model family containing knots is assumed, a design that stabilizes model estimation is necessary.

This study did not consider censoring, so DoE considering the censoring rate will be the subject of future simulation experiments. Currently, D-optimal designs were conducted for each true model family. It is necessary to examine the change in D-optimality criterion due to changes in sample size. Also, this study assumes the true model family is known and designed for each model family. Therefore, when the model family is unknown, the problem of experimental design for model family identification with good D-optimality for all model families will be a future challenge.

5 CONCLUSION AND FUTURE CHALLENGES

To evaluate the identification performance of EIC for accelerated lifetime models under more realistic settings, we designed experiments for model family identification using the D-optimality criterion. Assuming small l_1 and l_3 to consider problems where covariate space has few partitions with different slopes and previously linear laws are revealed as nonlinear, we identified model families $\mathscr{M}_0$, $\mathscr{M}_1$, and $\mathscr{M}_3$. To minimize the effect of parameter estimation on the selection of model families using EIC, we designed experiments by solving the integer programming problem with D-optimality as the objective function. Using these DoE, we evaluated the identification performance of EIC. As a result of the experimental design, for $\mathscr{M}_0$ and $\mathscr{M}_3$, designs with equal allocations could be used with some decrease in accuracy. However, for $\mathscr{M}_1$, designs with equal allocations did not stabilize estimation accuracy due to models containing knots not converging. The simulation results showed that for $\mathscr{M}_0$ and $\mathscr{M}_3$, there was no difference in identification accuracy of EIC between D-optimal designs and designs with equal allocations. For $\mathscr{M}_1$, D-optimal designs showed better identification accuracy than designs with equal allocations. This study assumes that each model family is designed for. Therefore, when the true model family cannot be assumed, it is better to identify model families using designs with equal allocations, though accuracy may slightly decrease.

In our DoE, we only considered complete data. Moreover, the designs assumed the three true models were known. Therefore, future challenges include DoE considering censoring and designing for unknown models with good D-optimality for all three model families.

REFERENCES

Akaike, H. (1973). Information theory and an extension of the maximum likelihood principle, In *Proceeding of 2nd International Symposium on Information Theory*:267–281.

Bagdonaviciius, V. & Nikulin, M. 2001. *Accelerated Life Models Modeling and Statistical Analysis*, Chapman and Hall/CRC.

Bagdonaviciius, V. and Levuliene, R. (2019). On accelerated life testing when the AFT model fails, *IEEE Transactions on Reliability* 68:1311–1319.

Efron, B. (1993). Bootstrap methods: Another look at the jackknife, *Annals of Statistics* 7: 1–26.

Escobar, L. A. and Meeker, W. Q. (1995). Planning accelerated life tests with two or more experimental factors, *Technometrics* 37: 411–427.

Escobar, L. A. and Meeker, W. Q. (2006). A review of accelerated test models, *Statistical Science* 21: 552–577.

Ishiguro, M., Sakamoto, Y. and Kitagawa, G. (1997). Bootstrapping log likelihood and EIC, an extension of AIC, *Annals of the Institute of Statistical Mathematics*. 49: 411–434.

Kececioglu, D. B. (2002). *Reliability and Life Testing Handbook*. Lancaster: DEStech Publications.

Konishi, S. and Kitagawa, G. (2008). *Information riteria and Statistical Modeling*, Springer Science and Business Media.

Kullback, S. and Leibler, R. A. (1951). On information and sufficiency, *Annals of Statistics* 22: 79–86.

Nelson, W. (2009) *Accelerated Testing Statistical Models, Test Plans, and Data Analysis*, John Wiley and Sons.

Park, C. and Padgett, W. J. (2005). Accelerated deterioration models for failure based on geometric Brownian motion and gamma processes, *Lifetime Data Analysis* 11: 511–527.

Park, J. W. and Yum, B. J. (1996). Optimal design of accelerated life tests with two stresses, *Naval Research Logistics* 43: 863–884.

Peck, D. S. and Zierdt, H. C. (1974). The reliability of semiconductor devices in bell system, *Proceedings of the IEEE* 2: 185–211.

Rigby, R. A. and Stasinopoulos, D. M. (2005). Generalized additive models for location, scale and shape, *Journal of the Royal Statistical Society: Series C* 54: 507–554.

Vonta, F. and Karagrigoriou, A. (2007). Variable selection strategies in survival models with multiple imputations, *Lifetime Data Analysis* 13: 295–315.

Wakimoto, T., Yamamoto, W. and Jin, L. (2023). Model selection for accelerated life test model using extended information criterion. *In 2023 5th International Conference on System Reliability and Safety Engineering*: 304–309

Wood, S. N. (2017). *Generalized Additive Models: An Introduction with R*, Second Edition, Chapman and Hall/CRC.

Recent Advances in Reliability and Maintenance Modeling – Okamura, Inoue & Xiao (Eds)

A dynamic predictive maintenance framework for multiple components based on remaining useful life prediction

Lubing Wang, Zhengbo Zhu & Xufeng Zhao
College of Economics and Management, Nanjing University of Aeronautics and Astronautics, Nanjing, China

ABSTRACT: The development of advanced sensor technologies has led to an increasing interest in component predictive maintenance. However, designing a predictive maintenance approach for multiple components is still a challenging problem as it requires effective integration of prediction and scheduling under resource constraints. This paper presents a dynamic predictive maintenance framework for multiple components that integrates Remaining Useful Life (RUL) prediction into maintenance plan. First, a hybrid deep learning model is designed for predicting the RULs of multiple components, which combines a one-dimensional convolutional neural network and a bidirectional long short-term memory network with temporal self-attention mechanism. Meanwhile, a Bayesian optimization method is proposed to further improve the RUL prediction performance of hybrid deep learning model. Then, a novel Mixed Integer Linear Programming (MILP) models is developed to handle multiple components predictive maintenance under constrained maintenance resources, which targets to minimize the maximum maintenance time for all components. The experimental results show that the proposed dynamic predictive maintenance framework can flexibly provide accurate RUL predictions and maintenance planning to reduce maintenance time.

Keywords: Predictive maintenance, remaining useful life, deep learning, mixed-integer linear programming

1 INTRODUCTION

In recent years, remaining useful life (RUL) prediction and predictive maintenance have played a crucial role in the prognostics and health management (PHM) of complex industrial systems. RUL refers to the time interval between the current time and the system failure [1]. Predictive maintenance is a proactive maintenance plan that can effectively predict possible hazards or failures in system operation. It is based on techniques such as condition monitoring, fault diagnosis, and condition prediction to determine the current operating conditions and future trends of the component and to develop an appropriate maintenance plan [2,3]. An effective component RUL prediction can alert decision-makers to take appropriate predictive maintenance decisions before a system failure occurs, reducing unplanned system downtime. Meanwhile, it can grasp the operating state of the component in time, reduce the maintenance cost of the system, and provide a reference for determining the corresponding maintenance decisions. Developed sensors and Internet of Things technologies create prerequisites for solving this problem by using data collected from multiple sensors as component degradation characteristics or health indicators. However, component RUL prediction is still challenging due to the complexity of operating conditions and environments faced by the component. Thus, it is essential to develop

DOI: 10.1201/9781003491309-28

an effective component RUL prediction method to support predictive maintenance decisions.

Currently, data-driven maintenance methods have been widely used in research for their great potential to improve operational efficiency and reduce maintenance costs compared to traditional model-based predictive maintenance methods. For example, Nguyen and Medjaher [4] proposed a dynamic predictive maintenance model based on prognostics information, which can effectively estimate the probability of component failure in different periods, and then make maintenance decisions by quickly evaluating the required maintenance costs in each period. Its purpose is to achieve effective maintenance of the component and order spare parts in time. Later, Chen *et al.* [5] developed a deep auto-encoder and a bidirectional long short-term memory network (Bi-LSTM) model to efficiently predict the health state and RUL of the component and make predictive maintenance decisions based on the prognostic results. Lu *et al.* [6] proposed a data-driven dynamic predictive maintenance strategy that utilizes online degradation data to predict the degradation of components, and continuously modifies the maintenance plan based on the prediction results, which aims to minimize the maintenance cost. Recently, Wang *et al.* [7] investigated a dynamic predictive maintenance model based on the component RUL prediction. This dynamic predictive maintenance model includes order, inventory, and maintenance decisions and considers the effect of system operating conditions on the predictive maintenance model.

Although the above study has taken the RUL prediction results to the maintenance planning, thus reducing the maintenance cost. However, there are still some unanswered questions. For example, the above studies have not considered the maintenance issues under maintenance resource constraints. Meanwhile, for critical components such as wind turbines, train engines, and aircraft engines, how to arrange for maintenance personnel to carry out maintenance is an issue that needs to be considered. This paper proposes a new multi-objective dynamic predictive maintenance framework for multiple components to address these issues. First, we propose a hybrid deep learning model to predict component RUL, combining a one-dimensional convolutional neural network (CNN), a Bi-LSTM with temporal self-attention mechanism, and a Bayesian optimization method. Unlike traditional methods that have to understand the degradation mechanism of components, this hybrid deep learning model can effectively predict components RUL through the sensor-collected time series data. Based on the multiple component RUL prediction results, we propose a novel mixed integer linear programming (MILP) model to handle multiple components predictive maintenance under constrained maintenance resources, which targets to minimize the maximum maintenance time for all components. The main novelties are as follows:

(i) A dynamic predictive maintenance framework integrates RUL prediction and maintenance plan is proposed, which provides a framework for optimizing multiple components predictive maintenance under resource constraints.
(ii) A powerful hybrid deep learning model is developed by combining Bayesian optimization method to achieve effective RUL prediction.
(iii) We consider the predictive maintenance problem for multiple components under resource constraints and develop a novel MILP model which aims to provide optimal maintenance plan and thus reduce equipment downtime.

The remainder of this paper proceeds as follows. Section 2 describes a hybrid deep learning model in detail. The predictive maintenance model for multiple components is described and established in Section 3. Section 4 analyzes the turbofan engine dataset from NASA and the experimental results illustrate the effectiveness of the proposed dynamic predictive maintenance framework. Finally, Section 5 discusses the conclusions.

2 HYBRID DEEP LEARNING MODEL

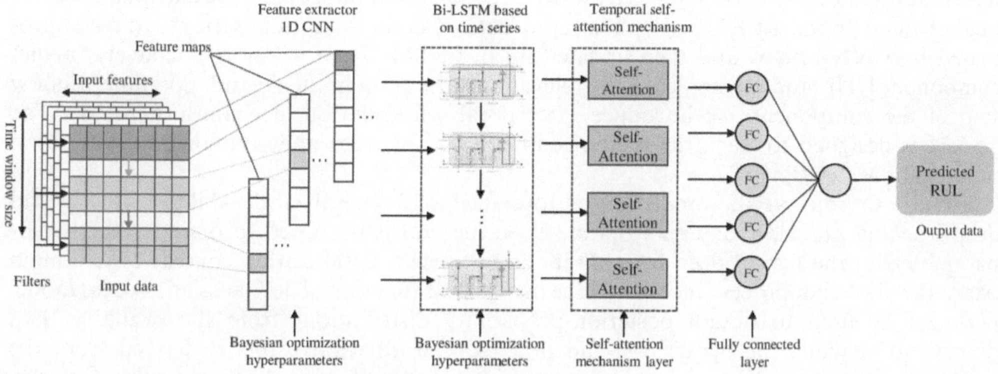

Figure 1. Structure of the proposed hybrid deep learning model.

In this section, a hybrid deep learning model is designed by combining CNN and Bi-LSTM models, TSAM, and Bayesian optimization hyperparameters, as illustrated in Figure 1. In Figure 1, CNN can analyze these sensor data and deeply extract feature information in the data to identify patterns related to the health status of the system. Bi-LSTM model can remember and connect previous information to the currently acquired data, capturing remote dependencies through several gates, thus improving the performance of the hybrid deep learning model. The TSAM further extracts the important information between the time window steps and features. Finally, Bayesian optimization hyperparameters are used to further improve the prediction performance of this hybrid deep learning model. In other words, the proposed hybrid deep learning model can fully utilize the capabilities of these techniques as mentioned above to enhance its predictive performance. The hybrid deep learning model consists of the following four main components.

First, we consider multi-dimensional data samples from multiple sensor detections X as input to CNN model:

$$X = \{x_1, x_2, \cdots, x_i\}, \tag{1}$$

where X is the number of samples and x_i is a data set for component operation status monitoring. We use the following formula to normalize x_i.

$$x_i' = (x_i - x_{\min})/(x_{\max} - x_{\min}), \tag{2}$$

where x_i' is the normalized data, and $x_{\min}$ and $x_{\max}$ are the minimum and maximum values in the data, respectively. The purpose of normalization is to compress the data into a range of 0 to 1, while reducing the impact of the data dimension on the prediction results. The normalized data is fed into the CNN model, and then the CNN model trains the data through convolutional layers, where each convolutional layer contains multiple convolutional kernels. CNN can effectively extract representative features hidden in monitoring data, but it is hard to learn information related to time series features. The monitoring data of component degradation usually contains a set of time series data and is inevitably related to the state of component degradation. To predict the degradation state of the component more effectively, we further introduce Bi-LSTM based on CNN model.

Bi-LSTM architecture is mainly composed of input gate, forget gate, and output gate. These gates regulate the flow of information within the network and decide which parts of the input should be remembered, forgotten, or output. Thus, Bi-LSTM can effectively capture long-term dependencies in input sequences and make more accurate predictions using time series data. Compared with a single LSTM, Bi-LSTM can learn time-related information in both forward and backward directions simultaneously, thus helping to better predict the component RUL. For the input data x_i' containing time series features is processed by CNN into x_t and then updated by Bi-LSTM. Then, to more effectively predict component RUL and further learn the relevant information in the degradation time window step of the component, we introduce a temporal self-attention mechanism (TSAM). The TSAM is designed to learn the relationship between features and component degradation time window steps.

Finally, to improve the prediction performance and generalization ability of the hybrid deep learning model, a Bayesian optimization method is proposed to optimize the hyperparameters in the hybrid deep learning model. Bayesian optimization consists of two main parts: the Gaussian process model and the acquisition function. The Gaussian process model $f(\mathbf{v})$ can be used to obtain posterior probability distributions from the available data observations, while the acquisition function uses an alternative model derived from the Gaussian process model to select the next point to be evaluated. The acquisition function consists of three main types: expected improvement (EI), probability of improvement, and the upper confidence bound. In this study, EI is chosen as the acquisition function. The optimal solution for the EI criterion is $\mathbf{v}^* = \arg\max_{\mathbf{v}\in\mathbf{V}} E[I(\mathbf{v})]$, where $\mathbf{v}$ is the hyperparameters to be optimized and $\mathbf{V}$ represents the hyperparameters search space.

3 PROBLEM DESCRIPTION AND FORMULATION

3.1 *Problem description*

In the data-driven predictive maintenance framework, component sensor monitoring points make the basic decision of whether to maintain or not based on the RUL predicted by the hybrid deep learning model. Maintenance operations are considered to be perfect operations by replacing components, so when is the best time to replace them? Meanwhile, components cannot be replaced when the system is on a mission. Therefore, we assume that predictive maintenance is performed only when the system is idle.

If the component is still operating during the running cycle, a maintenance decision is considered to be executed or not based on predicted RUL from the available monitoring data. Since maintenance cannot be performed during mission execution, it is necessary to know if the current RUL is lower than the mission cycle. If the answer is "yes" this implies that predictive maintenance must be performed because the component may fail while performing its mission. Thus, whether component i performs predictive maintenance is determined by the following equation:

$$Maintenance = \begin{cases} 1, & \widehat{RUL}_i \leq \Delta t, \\ 0, & others, \end{cases} \tag{3}$$

where $Maintenance = 1$ denotes that predictive maintenance is required, $Maintenance = 0$ denotes that predictive maintenance is not required, $\widehat{RUL}_i$ denotes the predicted RUL of component i, and Δt is a maintenance alarm threshold. If the component RUL is less than the maintenance alarm threshold, then the component is to be maintained, otherwise, a failure may occur. Given a maintenance alarm threshold Δt, when the number of n components RUL is less than the threshold Δt, then these components need to be maintained simultaneously. We consider these m maintenance teams performing predictive maintenance on n components. For each component i, it has a maintenance time t_i^m and can be maintained by only one maintenance team m once. In addition, maintenance team resources are limited due to maintenance costs. Each maintenance team may be required to maintain multiple components in a sequential relationship. Therefore, it is critical to organize maintenance teams to perform predictive maintenance and determine the optimal maintenance schedule based on the results of component RUL prediction. For this purpose, we develop a new MILP model to solve the predictive maintenance problem for multiple components. Before formulating the optimization model, we present a description of the relevant notations, as shown in Table 1 below.

Table 1. Description of parameters and variables.

Notation	Explanation
i,j	component index
m	maintenance team index
N	set of components, $N = \{1,2,\cdots,n\}$
V	set of components and maintenance depot, $V = \{0\} \cup N$, where 0 indicates maintenance depot
M	set of maintenance teams, $M = \{1,2,\cdots,m\}$
$\|M\|$	number of maintenance teams
r_{ij}^m	the time required for the maintenance team m to reach j from component i
t_i^m	maintenance time required for each maintenance team m to maintain component i
Q_m	maximum number of components that can be maintained by maintenance team m
L_i	the maintenance priority level of component i, which is used to depict the degradation severity of the component
B	a large positive number
Decision variables	
x_{ij}^m	$x_{ij}^m = 1$, if maintenance team m moves component i to j; otherwise, $x_{ij}^m = 0$
TC_i	the maintenance completion time of component i
$TC_{\max}$	the total maintenance completion time for all components

3.2 *Model formulation*

In order to efficiently perform predictive maintenance on multiple components, we develop a MILP model.

$$\min \ TC_{\max}, \tag{4}$$

$$\sum_{m\in M}\sum_{i\in V} x_{ij}^m = 1, \forall j \in N, i \neq j, \tag{5}$$

$$\sum_{j\in N} x_{0j}^m = 1, \forall m \in M, \tag{6}$$

$$\sum_{i\in N} x_{i0}^m = 1, \forall m \in M, \tag{7}$$

$$\sum_{i\in V} x_{ij}^m = \sum_{i\in V} x_{ji}^m, \forall m \in M, \forall j \in N, \tag{8}$$

$$\sum_{i\in V}\sum_{j\in V} x_{ij}^m \leq Q_m, \forall m \in M, \tag{9}$$

$$\sum_{m\in M}\sum_{i=1}^{|M|} x_{0i}^m = |M|, \tag{10}$$

$$x_{ij}^m = 0, \text{if } \widehat{RUL}_i \geq \widehat{RUL}_j \ \& \ L_i \leq L_j, \ \forall m \in M, \ \forall i \neq j \in N, \tag{11}$$

$$TC_i \geq \sum_{m \in M} x_{0i}^m (r_{0i}^m + t_i^m), \ \forall i \in N, \tag{12}$$

$$TC_j \geq TC_i + r_{ij}^m + t_j^m + B(x_{ij}^m - 1), \text{ if } TC_i + r_{ij}^m \geq \widehat{RUL}_j, \ \forall m \in M, \ \forall i,j \in N, j \geq i, \tag{13}$$

$$TC_j \geq \sum_{m \in M} \sum_{i \in V} x_{ij}^m (\widehat{RUL}_j + t_j^m), \text{ if } TC_i + r_{ij}^m < \widehat{RUL}_j, \ \forall j \in N, j \geq i, \tag{14}$$

$$TC_{\max} \geq TC_j, \ \forall j \in N, \tag{15}$$

$$\sum_{i \in V} x_{ij}^m \leq 1, \forall m \in M, \ \forall j \in N, \tag{16}$$

$$\sum_{i \in V} x_{ji}^m \leq 1, \forall m \in M, \ \forall j \in N, \tag{17}$$

$$TC_i \geq 0, \forall i \in N, \tag{18}$$

$$x_{ij}^m \in \{0, 1\}, \ \forall m \in M, \ \forall i,j \in V. \tag{19}$$

Objective function (4) is to minimize the total maintenance completion time for all components, thus reducing the downtime of the system. Constraint (5) ensures that each component must be maintained by only one maintenance team, and constraints (6) and (7) indicate that for any maintenance team should depart from maintenance depot 0 and return to it when the predictive maintenance task is completed. Constraint (8) represents a flow conservation constraint, which ensures that once a maintenance team m provides maintenance services for component j, then it must leave component j when the maintenance task is completed. Constraint (9) is a limit on the number of components allowed to be maintained per maintenance team, and its purpose is to ensure a rational allocation of maintenance team resources. Constraint (10) ensures that the first $|M|$ components with higher priority are prioritized to be maintained. Constraint (11) is developed to ensure that the maintenance sequence of each maintenance team respects the priority level of the components. Constraints (10) and (11) guarantee the maintenance priority order for components. Constraint (12) ensures that the maintenance completion time of each component cannot be less than its maintenance time and the time when the maintenance team arrives at the component. Constraint (13) implies that if two components are consecutively maintained by the same maintenance team, when the former component i satisfies $TC_i + r_{ij}^m \geq \widehat{RUL}_j$, the maintenance completion time of the latter component j is greater than the maintenance completion time of the former and the time of arrival at component j as well as its maintenance time. Meanwhile, constraint (13) is linearized using the big-M method. Constraint (14) shows that if two components are maintained consecutively by the same maintenance team, the maintenance time of the latter is not less than the sum of its maintenance time and $\widehat{RUL}_j$ when the former satisfies $TC_i + r_{ij}^m < \widehat{RUL}_j$. Constraints (13) and (14) integrate the component predicted RUL into the component predictive maintenance scheduling, thus enabling customized maintenance for components. Inequalities (16) and (17) limit the number of times a maintenance team can visit a component. If the component is maintained by maintenance team m, then maintenance team m visits it at most once, otherwise a travel cycle is generated. Constraints (18) and (19) are decision variable constraints.

4 EXPERIMENTAL VALIDATION

A turbofan engine component degradation dataset is adopted to show how the proposed dynamic predictive maintenance framework can be applied in engineering. The dataset is generated by the Commercial Modular Aero-Propulsion System Simulation (C-MAPSS) and simulates various degradation scenarios for homogeneous turbofan engine components. In

C-MAPSS dataset, the training dataset contains the entire lifecycle of each turbofan engine from startup to failure, while the test set contains only the data before the turbofan engine failure to predict the RUL of each turbofan engine. The C-MAPSS dataset is composed of data monitored by multiple sensors, and it has a total of 21 features. These sensors are located at different locations in the aircraft engine to monitor its operating conditions. The sensors labeled 1, 5, 6, 10, 16, 16, 18, and 19 do not affect the turbofan engine life degradation in FD001 of C-MAPSS dataset, which means that the data obtained from these 7 sensors have little or no effect on the aircraft engine RUL. Therefore, we have removed the data monitored by these 7 sensors from the training and test dataset in order to predict the aircraft engine RUL more efficiently. Finally, only the data monitored by 14 sensors are used for aircraft engine RUL prediction. For a detailed description of the C-MAPSS dataset see [8].

Based on the proposed hybrid deep learning model, we can predict the RUL of 100 turbofan engines in the FD001 dataset, as shown in Figure 2. From Figure 2, the proposed hybrid deep learning model can effectively predict the RUL of the turbofan engine. In particular, when the RUL of the turbofan engine is small, the prediction effect of the proposed hybrid deep learning model is better. This is because the smaller the RUL of the turbofan engine is, the more obvious its degradation features are, and the proposed prediction model is able to effectively capture these degradation features and improve its prediction performance. Meanwhile, Figure 3 corresponds to the number distribution of the predicted RUL interval for 100 turbofan engines in FD001, whose purpose is to carry out predictive maintenance of turbofan engines. According to Figure 3, it can be seen that when the maintenance alarm threshold $\Delta T = 10$, 9 turbofan engines need to be maintained in the RUL

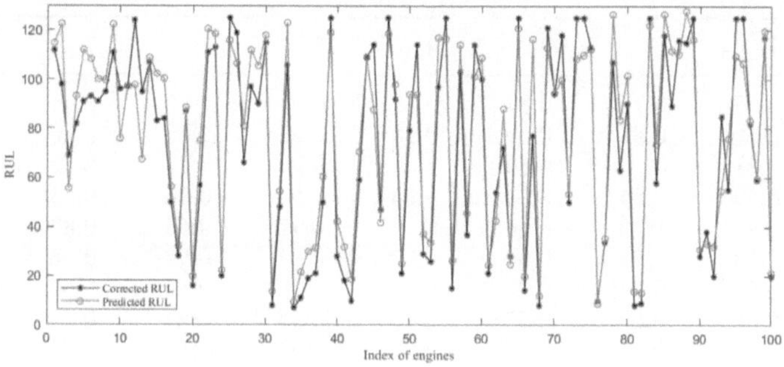

Figure 2. RUL prediction results on FD001.

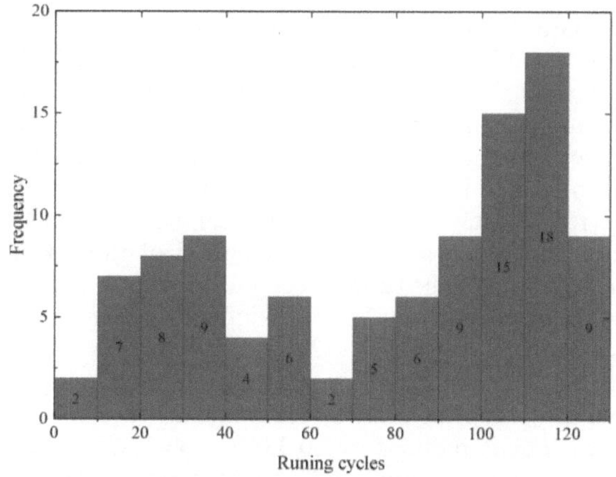

Figure 3. RUL prediction intervals for 100 aircraft engines in the FD001 test set.

intervals 30 to 40. For these turbofan engines, it is necessary to consider how to arrange for maintenance teams to maintain them and minimize their maintenance time, thereby reducing aircraft downtime. First, we give the predicted RULs for 9 turbofan engines in the RUL intervals 30 to 40 as shown in Table 2. Based on Table 2, w first preprocessed the data by subtracting the RUL by a multiple of the maintenance alarm threshold, i.e., $3\Delta T = 10$, which gives RUL as 0.92, 1.55, 1.89, 2.10, 2.23, 3.41, 3.81, 5.44, and 7.14. Furthermore, the required maintenance time for these 9 turbofan engines is randomly generated in [1.00, 2.00], and the time for the maintenance team to reach the turbofan engines is randomly generated in [0.50, 1.00], whose corresponding data are shown in Tables 3 and 4.

Table 2. Turbofan engine RUL prediction results.

Turbofan engines	1	2	3	4	5	6	7	8	9
$\widehat{RUL}$	30.92	31.55	31.89	32.10	32.23	33.41	33.81	35.44	37.14

Table 3. Maintenance time for turbofan engines.

Turbofan engines	1	2	3	4	5	6	7	8	9
Maintenance time	1.57	1.60	1.24	1.71	1.83	1.20	1.51	1.02	1.42

Table 4. Maintenance team arrival time at the turbofan engine.

V	0	1	2	3	4	5	6	7	8	9
0	0	0.63	0.72	0.87	0.66	0.95	0.76	0.94	0.76	0.90
1	0.63	0	0.61	0.59	0.90	0.58	0.50	0.56	0.72	0.76
2	0.72	0.61	0	0.54	1.00	0.74	0.77	0.62	0.94	0.95
3	0.87	0.59	0.54	0	0.64	0.50	0.90	0.53	0.63	0.58
4	0.66	0.90	1.00	0.64	0	0.92	0.50	0.83	0.70	0.91
5	0.95	0.58	0.74	0.50	0.92	0	0.99	0.86	0.57	0.75
6	0.76	0.50	0.77	0.90	0.50	0.99	0	0.64	0.62	0.60
7	0.94	0.56	0.62	0.53	0.83	0.86	0.64	0	0.54	0.94
8	0.76	0.72	0.94	0.63	0.70	0.57	0.62	0.54	0	0.72
9	0.90	0.76	0.95	0.58	0.91	0.75	0.60	0.94	0.72	0

Based on the above data, the commercial solver CPLEX 12.80 is used to solve the proposed MILP model exactly and the results are shown in Table 5. In Table 5, the first column N denotes the number of turbofan engines, i.e., the number of components; the second column M denotes the maintenance teams; and the third column Q denotes the number of turbofan engines that the maintenance team is allowed to maintain, where $Q = [N/M]^{-} + 1$. The fourth to sixth columns denote the objective function, the solution time, and the optimal maintenance planning, respectively. From Table 5, it can be seen that CPLEX can obtain an exact predictive maintenance program in a short period of time. For example, when the number of maintenance teams is 2, the turbofan engine predictive maintenance plans are 0-1-3-5-8-9-0 and 0-2-4-6-7-0. When the number of maintenance teams is 3, the turbofan engine predictive maintenance plans are 0-1-5-6-0, 0-2-4-9-0, and 0-3-7-8-0. Meanwhile, the number of maintenance teams has a large impact on the maintenance completion time of the turbofan engine. For example, when the number of maintenance teams is 3 compared to 2, the maintenance completion time is reduced by 10.38–8.56 = 1.82. In summary, the results in Table 5 and Figure 3 can provide decision makers with accurate RUL prediction results and optimal predictive maintenance schedules in a short period of time.

Table 5. Maintenance scheduling results for 9 aircraft engines.

N	*M*	*Q*	*obj*	CPU (s)	Planning
9	1	5	10.38	0.09	0-1-3-5-8-9-0
	2				0-2-4-6-7-0
9	1	4	8.56	0.06	0-1-5-6-0
	2				0-2-4-9-0
	3				0-3-7-8-0

5 CONCLUSIONS

This study has investigated a dynamic predictive maintenance framework for multiple components that integrates RUL prediction into maintenance plan. First, a hybrid deep learning model is designed for predicting the RULs of multiple components. Based on the predicted RUL of the component, we develop a MILP model that aim to minimize the maintenance completion time for all components. For MILP model, we first consider the effect of the predicted RUL on the component maintenance priority and then incorporate the component predicted RUL into the constructed MILP model. Finally, we illustrate the effectiveness of the dynamic predictive maintenance framework based on publicly available datasets from NASA. Experimental results show that our proposed dynamic predictive maintenance framework can effectively predict component RUL and implement maintenance, thus reducing maintenance time and fully utilizing other system performance.

ACKNOWLEDGMENTS

This work was supported by the Aeronautical Science Foundation of China under Grant 2022Z057052002 and the Postgraduate Research & Practice Innovation Program of Jiangsu Province under Grant KYCX23_0398.

REFERENCES

[1] Zhang J., Jiang Y., Wu S., *et al.* (2022). Prediction of remaining useful life based on bidirectional gated recurrent unit with temporal self-attention mechanism. *Reliability Engineering & System Safety*, 221: 108297.

[2] Zonta T., Da Costa C. A., da Rosa Righi R., *et al.* (2020). Predictive maintenance in the Industry 4.0: A systematic literature review[J]. *Computers & Industrial Engineering*, 150: 106889.

[3] Theissler A., Perez-Velazquez J., Kettelgerdes M., *et al.* (2021). Predictive maintenance enabled by machine learning: Use cases and challenges in the automotive industry[J]. *Reliability Engineering & System Safety*, 215: 107864.

[4] Nguyen K. T. P. and Medjaher K. (2019). A new dynamic predictive maintenance framework using deep learning for failure prognostics[J]. *Reliability Engineering & System Safety*, 188: 251–262.

[5] Chen C., Zhu Z. H., Shi J., *et al.* (2021). Dynamic predictive maintenance scheduling using deep learning ensemble for system health prognostics[J]. *IEEE Sensors Journal*, 21(23): 26878–26891.

[6] Lu B., Chen Z. and Zhao X. (2021). Data-driven dynamic predictive maintenance for a manufacturing system with quality deterioration and online sensors[J]. *Reliability Engineering & System Safety*, 212: 107628.

[7] Wang L., Zhu Z. and Zhao X. (2024). Dynamic predictive maintenance strategy for system remaining useful life prediction via deep learning ensemble method[J]. *Reliability Engineering & System Safety*: 110012.

[8] Saxena A., Goebel K., Simon D., *et al.* (2008). Damage propagation modeling for aircraft engine run-to-failure simulation[C]. *2008 International Conference on Prognostics and Health Management*. IEEE: 1–9.

Recent Advances in Reliability and Maintenance Modeling – Okamura, Inoue & Xiao (Eds)
© 2025 the Author(s), ISBN: 978-1-032-79098-5

Aero-engine remaining useful life prediction with missing data using diffusion model

Wei Wang, Zhiqiang Cai & Mingli Liu
School of Mechanical Engineering, Northwestern Polytechnical University, Xi'an, China
Ministry of Industry and Information Technology Key Laboratory of Industrial Engineering and Intelligent Manufacturing, Northwestern Polytechnical University, Xi'an, China

ABSTRACT: The use of deep learning for Remaining Useful Life (RUL) prediction has grown, bringing significant progress. However, a common obstacle in real-world applications is the challenge of dealing with missing data in condition monitoring data within test datasets. The absence of complete sensor data can severely impact the accuracy of model predictions. To address this issue, this study introduces a novel data-filling method named DiffFMD, which employs diffusion models. This method first utilizes a self-supervised learning strategy to train the diffusion models, effectively learning the distribution characteristics of the missing data in the test datasets. Subsequently, a denoising network is designed to integrate generative conditions and precisely capture the spatiotemporal correlations in monitoring data, thereby estimating noise levels. Experimental validation demonstrates that the proposed method effectively fills in the missing data and significantly enhances the accuracy of the RUL prediction model in test scenarios where data is incomplete.

Keywords: Remaining Useful Life prediction, Deep learning, Missing data, Diffusion model

1 INTRODUCTION

Aviation safety is a critical priority in the industry, with the reliability of aero-engines being central to preventing flight safety incidents. The application of Prognostics and Health Management (PHM) is key to enhancing aero-engine safety by facilitating ongoing monitoring and predictive maintenance [1]. A fundamental aspect of PHM, estimating the Remaining Useful Life (RUL) of equipment, allows for early intervention before performance drops to critical levels. Accurate RUL predictions help maintenance teams preemptively tackle potential issues, significantly reducing accident risks and enabling more efficient maintenance strategies, thereby saving on costs [2].

In recent years, with the advancement of deep learning, techniques such as Convolutional Neural Networks (CNNs) [3], Long Short-Term Memory networks (LSTMs) [4], and Attention mechanisms [5] have been widely applied in predicting RUL. These methods can automatically identify high-level features and complex patterns within the data, thereby achieving more accurate predictions. However, a key prerequisite for their successful application is the availability of complete monitoring data from operation to failure [6]. Incomplete monitoring data can greatly affect the accuracy of deep learning models in practical testing scenarios. Missing data is a frequent issue in real-world applications, often due to technical malfunctions, operational issues, or environmental conditions. Since complete data is essential for the predictive performance of these models, handling missing data in test datasets is a growing challenge.

DOI: 10.1201/9781003491309-29

Current literature seldom addresses the issue of missing monitoring data in testing scenarios, and existing solutions have not effectively met this challenge. This study introduces a novel data filling approach named DiffFMD, utilizing diffusion models to tackle the issue of missing monitoring data in test datasets. As a probabilistic generative model, diffusion models [7] offer a unique method of data generation compared to traditional Generative Adversarial Networks (GANs) [8] and Variational Autoencoders (VAEs) [9]. These models have gained attention for their exceptional performance in generating high-quality images, audio, and other data types [10,11]. By applying diffusion models to fill in condition monitoring data, our method can effectively match the distribution of missing data in the test datasets and generate high-quality, realistic samples. Specifically, this study employs a self-supervised learning approach to train the diffusion model, dividing the complete dataset into filling targets (simulating missing data in test datasets) and conditional data. In the iterative forward process, noise is gradually added to the filling targets, and then the original data is reconstructed and synthesized from Gaussian noise through a reverse denoising process. Furthermore, the study develops a denoising network that integrates conditional information and captures the spatiotemporal correlations of data during the denoising process, thereby accurately estimating the noise level. Experimental results demonstrate that our proposed method effectively fills missing data in testing scenarios and significantly improves the predictive performance of RUL prediction models on the filled data.

The remainder of this paper is organized as follows: Section 2 introduces problem definition and notations. Section 3 details the proposed method. Section 4 compares and analyzes experimental results. Section 5 concludes the paper.

2 PROBLEM DEFINITION AND NOTATIONS

The condition monitoring data collected by sensors often involve multiple variables, thus $X = \{x^{(i)} | i = 1, 2, \ldots, N_{train}\}$ can represent the monitoring data of N_{train} engines from the start of operation to the failure period. Here, the monitoring data of each engine is marked as $x^{(i)}$, which can be expressed by Equation (1):

$$x^{(i)} = \left\{x_1^{(i)}, x_2^{(i)}, \ldots, x_t^{(i)}, \ldots x_{L_i}^{(i)}\right\} \tag{1}$$

where $x_t^{(i)} = \left\{x_{t,j}^{(i)} | j = 1, 2, \ldots, N_{sensor}\right\}$ specifically refers to the data sample at time t in $x^{(i)}$, and N_{sensor} represents the number of sensors involved in the monitoring. L_i refers to the total number of monitoring time steps from the start of operation to failure of the engine i. Furthermore, $R = \{r^{(i)} | i = 1, 2, \ldots, N_{train}\}$ is used to represent the RUL value matching X, where $r^{(i)} = \left\{r_1^{(i)}, r_2^{(i)}, \ldots, r_t^{(i)}, \ldots r_{L_i}^{(i)}\right\}$. This study adopts the piecewise linear RUL function [3] to determine the RUL labels, as detailed in Equation (2):

$$r_t^{(i)} = \begin{cases} r_{\max}, & if \ \ t \leq t_{\max} \\ L_i - t, & if \ \ t > t_{\max} \end{cases} \tag{2}$$

where $r_{\max}$ represents the maximum RUL truncation value in the initial operation stage in the piecewise linear RUL function; $t_{\max}$ indicates the time step corresponding to this truncation value. To improve training efficiency and accelerate model convergence, we adopt the min-max normalization method [4] to process the original data, which can be described by Equation (3):

$$\widetilde{x}_{t,j}^{(i)} = \frac{x_{t,j}^{(i)} - x_{\min,j}}{x_{\max,j} - x_{\min,j}} \tag{3}$$

where $x_{t,j}^{(i)}$ refers to the monitoring data collected by the sensor j of engine i at the time step t, and the normalized data is represented as the corresponding $\tilde{x}_{t,j}^{(i)}$. $x_{\min,j}$ and $x_{\max,j}$ respectively represent the minimum and maximum values of the sensor j among all engines. Through this processing, the normalized monitoring data of all engines from operation to failure are marked as $\tilde{X} = \{\tilde{x}^{(i)} | i = 1, 2, \ldots, N_{train}\}$, where the monitoring data of each engine is represented as $\tilde{x}^{(i)}$, and the specific expression can be referred to Equation (4):

$$\tilde{x}^{(i)} = \left\{\tilde{x}_1^{(i)}, \tilde{x}_2^{(i)}, \ldots, \tilde{x}_t^{(i)}, \ldots \tilde{x}_{L_i}^{(i)}\right\} \tag{4}$$

3 METHODOLOGY

3.1 *DiffFMD framework*

This section elucidates how to fit the distribution of missing data in real condition monitoring data using DiffFMD, thereby filling missing data in test datasets. DiffFMD includes a forward process that incrementally adds noise to data and a reverse process that trains a denoising network at various noise levels to restore the original data. This method allows learning the data's probability distribution, ensuring the generated data's distribution closely matches that of the actual data, thus creating an effective model for generating condition monitoring data.

In practice, for the condition monitoring data in test datasets, if there are missing data, we cannot know the real data labels corresponding to these missing data. Therefore, it is impossible to complete the model's training process merely with the monitoring data in test datasets. To address this issue, we use a self-supervised training method that constructs data samples from the training set without missing data for the training process of DiffFMD. Without loss of generality, $\tilde{X}_0$ is used to represent the complete monitoring data in the training set after preprocessing. Firstly, the original complete data $\tilde{X}_0$ is divided into two parts: one part is the filling target $\tilde{X}_0^m$, used to simulate the missing data in this data sample; the other part is the conditional data $\tilde{X}_0^c$, which includes the missing data and serves as the generative conditional input during the model training process. Then, at time step t, by adding Gaussian noise ε to the filling target $\tilde{X}_0^m$, the noised data $\tilde{X}_t^m$ is obtained. At this point, the noised data $\tilde{X}_t^m$, the generative conditional data $\tilde{X}_0^c$, and the time step t constitute the input to the denoising network ε_θ, thus predicting the noise level added at time step t. The specific training process of DiffFMD is shown in Figure 1.

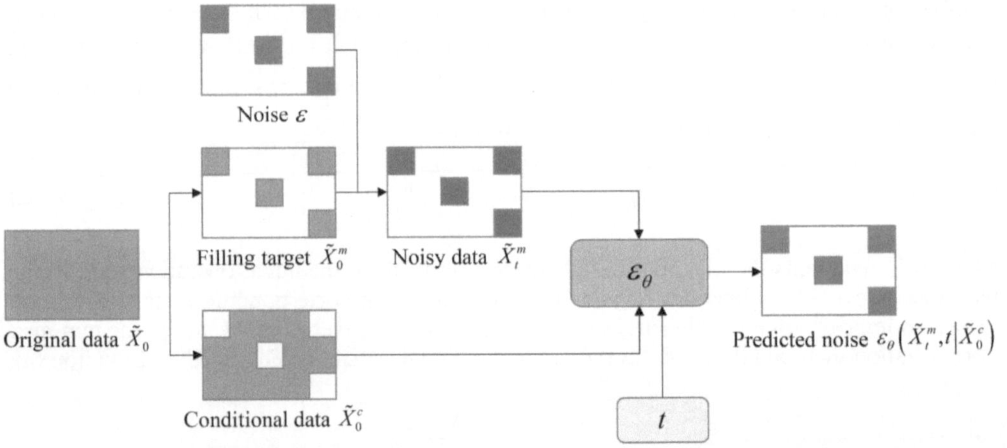

Figure 1. The training procedure of DiffFMD.

Specifically, the forward process is a Markov chain containing T time steps. The training process of the model aims to fit the distribution of the filling target $\widetilde{X}_0^m$; thus, we assume the distribution of the filling target $\widetilde{X}_0^m$ to be $q\left(\widetilde{X}_0^m\right)$. The forward diffusion process can be represented as Equation (5):

$$q\left(\widetilde{X}_{1:T}^m \middle| \widetilde{X}_0^m\right) = \prod_{t=1}^{T} q\left(\widetilde{X}_t^m \middle| \widetilde{X}_{t-1}^m\right) \tag{5}$$

The distribution of $\widetilde{X}_t^m$ can be expressed as Equation (6) [12]:

$$\widetilde{X}_t^m \sim q\left(\widetilde{X}_t^m | \widetilde{X}_0^m\right) = \mathcal{N}\left(\widetilde{X}_t^m; \sqrt{\bar{\alpha}_t}\widetilde{X}_0^m, (1-\bar{\alpha}_t)\mathrm{I}\right) \tag{6}$$

where $\alpha_t = 1 - \beta_t$, $\bar{\alpha}_t = \prod_{s=1}^{t} \alpha_s$. $\{\beta_t\}_{t=1}^{T}$ is an adjustable hyperparameter used to control the intensity of noise added during a single diffusion process.

After T steps of diffusion, the distribution of $\widetilde{X}_t^m$ will eventually be perturbed into a simple Gaussian distribution $\widetilde{X}_T^m$. Therefore, we can sample $\widetilde{X}_T^m$ from the Gaussian distribution $\mathcal{N}(0,\mathrm{I})$, and then denoise the noised data $\widetilde{X}_t^m$ through the reverse process to recover $\widetilde{X}_0^m$, thereby obtaining new data points sampled from the real data distribution. Here, a parameterized model $p_\theta\left(\widetilde{X}_{t-1}^m \middle| \widetilde{X}_t^m, \widetilde{X}_0^c\right)$ is used to approximate the distribution of the reverse process [7], which can be expressed as follows:

$$p_\theta\left(\widetilde{X}_{t-1}^m \middle| \widetilde{X}_t^m, \widetilde{X}_0^c\right) = \mathcal{N}\left(\widetilde{X}_{t-1}^m; \mu_\theta\left(\widetilde{X}_t^m, t \middle| \widetilde{X}_0^c\right), \Sigma_\theta\left(\widetilde{X}_t^m, t \middle| \widetilde{X}_0^c\right)\right) \tag{7}$$

$$\mu_\theta\left(\widetilde{X}_t^m, t \middle| \widetilde{X}_0^c\right) = \frac{1}{\sqrt{\alpha_t}}\left(\widetilde{X}_t^m - \frac{\beta_t}{\sqrt{1-\bar{\alpha}_t}}\varepsilon_\theta\left(\widetilde{X}_t^m, t \middle| \widetilde{X}_0^c\right)\right) \tag{8}$$

$$\Sigma_\theta\left(\widetilde{X}_t^m, t \middle| \widetilde{X}_0^c\right) = \sqrt{\widetilde{\beta}_t} \tag{9}$$

where $\widetilde{\beta}_t = (1 - \bar{\alpha}_{t-1}/1 - \bar{\alpha}_t)\beta_t$. ε_θ is the denoising network, a neural network model, which predicts the Gaussian noise ε by using the generative conditional data $\widetilde{X}_0^c$, the time step t, and the noised data $\widetilde{X}_t^m$ at that time step. The model structure of ε_θ will be detailed in Section 3.2. Then, the reverse generation process can be represented as Equation (10):

$$p_\theta\left(\widetilde{X}_{0:T}^m \middle| \widetilde{X}_0^c\right) = p_\theta\left(\widetilde{X}_T^m\right) \prod_{t=1}^{T} p_\theta\left(\widetilde{X}_{t-1}^m \middle| \widetilde{X}_t^m, \widetilde{X}_0^c\right) \tag{10}$$

Finally, we can train the reverse process by solving the following optimization problem [7]:

$$\min_\theta \mathcal{L}(\theta) := \min_\theta \widetilde{X}_{0,\varepsilon,t}^m \left[\left\|\varepsilon - \varepsilon_\theta\left(\sqrt{\bar{\alpha}_t}\widetilde{X}_0^m + \sqrt{1-\bar{\alpha}_t}\varepsilon, t \middle| \widetilde{X}_0^c\right)\right\|_2^2\right] \tag{11}$$

After the training is completed, we can fill in the missing data in test datasets. The specific method is as follows: for the condition monitoring data in test datasets containing missing data, we use these data as the generative conditional data $\widetilde{X}_0^c$, then start from $\widetilde{X}_T^m$, and gradually remove the Gaussian noise step by step, eventually obtaining the sampling result $\widetilde{X}_0^m$ from the data distribution $q\left(\widetilde{X}_0^m\right)$. Finally, by combining $\widetilde{X}_0^c$ and $\widetilde{X}_0^m$, we can obtain complete test set data.

3.2 *Denoising network architecture*

The accuracy of the denoising network ε_θ in predicting Gaussian noise at various diffusion time steps is crucial for the quality of the generated data. Based on this, we design a novel denoising network architecture, as shown in Figure 2. This model first utilizes a 1x1 convolutional neural network to expand the channel number of the input data $\widetilde{X}_t^m$ to N_c, followed by N_d feature extraction layers to learn diverse and complex features in a high-dimensional space. Within each feature extraction layer, a 2D Attention mechanism [13] is adopted to capture the complex spatiotemporal correlations of multivariate condition monitoring data. To enhance the model's convergence efficiency, residual connections are introduced within the feature extraction layers. Additionally, the model incorporates gated activation units [14] to integrate nonlinear information, ultimately achieving feature extraction through 1x1 convolution.

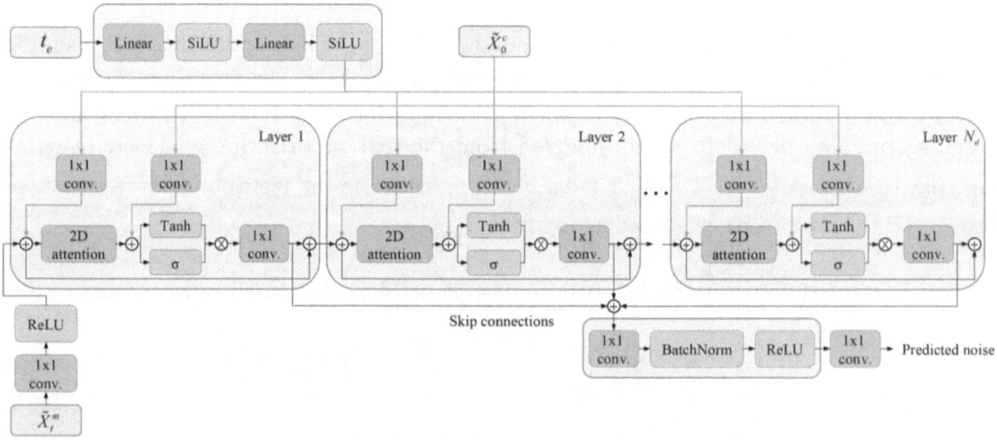

Figure 2. The architecture of denoising network.

In each feature extraction layer, in addition to the input noisy data $\widetilde{X}_t^m$, there are two additional inputs: the information of the diffusion time step t and the generation conditional information $\widetilde{X}_0^c$. To enable the denoising network to differentiate between various diffusion time steps, a time embedding scheme similar to the position encoding used in the Transformer [15] is adopted. Specifically, for each diffusion time step t, a dimension d encoding vector t_e is allocated to represent the embedding of that time step. For $i = 0, 1, \ldots, d/2 - 1$, the encoding vector's calculation follows the form $t_e(i) = \sin\left(10^{4i/(d/2-1)}\ t\right)$, $t_e(i + d/2) = \cos\left(10^{4i/(d/2-1)}\ t\right)$. Subsequently, the encoding vector t_e is processed through two fully connected layers utilizing the SiLU activation function, whose parameters are shared across all feature extraction layers. The processed time step embedding is then fed into each feature extraction layer and, through 1x1 convolution, added to the input of the 2D Attention as the time step embedding of that layer. For the generation conditional information $\widetilde{X}_0^c$, it is similarly fed into each feature extraction layer and, after being processed by 1x1 convolution, merged with the output of the 2D Attention to form the generation conditional embedding. Next, the parameterized skip connections of all feature extraction layers are summed, and this sum is then processed through a 1x1 convolution, followed by batch normalization and a ReLU activation layer. Finally, another 1x1 convolutional neural network adjusts the output channel number to the predetermined level, generating the final predicted noise.

4 EXPERIMENTS

4.1 *Data description*

This study assesses the performance of DiffFMD in addressing missing data with the C-MAPSS dataset [16]. It comprises four subsets, categorized into two groups based on variations in operating conditions and failure modes: the first group (specifically, FD001 and FD003) follows a single data distribution pattern, while the second (namely, FD002 and FD004) exhibits multiple data distribution characteristics. Given that DiffFMD is not designed to adapt to multiple distribution characteristics, we select the FD001 and FD003 subsets for experimental analysis. Each subset includes 100 engines for training and 100 for testing, all monitored by 21 sensors. However, data from some sensors remain unchanged during operation and are not useful for predicting the RUL of engines. Thus, only data from 14 sensors [3] that offer relevant information are used for model training. Equation (2) is used to calculate RUL labels, and Equation (3) for data normalization. The monitoring data is then segmented by a sliding window over time [2], with window size L and step size 1, to enhance training effectiveness.

4.2 *Implementation details*

The experimental environment is set up on a Linux system, equipped with an NVIDIA GeForce GTX 3090 GPU. During the model training, the Adam optimizer is used to optimize the model parameters. In the experiments, the initial truncation threshold for the piecewise linear RUL objective function is set to $r_{\max} = 130$, and the size of the sliding window L is set to 30. The time step T of the diffusion process is set to 50, and the variance of Gaussian noise linearly increases from $\beta_1 = 0.0001$ to $\beta_T = 0.5$. The denoising network configuration includes: the number of channels $N_c = 64$, the number of feature extraction layers $N_d = 4$, and the dimension of the encoding vector for diffusion time steps $d = 128$.

We employ two metrics, Root Mean Square Error (RMSE) and Scoring Function (Score) [16], to evaluate the model's prediction performance on RUL. The specific formula of RMSE is provided in Equation (12):

$$RMSE = \sqrt{\frac{1}{N_{train}} \sum_{i=1}^{N_{train}} \left(\hat{r}_t^{(i)} - r_t^{(i)} \right)^2} \tag{12}$$

The method for calculating the Score is detailed in Equation (13):

$$Score = \sum_{i=1}^{N_{train}} s_i, s_i = \begin{cases} e^{-\frac{\hat{r}_t^{(i)} - r_t^{(i)}}{13}} - 1, \hat{r}_t^{(i)} - r_t^{(i)} < 0 \\ e^{\frac{\hat{r}_t^{(i)} - r_t^{(i)}}{10}} - 1, \hat{r}_t^{(i)} - r_t^{(i)} \geq 0 \end{cases} \tag{13}$$

where $r_t^{(i)}$ represents the actual RUL value of the engine i at time step t; $\hat{r}_t^{(i)}$ is the corresponding prediction value. For both RMSE and Score, a lower value indicates better prediction performance of the model.

4.3 *Experimental results and performance analysiss*

To investigate whether DiffFMD can effectively fill in missing data in test datasets, we use the test set filled with DiffFMD to predict the RUL. Specifically, for the test data in the C-MAPSS dataset, we construct different scenarios of data missing by setting different missing ratios (20%, 40%, 60%, 80%). These missing data are achieved by setting a certain proportion of data values in the original dataset to zero, thereby simulating data missing. Subsequently, DiffFMD is applied to fill in the missing values in the test datasets with different missing ratios, and the prediction model is used to predict the RUL of the filled data, thereby verifying the effectiveness of DiffFMD in processing filled data.

In this study, we employ two different prediction models to predict the RUL of engines, specifically, a method based on LSTM and a method based on Transformer. Both methods reference existing RUL prediction techniques and utilize LSTM [4] and Transformer [5] as their core frameworks, respectively. Given that some models were proposed earlier, to achieve better prediction effects, we optimize and upgrade the structure and parameters of these models, and introduce more advanced activation functions and normalization methods to enhance prediction performance. The specific configuration details are described in the Appendix. To ensure the fairness of experimental results, all models adopt a unified data processing procedure (covering data preprocessing and segmenting methods), the same training strategy (including optimizer selection, learning rate adjustment mechanism, and parameter initialization techniques). At the same time, to ensure the consistency of experimental conditions, the training and testing of all models are conducted under the same hardware configuration.

Table 1. Comparison results of the LSTM-based method.

	RMSE		Score	
Data type	FD001	FD003	FD001	FD003
Complete	13.79	13.75	351.82	374.75
20% missing	30.55	38.79	13099.20	47351.75
20% missing + DiffFMD	**16.41**	**16.03**	**667.11**	**600.73**
40% missing	40.72	53.49	22614.47	278812.50
40% missing + DiffFMD	**19.61**	**18.79**	**1134.32**	**1135.58**
60% missing	49.48	66.58	481717.96	1890327.14
60% missing + DiffFMD	**25.07**	**22.66**	**3612.90**	**2501.22**
80% missing	56.48	68.06	770255.27	2249198.57
80% missing + DiffFMD	**31.95**	**30.33**	**8581.35**	**7159.68**

The comparison results shown in Table 1 reveal the performance of the LSTM-based method when dealing with test sets with different proportions of missing data. The analysis of this table shows that the prediction model achieves its optimal performance in the absence of missing data. However, as the proportion of missing data increases from 20% to 80%, the RMSE and Score of the two datasets (FD001 and FD003) show significant deterioration, highlighting the importance of data integrity in maintaining prediction accuracy. Notably, in all cases of missing data, using DiffFMD to fill in the data significantly improves the model's predictive ability. Taking a 20% missing ratio as an example, after using DiffFMD, the RMSE of FD001 decreases from 30.55 to 16.41, and the RMSE of FD003 also decreases from 38.79 to 16.03. Even at higher missing ratios, the significant improvement in the performance of the prediction model remains consistent. This finding indicates that DiffFMD can effectively fit the true distribution of missing data, generate high-quality data samples, effectively fill in missing data in test datasets, and greatly reduce the negative impact of data incompleteness on prediction accuracy.

Table 2 presents the prediction results of the Transformer-based method on test sets with different data missing ratios. These results indicate that under the condition of complete data, the baseline performance of the Transformer-based method is slightly better than that of the LSTM-based method. However, as the missing ratio increases, the performance of the Trans-former-based method also shows a significant downward trend, especially in the

Table 2. Comparison results of the Transformer-based method.

Data type	RMSE		Score	
	FD001	FD003	FD001	FD003
Complete	12.60	13.53	271.50	314.52
20% missing	26.07	35.70	5234.31	127987.19
20% missing + DiffFMD	**15.52**	**15.90**	**571.80**	**634.92**
40% missing	34.82	49.78	90251.35	369959.58
40% missing + DiffFMD	**18.91**	**19.63**	**1043.40**	**1678.98**
60% missing	43.25	62.72	146885.18	1507361.34
60% missing + DiffFMD	**25.01**	**24.91**	**3499.46**	**5153.41**
80% missing	52.69	67.93	290739.78	2064333.15
80% missing + DiffFMD	**31.98**	**33.24**	**10099.48**	**15256.33**

absence of the DiffFMD data filling technique. When the missing ratio reaches 80%, the RMSEs for FD001 and FD003 deteriorate to 52.69 and 67.93, respectively, while the Scores decrease even more significantly, a trend similar to that of the LSTM-based method. After introducing DiffFMD for data filling, the performance of the Transformer-based method significantly improves. At a 20% data missing ratio, the RMSEs for FD001 and FD003 are reduced from 26.07 and 35.70 to 15.52 and 15.90, respectively. This significant improvement in performance persists at higher missing ratios, further validating the effectiveness of DiffFMD in data filling.

5 CONCLUSIONS

This study introduces DiffFMD, a novel method using diffusion models to fill missing data in test datasets. It starts by training the model through a self-supervised approach, splitting a complete dataset into a target for filling and another for conditional generation. The technique involves adding noise to the target and then removing it in a reverse process using Gaussian noise to restore the original data. The research also presents a denoising network that leverages conditional information and understands spatiotemporal relationships to accurately estimate noise levels. Results show that DiffFMD effectively fills missing data in test datasets and significantly enhances the accuracy of RUL prediction models using this data.

ACKNOWLEDGMENT

This work was supported by the National Natural Science Foundation of China [72271200, 72231008] and the Distinguished Young Scholar Program of Shaanxi Province [2023-JQ-JC-10].

REFERENCES

[1] Zhao K., Jia Z., Jia F., and Shao H., (2023). Multi-scale integrated deep self-attention network for predicting remaining useful life of aero-engine, *Engineering Applications of Artificial Intelligence*, vol. 120, p. 105860.

[2] Zhang X., Sun J., Wang J., Jin Y., Wang L., and Liu Z., (2023). Paoltransformer: Pruning-adaptive optimal lightweight transformer model for aero-engine remaining useful life prediction,' *Reliability Engineering & System Safety*, vol. 240, p. 109605.

[3] Li X., Ding Q., and Sun J.-Q., (2018). Remaining useful life estimation in prognostics using deep convolution neural networks, *Reliability Engineering & System Safety*, vol. 172, pp. 1–11.

[4] Zheng S., Ristovski K., Farahat A., and Gupta C., (2017). Long short-term memory network for remaining useful life estimation, in *2017 IEEE International Conference on Prognostics and Health Management (ICPHM)*, pp. 88–95, IEEE.
[5] Chen D., Hong W., and Zhou X., (2022). Transformer network for remaining useful life prediction of lithium-ion batteries, *IEEE Access*, vol. 10, pp. 19621–19628.
[6] Zio E., (2022). Prognostics and health management (phm): Where are we and where do we (need to) go in theory and practice, *Reliability Engineering & System Safety*, vol. 218, p. 108119.
[7] Ho J., Jain A., and Abbeel P., (2020). Denoising diffusion probabilistic models, *Advances in Neural Information Processing Systems*, vol. 33, pp. 6840–6851.
[8] Goodfellow I., Pouget-Abadie J., Mirza M., Xu B., Warde-Farley D., Ozair S., Courville A., and Bengio Y., (2020). Generative adversarial networks, *Communications of the ACM*, vol. 63, no. 11, pp. 139–144.
[9] Kingma D. P. and Welling M., (2013). Auto-encoding variational bayes, *arXiv Preprint arXiv:1312.6114.*
[10] Croitoru F.-A., Hondru V., Ionescu R. T., and Shah M., (2023). Diffusion models in vision: A survey, *IEEE Transactions on Pattern Analysis and Machine Intelligence.*
[11] Lin Z., Gong Y., Shen Y., Wu T., Fan Z., Lin C., Duan N., and Chen W., (2023). Text generation with diffusion language models: A pre-training approach with continuous paragraph denoise, in *International Conference on Machine Learning*, pp. 21051–21064, PMLR.
[12] Luo C., (2022). Understanding diffusion models: A unified perspective, *arXiv preprint arXiv:2208.11970.*
[13] Tashiro Y., Song J., Song Y., and Ermon S., (2021). Csdi: Conditional score-based diffusion models for probabilistic time series imputation, *Advances in Neural Information Processing Systems*, vol. 34, pp. 24804–24816.
[14] Oord A. v. d., Dieleman S., Zen H., Simonyan K., Vinyals O., Graves A., Kalchbrenner N., Senior A., and Kavukcuoglu K., (2016). Wavenet: A generative model for raw audio, *arXiv preprint arXiv:1609.03499.*
[15] Vaswani A., Shazeer N., Parmar N., Uszkoreit J., Jones L., Gomez A. N., Kaiser Ł., and Polosukhin I., (2017). Attention is all you need, *Advances in Neural Information Processing Systems*, vol. 30.
[16] Saxena A., Goebel K., Simon D., and Eklund N., (2008). Damage propagation modeling for aircraft engine run-to-failure simulation, in *2008 International Conference on Prognostics and Health Management*, pp. 1–9, IEEE.

APPENDIX

Table 3. Specific configuration of the LSTM-based method.

Layer	Network parameters
1	LSTM (hidden_size = 64, bias = True, bidirectional = False), Dropout (p = 0.1)
2	LSTM (hidden_size = 64, bias = True, bidirectional = False), Dropout (p = 0.1)
3	Linear (units = 16, bias = True), Activation (ReLU), Dropout (p = 0.1)
4	Linear (units = 8, bias = True), Activation (ReLU), Dropout (p = 0.1)
5	Linear (units = 1, bias = True)

Table 4. Specific configuration of the Transformer-based method.

Layer	Network parameters
1	Linear (units = 64, bias = True)
2	Positional Encoding (d model = 64), Dropout (p = 0.1)
3	Transformer Encoder (num layers = 2, d model = 64, nhead = 8, dim feedforward = 512), Dropout (p = 0.1)
4	Linear (units = 16, bias = True), Activation (ReLU), Dropout (p = 0.1)
5	Linear (units = 8, bias = True), Activation (ReLU), Dropout (p = 0.1)
6	Linear (units = 1, bias = True)

Recent Advances in Reliability and Maintenance Modeling – Okamura, Inoue & Xiao (Eds)
© 2025 the Author(s), ISBN: 978-1-032-79098-5

Strategy of capital investment decision-making by integrating engineering analysis to financial analysis at the corporate level for the electricity company sustainability

D.B.K. Winarno & H. Nugraha
PT PLN (Persero), Jakarta, Indonesia

ABSTRACT: The electricity company is currently having several challenges, one of which is the energy transition to achieve net zero emission in 2060. Therefore, the electricity company must demonstrate the ability to increase profit through the electricity company's financial indicator Return on Assets (RoA) at least 4%. In the previous research, several methods were developed to solve the problems, namely by optimizing asset interventions. However, these methods cannot be applied in the electricity company because optimization was only performed at the equipment level and/or at the power generation level. This is because assets in the power generation can also generate revenue, which can be one of the considerations in the optimization. Therefore, the electricity company seeks to increase asset productivity by prioritizing asset intervention at the corporate level and using the economic lifetime method. The prioritization is carried out based on key performance indicators (KPIs) of corporate, the power system priority index based on prediction of the power system risk cost, prediction of the highest equivalent annualized profit with combinations of all intervention scenarios (with the intervention scenario being carried out now or postponed in the following years), all compliance criteria (environmental, regulatory, and safety), and prediction of corporate capital cost limit per year.

1 INTRODUCTION

In an organization (an electricity company), business objectives or visions are contained in the company's long-term plan which is published every 5 (five) years. To achieve the business objectives for the next 5 (five) years, strategic goals and strategic enablers are needed. Strategic enablers consist of 4 (four) enablers, namely organization and people, technology advancement, financial sustainability, and national development. One of the strategic objectives of financial sustainability is to increase company profitability, one of the indicators is return on assets (RoA). The company must be able to prioritize capital costs to increase RoA.

In these conditions, decisions regarding asset management become critical. This means that the electricity company needs to improve its electricity business planning capabilities, improve investment accuracy, and understand how parameters of economic growth relate to electricity demand. Therefore, the electricity company, especially top management, must understand the role of asset management in improving the quality of business decision-making, customer satisfaction, and be able to implement good practices from asset management and asset management systems effectively within the organization. Decision-making is carried out to produce optimal activities, so that organizational objectives are achieved. The decision-making consists of: (a) strategic asset management plans (SAMPs) decision-making, (b) asset

DOI: 10.1201/9781003491309-30

management plans decision-making, which consists of capital investment decision-making and operation & maintenance (O&M) decision-making, and (c) implementation of asset decision-making management plans (delivery of plans). In making decisions, asset management plans are divided into 3 (three) levels, namely: (a) individual assets, (b) asset system, and (c) portfolio asset.

In the previous research, optimizing asset intervention at the individual asset level (component / equipment level) were developed by carrying out maintenance strategy (Arthur *et al.* 2018), (Esselman T. *et al.* 2012), (EPRI 2016), (EPRI 1998), (Sliter and George E. 2003), (Sabouhi H. *et al.* 2016), (Wibawa A. *et al.* 2019), (Zarei S. and Kajuei P.G. 2017), operation strategy (Fu, C.Z. *et al.*, 2015), (Wibawa A. *et al.* 2019), and fuel strategy (Wibawa A. *et al.* 2019). However, these methods cannot be applied in the electricity company because the optimization was only performed at the individual asset level.

Methods were also developed optimizing asset intervention at the asset system level (system / power generation level) by carrying out maintenance strategy (EPRI 1998), (EPRI 2016), (Raghawan S. and Chowdhury b. 2012), (Sabouhi, H. *et al.* 2016), (Zarei S. and Kajuei P.G. 2017), operation strategy (Fu C.Z. *et al.* 2015), (Hübela M. 2017), (Luo J. *et al.* 2016), (Mechleri E. *et al.* 2017), (Zhang C. 2015), dan fuel strategy (Li M. *et al.* 2010), (Munir, S. *et al.*, 2011), (Ploumen P. *et al.* 2011), (Stover B. *et al.* 2011) (Xia J. *et al.* 2014), (Xu C. *et al.* 2016). However, these methods cannot be applied in the electricity company because the optimization was only performed at the asset system level and did not make revenue as one of the considerations.

In addition, there were also prioritizing asset intervention at the asset portfolio level (structure / corporate level) by carrying out maintenance strategy (EPRI 1998), (EPRI 2016), operation strategy (Xu J. *et al.* 2017), (Xin T. *et al.* 2019). However, these methods cannot be applied in the electricity company because the prioritization did not make revenue as one of the considerations.

Based on the previous research, it was necessary to prioritize asset intervention at the asset portfolio level by making revenue as one of the considerations. The prioritization was carried out based on: (a) key performance indicators (KPIs) of corporate, (b) the system priority index, (c) prediction of the highest equivalent annualized profit with combinations of all intervention scenarios (with the intervention scenario being carried out now or postponed in the following years), (d) all compliance criteria, and (e) prediction of corporate capital cost limit per year. Before prioritizing asset intervention at the asset portfolio level, the asset must be optimized first at the individual asset level and the asset system level respectively. Therefore, only the most optimal asset intervention will be optimized at the portfolio level so that it will increase cost efficiency and asset productivity.

2 CAPITAL INVESTMENT DECISION-MAKING AT THE INDIVIDUAL ASSET LEVEL (THE EQUIPMENT LEVEL)

2.1 *Current condition of equipment*

This describes the realizations of technical and financial conditions of equipment which depend on the scope of project, which can be seen in Table 1. For example, if perform a replacement project, then it is only necessary the realizations of technical condition of equipment and power generation and financial condition. The parameters used to inform the technical condition of equipment since the commercial operation date (COD) until now are: (a) failure rate of equipment obtained from enterprise asset management (EAM), (b) failure distribution of equipment, and (c) downtime distribution of equipment. Failure distribution and downtime distribution are calculated by using the reliability software. The parameters will be used to calculate the predictions of technical condition of equipment, which can be seen in section 2.2.

Table 1. Realization and prediction to scope of project.

Scope of Project	Realization and Prediction				
	Equip.	Power Gen.	Power System	Tech. Support	Financ.
Replacement	●	●			●
Uprating	●	●			●
Upgrading	●	●			●
Modernization	●	●			●
Periodic Maintenance		●			●
Relocation		●	●	●	●
Disposal		●	●	●	●
New Power Gen Construc.			●	●	

● = must be fulfilled.

The parameters used to inform the financial condition since construction until now are: (a) capital cost; (b) O&M cost, (c) equivalent annualized capital cost, (d) equivalent annualized O&M cost, and (e) equivalent annualized cost. The equivalent annualized costs are calculated by using Equation 1.

$$EAC = NPV/A_{t,r} = NPV/\left(\left(1 - 1/(1+r)^{T}\right)/r\right) \tag{1}$$

where EAC = Equivalent Annualized Cost (USD/year); NPV = Net Present Value (USD); $A_{t,r}$ = annuity factor; r = discount rate (%); and T = year.

2.2 *Operational study*

This describes the predictions of technical condition of equipment until its end of life from: (a) business as usual (BaU), (b) main project, and (c) alternative projects. The parameters used to inform the technical condition are: (a) failure rate of equipment, (b) failure distribution, and (c) downtime distribution. In replacement project, the parameters are obtained from the realization.

2.3 *Financial study*

This describes the predictions of the financial condition of equipment until its end of life from: (a) BaU, (b) main project, and (c) alternative projects. The parameters used to inform the financial condition are: (a) capital cost, (b) operation cost calculated by using Equation 2, (c) maintenance cost calculated by using Equations 4–5, (d) equivalent annualized capital cost, (e) equivalent annualized O&M cost, (f) equivalent annualized disposal cost, (g) equivalent annualized cost.

$$O_t = OC \times MTBF \times (\lambda + n) \times T \tag{2}$$

$$MTBF = (8760 - (\lambda \times MDT))/\lambda \tag{3}$$

$$M_r = (\lambda \times M_{CR}) + (CR_{RoM} \times (EFDH + FOH)) + ((EFDH + FOH) \times \text{Asset Criticality}) \tag{4}$$

$$M_i = (n \times M_{CS}) + (CR_{RoM} \times (EFDH + POH)) + ((EFDH + POH) \times \text{Asset Criticality}) \tag{5}$$

$$\text{Asset Criticality} = (ENS/1\text{ hour} \times (7\% \times BPP_{SYST})) + ((DMN - ENS/1\text{ hour}) \times \text{Extra Fuel Cost}) \tag{6}$$

$$\text{Extra Fuel Cost} = \text{Marginal Cost} - BPP_{KIT} \tag{7}$$

where O_t = operation cost (USD); OC = operation labour cost (USD/hour); λ = failure rate (1/year); MTBF = Mean Time Between Failure (hours); MDT = Mean Downtime (hours); n = inspection rate (1/year); M_r = repair cost (USD); M_i = inspection cost (USD); M_{CR} = average material cost per failure rate (USD.year); M_{CS} = average material cost per inspection rate (USD.year); CR_{RoM} = average service cost per failure rate (USD.year); EFDH = Equivalent Forced Derated Hours (hours); EPDH = Equivalent Planned Derated Hours (hours); FOH = Forced Outage Hours (hours); POH = Planned Outage Hours (hours); ENS = Energy not Served (kW); DMN = net capacity (kW); BPP_{SYST} = production cost of this power system (USD/kWh); and BPP_{KIT} = production cost of this power generation (USD/kWh).

2.4 *Risk study*

This describes the predictions of risk cost of equipment until its end of life from: (a) BaU, (b) main project, and (c) alternative projects. The equipment risk cost is calculated using Equation 8.

$$\text{Equipment Risk Cost} = (\lambda \times M_{CR}) + (CR_{RoM} \times (EFDH + FOH)) + ((EFDH + FOH) \times \text{Asset Criticality}) + \Delta F_t \quad (8)$$

$$\Delta F_t = F_{t.\text{scenario}} - F_{t.\text{improvement}} \quad (9)$$

2.5 *Optimization*

This describes capital investment decision-making at the equipment level based on comparison of scenarios: (a) BaU, (b) main project, and (c) alternative projects. The project optimization method is carried out based on: (a) economic lifetime of equipment (the minimum cost point of the equipment until its end of life), (b) prediction of the lowest equivalent annualized cost of equipment (ISO 55010 2019), and (c) prediction of the equipment risk cost based on risk appetite statements. The most optimal project will be analyzed at the asset system level. In the case study of equipment replacement, the project optimization can be seen in Table 2.

Table 2. Optimization at the equipment level.

	Prediction		
Parameter	BaU	Main Project	Alternative Project
EAC (USD)	28,633.44	25,220.30	26,061.47
Total of Equipment Risk Cost (USD)	177,758.22	158,588,52	165,781.50
	Yes ~~/ No *~~	Yes ~~/ No *~~	Yes ~~/ No *~~
Optimization	~~BaU /~~ Main Project ~~/ Alternative … *~~		

3 CAPITAL INVESTMENT DECISION-MAKING AT THE ASSET SYSTEM LEVEL (THE POWER GENERATION LEVEL)

3.1 *Current condition of power generation and power system*

This describes the realizations of technical and financial conditions of power generation and technical of power system and technical supporting which depend on the scope of project, which can be seen in Table 1. The parameters used to inform technical condition of power generation since COD until now are: (a) net plant heat rate (NPHR), (b) NPHR distribution calculated by using the reliability software, (c) failure rate of power generation, (d) EFDH and FOH, (e) equivalent availability factor (EAF), and (f) capacity factor (CF). NPHR, failure rate, EFDH, FOH, EAF, and CF are obtained from performance report.

The parameters used to inform technical condition of power system since the power generation's COD until now are: (a) failure rate of power system, (b) FOH, (c) power flow, (d) short circuit, (e) system stability, (f) power quality, and (g) load (dispatching). Failure rate and FOH are obtained from performance report. Power flow, short circuit, system stability, power quality, and load are calculated by using the power system analysis software.

The parameters used to inform technical condition of technical supporting since the power generation's COD until now are: (a) demand, (b) energy potential, (c) topography, (d) soil investigation, (e) hydrology/hydro-oceaonography, (f) Climatology, and (g) seismography.

The parameters used to inform the financial condition since construction until now are: (a) capital cost; (b) O&M cost, (c) fuel cost, (d) revenue, (e) equivalent annualized capital cost, (f) equivalent annualized O&M cost, (g) equivalent annualized fuel cost, (h) equivalent annualized cost, (i) equivalent annualized revenue calculated by using Equation 1, and (j) equivalent annualized profit.

3.2 *Operational study*

This describes the predictions of technical condition of power generation, power system, and technical supporting until power generation's end of life from: (a) feasibility study (FS), (b) BaU, (c) main project, and (d) alternative projects. The parameters used to inform technical condition of power generation are: (a) NPHR, (b) NPHR distribution, (c) failure rate of power generation, (d) EFDH and FOH, (e) EAF, and (f) CF. In reliability project, NPHR and NPHR distribution are obtained from the realization. Meanwhile in efficiency project, NPHR is calculated by using Equation 10. EFDH, FOH, and EAF are calculated by using the reliability software.

The parameters used to inform technical condition of power system are: (a) failure rate of power system, (b) FOH, (c) power flow, (d) short circuit, (e) system stability, (f) power quality, and (g) load. Failure rate and FOH are calculated by using the reliability software. Power flow, short circuit, system stability, power quality, and load are calculated by using the power system analysis software.

The parameters used to inform technical condition of technical supporting are: (a) demand, (b) energy potential, (c) topography, (d) soil investigation, (e) hydrology/hydro-oceaonography, (f) climatology, and (g) seismography.

$$\text{NPHR} = (\text{Coal Flow} \times \text{HHV})/(\text{GGO} - \text{Aux}) \tag{10}$$

where NPHR = Net Plant Heat Rate (kcal/kWh); HHV = higher heating value of coal (kcal/kg); GGO = Gross Generator Output (kWh); and Aux = auxiliary power (kWh).

3.3 *Financial study*

This describes the predictions of financial condition of power generation until its end of life from: (a) FS, (b) BaU, (c) main project, and (d) alternative projects. The parameters used to inform the financial condition are: (a) capital cost, (b) O&M cost, (c) fuel cost calculated by using Equations 11–12, (d) revenue calculated by using Equations 13–14, (e) equivalent annualized capital cost, (f) equivalent annualized O&M cost, (g) equivalent annualized fuel cost, (h) equivalent annualized disposal cost, (i) equivalent annualized cost, (j) equivalent annualized revenue, and (k) equivalent annualized profit.

$$F_t = (C_f \times \text{NPHR} \times (\text{GGO} - \text{Aux}) \times \text{EAF})/\text{HHV} \quad \text{if EAF} < \text{CF} \tag{11}$$

$$F_t = (C_f \times \text{NPHR} \times (\text{GGO} - \text{Aux}) \times \text{CF})/\text{HHV} \quad \text{if EAF} \geq \text{CF} \tag{12}$$

$$TR = \text{Electric Price} \times (GGO - Aux) \times EAF \quad \text{if } EAF < CF \tag{13}$$

$$TR = \text{Electric Price} \times (GGO - Aux) \times CF \quad \text{if } EAF \geq CF \tag{14}$$

where F_t = fuel cost (USD); C_f = coal unit price (USD/kg); EAF = Equivalent Availability Factor (%); CF = Capacity Factor (%); and TR = total revenue (USD).

3.4 *Risk study*

This describes the predictions of risk cost of power generation and power system until power generation's end of life from: (a) FS, (b) BaU, (c) main project, and (d) alternative projects. The power generation risk cost is calculated using Equation 8.

3.5 *Optimization*

This describes capital investment decision-making at the power generation level based on comparison of scenarios: (a) FS, (b) BaU, (c) main project, and (d) alternative projects. The project optimization method is carried out based on: (a) economic lifetime of power generation (the maximum profit point of the power generation until its end of life), (b) KPIs of power generation, (c) the equipment priority index based on prediction of the equipment risk cost which can be seen in Table 3, (d) prediction of the highest equivalent annualized profit of power generation, (e) predictions of the power generation risk cost and the power system risk cost based on risk appetite statements, (f) predictions of power flow, short circuit, system stability, power quality, dan load based on grid code, and (g) predictions of energy potential, topography, soil investigation, hydrology/hydro-oceaonography, climatology, and seismography. The most optimal project will be analyzed at the asset portfolio level. In the case study of equipment replacement, the project optimization can be seen in Table 4.

Table 3. Equipment priority index (PI) in 2023.

Equipment Names	Year	Risk Cost (USD)
Mill Seal Air Cold Primary Air Duct	2023	2,971.26
Coal Feeder Seal Air Damper	2023	2,841.97
Seal Air Fan Outlet Damper	2023	2,787.55
Mill Seal Air Header	2023	2,670.48
Mill Seal Air Fan Inlet Duct	2023	1,857.49
Mill Seal Air Fan Outlet Duct	2023	1,857.49
Coal Feeder Seal Air Duct	2023	1,857.49

Table 4. Optimization at the power generation level.

	Prediction		
Parameter	BaU	Main Project	Alternative Project
Key Performance Indicator (KPI)			
EFDH + FOH	1,277.31	1,016.83	1,534.56
	Yes ~~/ No *~~	Yes ~~/ No *~~	Yes ~~/ No *~~
EAF	74.46	77.43	82.48
	Yes ~~/ No *~~	Yes ~~/ No *~~	Yes ~~/ No *~~

(*continued*)

Table 4. Continued

Parameter	Prediction		
	BaU	Main Project	Alternative Project
EA Profit (USD x 1,000)	−121,923.87	36,587.25	7,870.27
Total Risk Cost of Power Gen. (USD x 10,000)	9,660.15	8,618.39	9,772.96
	Yes ~~/ No *~~	Yes ~~/ No *~~	Yes ~~/ No *~~
Total Risk Cost of Power System (USD)	–	–	–
Power Flow	–	–	–
Short Circuit	–	–	–
System Stability	–	–	–
Power Quality	–	–	–
Load	–	–	–
Energy Potential	–	–	–
Topography	–	–	–
Soil Investigat.	–	–	–
Hydrology / Hydro-oceaon.	–	–	–
Climatology	–	–	–
Seismography	–	–	–
Optimization	~~BaU /~~ Main Project ~~/ Alternative . . . *~~		

4 CAPITAL INVESTMENT DECISION-MAKING AT THE ASSET PORTFOLIO LEVEL (THE CORPORATE LEVEL)

Capital investment decision-making at the asset portfolio level is the decision-making at the corporate level. The decision-making is carried out based on a comparison of all projects in corporate (not only projects in the power generation). The project prioritization method is carried out based on: (a) KPIs of corporate, (b) the system priority index based on prediction of the system risk cost (structure, system, and component in power system, system is power generation, transformer, etc), (c) prediction of the highest equivalent annualized profit with combinations of all intervention scenarios, (d) all compliance criteria (environmental, regulatory, and safety), and (e) prediction of corporate capital cost limit per year. The project prioritization can be seen in Tables 5 and 6.

Table 5. Prioritization at the corporate level.

Parameter	Prediction		
	Project 1	Project 2	Project 3
Equivalent Annualized Profit (EA Profit, USD)			
Postp. = 0	36,587,250.57	36,125,188.95	35,442,342.08
Postp. = 1	34,110,610.75	35,838,051.87	34,858,983.19
Postp. = 2	33,742,917.03	35,128,552.17	34,612,519.75
Environmental			
Postp. = 0	Yes ~~/ No *~~	Yes ~~/ No *~~	Yes ~~/ No *~~
Postp. = 1	Yes ~~/ No *~~	Yes ~~/ No *~~	Yes ~~/ No *~~
Postp. = 2	Yes ~~/ No *~~	Yes ~~/ No *~~	Yes ~~/ No *~~
Regulatory			
Postp. = 0	Yes ~~/ No *~~	Yes ~~/ No *~~	Yes ~~/ No *~~
Postp. = 1	Yes ~~/ No *~~	Yes ~~/ No *~~	Yes ~~/ No *~~
Postp. = 2	Yes ~~/ No *~~	Yes ~~/ No *~~	Yes ~~/ No *~~
Safety			

(continued)

Table 5. Continued

Parameter	Prediction		
	Project 1	Project 2	Project 3
Postp. = 0	Yes ~~/ No *~~	Yes ~~/ No *~~	Yes ~~/ No *~~
Postp. = 1	Yes ~~/ No *~~	Yes ~~/ No *~~	Yes ~~/ No *~~
Postp. = 2	Yes ~~/ No *~~	Yes ~~/ No *~~	Yes ~~/ No *~~
Prioritization	Combination III Project 2 will be postponed for 1 (one) year (The scenarios combination can be seen in Table 6)		

Table 6. Project scenarios combination.

Project Names	Postp. = 0	Postp. = 1	Postp. = 2
Equivalent Annualized Profit – Combination I			
Project 1 (USD)	36,587,250.57	–	–
Project 2 (USD)	36,125,188.95	–	–
Project 3 (USD)	35,442,342.08	–	–
EA Profit (USD)	108,154,781.60		
KPI	Yes ~~/ No *~~		
Capital Cost Ltd	~~Yes /~~ No *		
Equivalent Annualized Profit – Combination II			
Project 1 (USD)	36,587,250.57	–	–
Project 2 (USD)	36,125,188.95	–	–
Project 3 (USD)	–	34,858,983.19	–
EA Profit (USD)	107,571,422.71		
KPI	Yes ~~/ No *~~		
Capital Cost Ltd	Yes ~~/ No *~~		
Equivalent Annualized Profit – Combination …			

5 CONCLUSION

In the electricity company, robust capital investment decision-making at the asset portfolio level is very important. Identify and build on the information needed to build a robust plan, including: (a) understand the relationship between stakeholder needs, including capacity and functionality, by using assets to best meet stakeholder needs, (b) understand the cost of assets throughout their life cycle, (c) understand the relationship between maintenance and asset reliability/performance, (d) understand what is required in terms of design and operation of the asset to meet regulatory and other requirements, (e) use the decision-making to determine the type and frequency of maintenance to be performed throughout the asset's life cycle and the performance characteristics that will identify when an asset has reached the end of its economic life.

The main output of this activity is the identification of maintenance strategies for all types of assets. Also included in the scope of this activity group is decision-making on capital investment, which is the basis for identifying the need for developing new assets, the optimal timing for replacement or upgrading of capabilities and providing strong justification for the selected options so that it will increase cost efficiency and asset productivity.

REFERENCES

Arthur, Melani, A. H. A., Murad, C. A., Netto, A. C. (2018). Criticality-based maintenance of a coal-fired power plant, *Energy*.

Electric Power Research Institute (EPRI). (1998). *Nuclear Plant Life Cycle Management Implementation Guide*. EPRI-106109. Palo Alto.

Electric Power Research Institute (EPRI). (2016). *Integrated Life Cycle Management*. EPRI-3002008005.

Esselman, T., Bruck, P., Menger, C. (2012). Integrated Life Cycle Management: A Strategy for Plants to Extend Operating Lifetimes Safely with High Operational Reliability. 1–8. IAEA.

Fu, C. Z., Anantharaman, R., Jordal, K., Gundersen, T. (2015). Thermal efficiency of coal fired power plant: from theoretical to practical assessment. *Journal of Energy Conversion and Management*. 530–544.

Hübela, M., Meinked, S., Andrénb, M. T., Wedding, C., Nockea, J., Gierowa, C., Hassela, E., Funkquistc, J. (2017). Modelling and Simulation of a Coal-Fired Power Plant for Start-Up Optimisation. *Applied Energy*. 319–331.

International Organisation for Standardization (ISO) 55010. (2019). Guidance on the Alignment of Financial and Non-Financial Function in Asset Management.

Li, M., Rao, A. D., Brouwer, J., Samuelsen, G. S. (2010). Design of highly efficient coal-based integrated gasification fuel cell power plants. *Journal of Power Sources*. 5707–5718.

Luo, J., Wu, L., Wan, W. (2016). Optimization of the exhaust gas oxygen content for coal fired power plant boiler. *The 8th International Conference on Applied Energy* (ICAE).

Mechleri, E., Fennell, P. S., Dowell, N. M. (2017). Flexible operation strategies for coal and gas-CCS power stations under the UK and USA markets. *Energy Procedia*. 6543–6551.

Munir, S., Nimmo, W., Gibbs, B. M. (2011). The Effect of Air Staged, Co-combustion of Pulverized Coal and Biomass Blends on NOx Emissions and Combustion Efficiency.

Ploumen, P., Stienstra, G., Kamphuis, H.. (2011). Reduction of CO2 emissions of coal fired power plants by optimizing steam water cycle. *Energy Procedia*. 2074–2081.

Raghawan, S., Chowdhury, B. (2012). *Developing Life Cycle Management Plant for Power Plant Components*. Conference: North American Power Symposium (NAPS).

Sabouhi, H., Firuzabad, M. F., Dehghanian, P. (2016). Identifying critical components of combined cycle power plants for implementation of reliability centered maintenance. *Journal of Power and Energy Systems*.

Sliter, George, E. (2003). Life Cycle Management in the US Nuclear Power Industry, *Transactions of the 17th International Conference on Structural Mechanics in Reactor Technology* (SMiRT 17). 1–8. Prague.

Stover, B., Bergins, C., Klebes, J. (2011). Optimized post combustion carbon capturing on coal fired power plants. *Energy Procedia*. 1637–1643.

Wibawa, A., Ichsani, D., Yuniarto, N. (2019). Power plant life cycle cost management framework: a literature review. *Journal of Physics: Conference Series, the International Conference on Sciences and Technology Applications*.

Xia, J., Chen, G., Tan, P., Zhang, C. (2014). An Online case-based reasoning system for coal blends combustion optimization of thermal power plant. *Electrical Power and Energy System*. 299–331.

Xin, T., Xu, C., Liu, W., Xu, G., Yang, Y. (2019). Thermodynamic analysis of a novel oxy-combustion supercritical carbon dioxide power cycle integrating solar-driven coal gasification. *Energy Procedia*. 1962–1967.

Xu, C., Xu, G., Zhao, S., Dong, W., Zhou, L., Yang, Y. (2016). A theoretical investigation of energy efficiency improvement by coal pre-drying in coal fired power plant. *Journal Energy Conversion and Management*. 580–588.

Xu, J., Gu, Y., Chen, D., Li, Q. (2017). Data mining based plant-level load dispatching strategy for the coal-fired power plant coal-saving: a case study. *Applied Thermal Engineering*. 553–559.

Zarei, S., Kajuei, P. G. (2017). Evaluation of various maintenance strategies for reliability assessment of thermal power plants. *Indonesian Journal of Electrical Engineering and Computer Science*. 617–624.

Zhang, C. (2015). A software for optimizing the thermal power plant operation under environmental constraints. *International Conference on Mechatronics, Materials, Chemistry and Computer Engineering (ICMMCCE)*.

Recent Advances in Reliability and Maintenance Modeling – Okamura, Inoue & Xiao (Eds)
© 2025 the Author(s), ISBN: 978-1-032-79098-5

A note on block type policies with Markovian opportunity arrival process

Jing Wu, Junjun Zheng, Hiroyuki Okamura & Tadashi Dohi
Graduate School of Advanced Science and Engineering, Hiroshima University, Kagamiyama, Higashi-Hiroshima, Japan

Cunhua Qian
School of Economics and Management, Nanjing Tech University, Nanjing, China

ABSTRACT: This paper considers two important opportunity-based block type models under the assumption that the arrival of opportunities obeys a stochastic point process. For more details, the block type models under the replacement first and replacement last disciplines are formulated in the environment where the opportunity process obyes a Markovian Arrival Process (MAP). Replacement first is the preventive replacement is performed at the prescheduled time T or the opportunity point Y_N, whichever comes first. Conversely, Replacement last is the preventive replacement is performed at the prescheduled time T or the opportunity point Y_N, whichever comes last. The MAP is one of the general point processes and involves a Poisson process and some renewal process classes. In this paper, the expected costs in the steady state under replacement first and replacement last disciplines are formulated.

Keywords: Block type, Minimal repair, Opportunity, Replacement first, Replacement last, Markovian arrival process

1 INTRODUCTION

As important components within industrial manufacturing, production equipment system commonly undergoes failures over time due to prolonged operation and frequent usage. Therefore, these outstanding managers often try to carry out a lot of simply and effective preventive replacement policies to reduce maintenance costs and enhance system reliability. In a repairable system, the block type policies are extensively applied because of its practicality and cost-effectiveness. In general, block type policies include block replacement, 'no replacement at failure' (Nakagawa 2005) and minimal repair policies. In this paper, we focus exclusively on the minimal repair policy, as the function expressions for these types of policies are very similar. In the context of reliability, the minimal repair policy is that a minimal repair is performed at each failure and the preventive replacement is carried out at a fixed times nT $(n = 1, 2, \ldots)$. Indeed, a minimal repair can return a faulty system to the state it was in just before the failure occurred, without altering its failure rate.

Over the past six decades, various minimal repair policies have been extensively proposed and developed. Among these policies, periodic replacement policy with minimal repair is one of the most well-known minimal repair policies. Barlow and Hunter (1960) firstly proposed the definition of minimal repair policy. Based on this study (Barlow and

 DOI: 10.1201/9781003491309-31

Hunter 1960), some important modified minimal repair policies were studied to improve its performance in real industrial problem. For examples, Makabe and Morimura (1963) introduced a classical minimal repair policy in which preventive replacement is made at *Nth* failure. Muth (1977) proposed another somewhat different minimal repair policy where the preventive replacement is carried out at the first failure after the fixed time nT. Phelps (1981) systematically compared and analyzed these two classical modified minimal repair models in (Makabe and Morimura 1963; Muth 1977). Its result indicated that former one is better than the latter one in terms of long-run average cost rate at condition of increasing failure rate. Besides these models, Dohi *et al.* (2001) considered the optimal periodic maintenance with minimal repair policy under an intermittently used environment. Liu and Wang (2022) took account of the non-negligible repair time into minimal repair model. Cha and Finkelstein (2024) studied the optimal preventive maintenance policy for the constrained multi-attempt minimal repair.

For more convenience in preventive maintenance, the random arrival of opportunities is considered. In reliability theory, an opportunity means the special maintenance time point or spare parts at the lower price. The classical opportunity-based preventive replacement models include the opportunity-based age models (Dekker and Dijkstra 1992 and Dekker and Smeitink 1994) and the opportunity-based block model (Dekker and Smeitink 1991). In recent, Wang *et al.* (2023) proposed an opportunistic condition-based maintenance policy for electrical distribution systems and the Monte Carlo simulation framework for maintenance decision-making. Another promising avenue of research within opportunity-based replacement models is to integrate them with age replacement models. The important models are the replacement first and replacement last (Zhao and Nakagawa 2012 and Zhao *et al.* 2014). More concretely, replacement first is described that the preventive replacement is done at a random opportunity point Y (working cycle) or the prescheduled time T, whichever occurs first. Replacement last is that the preventive replacement is done at a random opportunity point Y (working cycle) or the prescheduled time T, whichever occurs last. This study was supported by (Dohi and Okamur 2020; Zhang 2021 and Wu *et al.* 2024). Especially, Wu *et al.* (2024a) formulated the replacement first and replacement last models in discrete time by introducing the concept of replacement option priority.

In the most preventive replacement models, the authors applied the Poisson process or renewal process to describe the process of opportunity arrivals. For more details, in the Poisson process, opportunities arrive with inter-arrival times follows an exponential distribution characterized by a constant rate. In contrast, the renewal process extends beyond the constraints of exponential distribution by a more general distribution of inter-arrival times for opportunities. In other words, the renewal process is a generalized version of the Poisson process, although it does not account for correlations between successive inter-arrival times. As we know, the opportunity arrival process may have a more complex mechanism in some practical scenarios. Zheng *et al.* (2021) have formulated the age replacement models with Markovian opportunity process (MAP). In this paper, we discuss the opportunity-based minimal replacement models where the arrivals of opportunities obey Markovian arrival process. We first revisit the standard minimal repair model and the minimal repair models under replacement first and last disciplines (Zhao and Nakagawa 2012 and Wu *et al.* 2024b). Then we formulate the opportunity-based minimal repair models with MAP. Under this environment, the inter-arrival time of opportunities follows a phase-type (PH) distribution and the correlation between inter-arrival times is taken into account. This paper is arranged as follows. Section 2 introduces the formulation of the standard minimal repair model (Barlow and Hunter 1960), the opportunity-based minimal repair models under replacement first and last (Zhao and Nakagawa 2012 and Wu *et al.* 2024b). In Section 3, the formulation of the long-run average costs for these models under the assumption that the arrival of opportunities follows a MAP are presented. Finally, in Section 4, this paper is concluded with some remarks.

2 MINIMAL REPAIR MODELS

2.1 *Standard minimal repair model*

First, we formulate the standard minimal repair model. In standard minimal repair model, a failed item is repaired minimally with replacement cost c_M during the replacement interval T, and the prescheduled replacement with preventive replacement cost c_P for the non-failed item is performed at T, $2T$, $3T$, $\cdots$. Let $EC_S(T)$ denote the long-run average cost in the steady state during one cycle. Then we have (Barlow and Hunter 1960; Zhao and Nakagawa 2012 and Wu *et al.* 2024b)

$$EC_S(T) = \frac{c_M H(T) + c_P}{T}, \tag{1}$$

where $H(T) = \int_0^T h(t)dt$ and $h(t) = f(t)/\bar{F}(t)$ are called cumulative failure rate function and failure rate function.

The problem is to determine the optimal preventive replacement time T^* minimizing $EC_S(T)$.

Theorem 1: Suppose that $h(t)$ is the failure rate of $F(t)$. Then the sufficient condition for the existence of the optimal replacement time which minimizes the long-run average cost is

- $h(t)$ is an increasing function of t.

Proof. The purpose is to find the optimal preventive replacement time T^* minimizing $EC_S(T)$. Thus, we first differentiate $EC_S(T)$ and set it equal 0:

$$c_M \int_0^T [h(T) - h(t)]dt - c_P = 0 \tag{2}$$

Let $q(T)$ denote the left hand of Equation (2). If the failure rate function $h(t)$ is increasing, then $q(T)$ is increasing. Since

$$q(0) = -c_P < 0, \tag{3}$$

and if

$$q(\infty) = c_M \int_0^\infty [h(\infty) - h(t)]dt - c_P > 0, \tag{4}$$

there exists a unique and finite T^* which satisfies Eq. (2). At this case, the related minimum expected cost in steady state is given by $EC_S(T^*) = c_M h(T^*)$; otherwise, the optimal preventive replacement time $T^* \to \infty$.

2.2 *Replacement first and last disciplines*

First, we formulate the replacement first model. Suppose that the unit is preventively replaced at a prescheduled preventive replacement time T ($0 < T \leq \infty$) or at a rand opportunity point $T_N = Y_1 + \ldots + Y_N$, whichever occurs first and undergoes minimal repair at each failure between replacements. The probability that the non-failed unit is replaced at T_N is $G^{(N)}(T)$, and the probability that it is replaced at time T is $1 - G^{(N)}(T) = \overline{G}^{(N)}(T)$. Then, the expected cost in the steady state is given by

$$EC_F(T, N) = \frac{C_F(T, N)}{L_F(T, N)}, \tag{5}$$

where the total expected cost during one cycle is

$$C_F(T,N) = c_M \int_0^T \overline{G}^{(N)}(t)h(t)dt + c_Y G^{(N)}(T) + c_P \overline{G}^{(N)}(T)$$

$$= c_P + c_M \int_0^T \overline{G}^{(N)}(t)h(t)dt - (c_P - c_Y)G^{(N)}(T), \quad (6)$$

and the mean time to replacement during one cycle is

$$L_F(T,N) = T\overline{G}^{(N)}(T) + \int_0^T tg^{(N)}(t)dt = \int_0^T \overline{G}^{(N)}(t)dt. \quad (7)$$

The aim is to determine the optimal pair of (T_F^*, N_F^*) minimizing $EC_F(T,N)$.

Next, we consider the replacement last model. We suppose that the unit is replaced at a prescheduled preventive replacement time T ($0 < T \leq \infty$) or at a random opportunity point $T_N = Y_0 + Y_1 + \ldots + Y_N$, whichever occurs last and undergoes minimal repair at each failure between replacements. Then, we formulate the preventive block replacement model under RL discipline. In RL discipline, the expected cost in the steady state is given by

$$EC_L(T,N) = \frac{C_L(T,N)}{L_L(T,N)}, \quad (8)$$

where the total expected cost during one cycle is

$$C_L(T,N) = c_M \left[H(T) + \int_T^\infty \overline{G}^{(N)}(t)h(t)dt\right] + c_P G^{(N)}(T) + c_Y \overline{G}^{(N)}(T)$$

$$= c_P + c_M \left[H(T) + \int_T^\infty \overline{G}^{(N)}(t)h(t)dt\right] - (c_P - c_Y)\overline{G}^{(N)}(T). \quad (9)$$

and the mean time to replacement during one cycle is

$$L_L(T,N) = TG^{(N)}(T) + \int_T^\infty tdG^{(N)}(t) = T + \int_T^\infty \overline{G}^{(N)}(t)dt. \quad (10)$$

The purpose is to determine the optimal pair of (T_L^*, N_L^*) minimizing $EC_L(T,N)$. Zhao and Nakagawa (2012) discussed the above models at the condition that $c_P = c_Y$. Wu *et al.* (2024b) further discussed the above models with and without discounting.

3 MINIMAL REPAIR MODELS UNDER MARKOVIAN OPPORTUNITIES

3.1 *Markovian arrival process*

Suppose that the arrival stream of opportunities follows a MAP. The MAP is an extension of the Poisson process where the arrival rate varies according to an m-state continuous-time Markov chain (CTMC), and it is well-known that MAP is dense for any point process. In other words, MAP can approximate any point process with any precision.

Let $N(t)$ be the cumulative number of arrivals following a MAP by time t. When the underlying CTMC is given by $J(t)$ on the state space $\{1, 2, \ldots, m\}$, the MAP is represented by the two-dimensional point process $(N(t), J(t))$ which is a CTMC with the following infinitesimal generator:

$$\mathbf{Q}_{\mathrm{MAP}} = \begin{pmatrix} \mathbf{D}_0 & \mathbf{D}_1 & & \\ & \mathbf{D}_0 & \mathbf{D}_1 & \\ & & \ddots & \ddots \end{pmatrix}, \tag{11}$$

where $\mathbf{D}_0$ is an m-by-m infinitesimal generator for phase process without the occurrence of arrivals and $\mathbf{D}_1$ is an m-by-m transition rate matrix when an arrival occurs. Also, the state of $J(t)$ is called the phase of MAP. More specifically, for the MAP with m phases,

$$\mathbf{D}_0 = \begin{pmatrix} -q_{1,1} & q_{1,2} & \cdots & q_{1,m} \\ q_{2,1} & -q_{2,2} & \cdots & q_{2,m} \\ \vdots & \vdots & \ddots & \vdots \\ q_{m,1} & q_{m,2} & \cdots & -q_{m,m} \end{pmatrix}, \tag{12}$$

$$\mathbf{D}_1 = \begin{pmatrix} \lambda_{1,1} & \lambda_{1,2} & \cdots & \lambda_{1,m} \\ \lambda_{2,1} & \lambda_{2,2} & \cdots & \lambda_{2,m} \\ \vdots & \vdots & \ddots & \vdots \\ \lambda_{m,1} & \lambda_{m,2} & \cdots & \lambda_{m,m} \end{pmatrix}, \tag{13}$$

where $q_{i,j(i \neq j)}$ represents the transition rate from phase i to phase j, $\lambda_{i,j}$ represents the arrival rate when the phase transits from i phase to j phase, and $q_{i,i} = \sum_{j=1, j \neq i}^{m} q_{i,j} + \sum_{j=1}^{m} \lambda_{i,j}$.

Since the infinitesimal generator of phase process $J(t)$ becomes $\mathbf{Q} = \mathbf{D}_0 + \mathbf{D}_1$, the state probability vector of phase at time t is obtained by solving the following differential equation (Kolmogorov forward equation). Let $\pi(t)$ be the state probability vector of phase at time t, i.e., $\pi(t) = [P(J(t) = i)]$, and we have

$$\frac{d}{dt}\pi(t) = \pi(t)\mathbf{Q}. \tag{14}$$

The above differential equation can be solved when the initial condition $\mathbf{p}(0)$ is given. Alternatively, the state probability vector of phase can be expressed with the matrix exponential form:

$$\pi(t) = \pi(0)\exp(\mathbf{Q}t). \tag{15}$$

Also, the inter-arrival time distribution of MAP is represented by a PH distribution. Concretely, if the state probability vector of phase at a time s is given by $\mathbf{p}_s$, then the time to the next arrival in the steady state Y follows the PH distribution whose c.d.f. and p.d.f. are

$$P(Y \leq x) = 1 - \pi_s \exp(\mathbf{D}_0 x)\mathbf{1}, \tag{16}$$

$$f_Y(x) = \pi_s \exp(\mathbf{D}_0 x)\xi, \tag{17}$$

where $\mathbf{1}$ is a column vector whose elements are 1 and $\xi = \mathbf{D}_1\mathbf{1}$. Also, the expected time interval of opportunities in the steady state becomes

$$\mathrm{E}[Y] = \pi_s(-\mathbf{D}_0)^{-1}\mathbf{1}, \tag{18}$$

and further the coefficient of variation (CV) of Y is given by (for more details see)

$$\mathrm{CV}_Y = \frac{\sqrt{2\pi_s(-\mathbf{D}_0)^{-2}\mathbf{1} - \left(\pi_s(-\mathbf{D}_0)^{-1}\mathbf{1}\right)^2}}{\pi_s(-\mathbf{D}_0)^{-1}\mathbf{1}}. \tag{19}$$

3.2 *Model description*

Suppose that the arrival stream of opportunities follows a MAP with $\mathbf{D}_0$ and $\mathbf{D}_1$. Without loss of generality, the internal states of MAP consist of m phases, i.e., $\mathbf{D}_0$ and $\mathbf{D}_1$ are m-by-m matrices. Let X be a failure time of unit from $t = 0$, and $F(t)$ and $f(t)$ are the c.d.f. and p.d.f. of X. Also, c_Y, c_P, and c_M are the unit's replacement costs at which the unit is replaced by an opportunity, the scheduled time, and a failure. In this modeling, we consider the following two cases for the interaction between the phase process of opportunities and unit replacement.

- Case 1 (Independent case): The phase of opportunity is independent of unit replacement.
- Case 2 (Dependent case): The phase of opportunity is i renewal when a unit is replaced.

We formulate the long-run expected cost under replacement first and last disciplines with Markovian opportunity process. Let $P_F(T,N)$ and $P_L(T,N)$ be the transition probability matrices of embedded DTMCs for replacement first and last models in Case 1. Then, we have

$$P_F(T,N) = \int_0^T g^{(N)}(t)dt + \overline{G}^{(N)}(T). \tag{20}$$

$$P_L(T,N) = \int_0^T g^{(N)}(t)exp(\mathbf{D}_0(T-t))dt + \overline{G}^{(N)}(T)\int_0^\infty exp(\mathbf{D}_0 t)\mathbf{D}_1 dt. \tag{21}$$

where $\overline{\mathbf{G}}^{(N)}(x)$ and $\mathbf{g}^{(N)}(x)$ are m-by-m matrices corresponding to complementary c.d.f. and p.d.f. of the arrival time of the N-th opportunity. Since the convolution of PH distribution can be expressed by expanding the number of phases, they are given by

$$\overline{\mathbf{G}}^{(N)}(x) = \mathbf{A}_N \exp(\mathbf{Q}_N x)\mathbf{B}_N, \tag{22}$$

$$\mathbf{g}^{(N)}(x) = \mathbf{A}_N \exp(\mathbf{Q}_N x)\mathbf{C}_N. \tag{23}$$

In the above, $\mathbf{Q}_N$ is an infinitesimal generator of underlying CTMC in MAP, and is an N-by-N sub-block matrix of $\mathbf{Q}_{\mathrm{MAP}}$ in Equation (11), i.e.,

$$\mathbf{Q}_N = \left.\begin{pmatrix} \mathbf{D}_0 & \mathbf{D}_1 & & & \\ & \mathbf{D}_0 & \mathbf{D}_1 & & \\ & & \ddots & \ddots & \\ & & & \mathbf{D}_0 & \mathbf{D}_1 \\ & & & & \mathbf{D}_0 \end{pmatrix}\right\}\text{Nblocks}. \tag{24}$$

Also, $\mathbf{A}_N$, $\mathbf{B}_N$, and $\mathbf{C}_N$ are

$$\mathbf{A}_N = \underbrace{(\mathbf{I} \quad \mathbf{O} \cdots \quad \mathbf{O})}_{\text{Nblocks}}, \tag{25}$$

$$\mathbf{B}_N = \left.\begin{pmatrix} \mathbf{I} \\ \vdots \\ \mathbf{I} \end{pmatrix}\right\} \text{Nblocks}, \quad \mathbf{C}_N = \left.\begin{pmatrix} \mathbf{O} \\ \vdots \\ \mathbf{O} \\ \mathbf{D}_1 \end{pmatrix}\right\} \text{Nblocks}, \tag{26}$$

where $\mathbf{I}$ and $\mathbf{O}$ are m-by-m identity and zero matrices. The probability vector of phase at the beginning of a cycle is obtained by solving the stationary vector for $\mathbf{P}_F(T,N)$ and $\mathbf{P}_L(T,N)$.

Let $G_{\text{PH}}^{(N)}(t|T)$ be the c.d.f. of the arrival time of the N-th opportunity for a given probability vector at the beginning of a cycle π_0, i.e., $G_{\text{PH}}^{(N)}(t|T) = 1 - \int_0^t \pi_0 \mathbf{G}^{(N)}(s)\mathbf{1}ds$. By substituting $G_{\text{PH}}^{(N)}(t|T)$ into $G^{(N)}(t)$, we can get the long-run average costs for replacement first and last disciplines under Markovian opportunity process:

$$EC_F(T,N) = \frac{C_F(T,N)}{L_F(T,N)}, \tag{27}$$

$$L_F(T,N) = \int_0^T \overline{G}_{\text{PH}}^{(N)}(x|T)dx, \tag{28}$$

$$\begin{aligned} C_F(T,N) &= c_P + c_M F \int_0^T \overline{G}_{\text{PH}}^{(N)}(x|T)dH(x) \\ &\quad -(c_P - c_Y)G_{\text{PH}}^{(N)}(T|T). \end{aligned} \tag{29}$$

and

$$EC_L(T,N) = \frac{C_L(T,N)}{L_L(T,N)}, \tag{30}$$

$$L_L(T,N) = T + \int_T^\infty \overline{G}_{\text{PH}}^{(N)}(x|T)dx, \tag{31}$$

$$\begin{aligned} C_L(T,N) &= c_P + c_M \left[H(T) + \int_T^\infty \overline{G}_{\text{PH}}^{(N)}(x|T)dH(x) \right] \\ &\quad -(c_P - c_Y)\overline{G}_{\text{PH}}^{(N)}(T|T). \end{aligned} \tag{32}$$

In Case I, π_0 should be computed from the embedded DTMC. By using $P_F(T,N)$, the probability vector of phase at the beginning of a cycle in steady state can be obtained by solving the following equations

$$\pi_0 = \pi_0 \pi_F(T,N), \quad \pi_0 \mathbf{1} = 1. \tag{33}$$

It should be noted that π_0 depends on T.

In Case II, we obtain the long-run average cost by replacing the stationary vector π_0 to the initial vector for the phase process. It is straightforwardly to find that the initial probability vector $\pi_0 = [\pi_0(1), \ldots, \pi_0(m)]$ is directly provided as the initial vector of phase process. If $\mathbf{p}_0$ in Case II is given by π_c such that $\pi_c \mathbf{Q} = \mathbf{0}$, Y is independent of T.

4 CONCLUDING REMARKS

This paper examines optimal minimal repair policies within the framework of Markovian arrival processes for opportunistic events. In past studies, it has been common to model the arrival of opportunities using renewal process theory. By adopting the MAP, we are able to relax this assumption. Additionally, the paper presents explicit formulas for calculating long-run average costs through the application of the Markov renewal theorem. It is worth noting that PH distributions can accurately approximate any point process for opportunities with high precision. Leveraging these techniques, one can develop system maintenance plans based on minimal repair strategies derived from empirical data.

In future work, we intend to extend the application of PH/MAP techniques to shock models. In our current model, the occurrence of opportunities is essentially independent of failure time. However, in the context of shock models, the number of shocks is strongly correlated with failure time. Furthermore, we aim to utilize PH/MAP techniques to analyze the correlation between missions in preventive replacement models with multiple missions.

REFERENCES

Barlow, R. E. and Hunter, L. C. 1960. Optimum preventive maintenance policies. *Operations Research*, 8 (1): 90–100.

Cha, J. H. & Finkelstein, M. 2024. Preventive maintenance for the constrained multi-attempt minimal repair. *Reliability Engineering & System Safety*, 243: 109899.

Dekker, R. and Dijkstra, M. C. (1992). Opportunity-based age replacement: Exponentially distributed times between opportunities. *Naval Research Logistics*, 39(2): 175–190.

Dekker, R. and Smeitink, E. (1994). Preventive maintenance at opportunities of restricted duration. *Naval Research Logistics*, 41(3): 335–353.

Dohi, T. and Okamura, H. (2020). Failure-correlated opportunity-based age replacement models. *International Journal of Reliability, Quality and Safety Engineering*, 27(02), 2040008.

Dohi, T., Kaio, N., and Osaki, S. (2001). Optimal periodic maintenance strategy under an intermittently used environment. *IIE Transactions*, 33(12): 1037–1046.

Liu, P. and Wang, G. (2022). Minimal repair models with non-negligible repair time. *Reliability Engineering & System Safety*, 217: 108046.

Makabe, H. and Morimura, H. (1963). On some preventive maintenance policies. *Journal of Operations Research Society of Japan*, 6: 17–47.

Muth, E. J. (1977). An optimal decision rule for repair vs replacement. *IEEE Transactions on Reliability*, 26(3): 179–181.

Nakagawa T. (1979). A summary of block replacement policies. *RAIRO-Operations Research-Recherche Operationnelle*, 13: 351–361.

Phelps, R. I. (1981). Replacement policies under minimal repair. *Journal of the Operational Research Society*, 32(7): 549–554.

Wu, J., Qian, C. and Dohi, T. 2024a. Optimal opportunity-based age replacement policies in discrete time. *Reliability Engineering & System Safety*, 241, 109587.

Wu, J., Zheng, J., Okamura, H., and Dohi, T. (2024b). Opportunistic Block Type Replacement Policies With and Without Discounting. Under submission

Zhang, Q., Yao, W., Xu, P. and Fang, Z. (2023). Optimal age replacement policies of mission-oriented systems with discounting. *Computers & Industrial Engineering*, 177: 109027.

Zhao, X., Nakagawa, T. and Zuo, M. J. (2014). Optimal replacement last with continuous and discrete policies. *IEEE Transactions on Reliability*, 63(4): 868–880.

Zheng, J., Okamura, H., and Dohi, T. (2021). Age replacement with Markovian opportunity process. *Reliability Engineering & System Safety*, 216, 107949.

Recent Advances in Reliability and Maintenance Modeling – Okamura, Inoue & Xiao (Eds)
© 2025 the Author(s), ISBN: 978-1-032-79098-5

Exploring the dependency relationship between software bug prediction and information criteria

Jingchi Wu, Tadashi Dohi & Hiroyuki Okamura
Hiroshima University, Hiroshima, Japan

ABSTRACT: The information criteria are often used for model selection to determine the best fit model to the observed data, but do not guarantee the best prediction model for the future (unknown) data. The present article concerns the dependency relationship between software bug prediction and information criteria, and investigates the applicability of software reliability models selected by the information criteria through empirical experiments with actual software bug count data.

Keywords: software reliability, non-homogeneous Poisson process, bug prediction, information criteria, empirical study

1 INTRODUCTION

Assessment of quantitative software reliability is an important issue for software test engineers and software project managers, where the quantitative software reliability is defined as the probability that software failures caused by software faults do not occur over a period of time. While software bug-count processes are considered as non-decreasing integer-valued stochastic processes, non-homogeneous Poisson process (NHPP)-based software reliability models (SRMs) have been frequently used (Lyu 1996). Goel and Okumoto (1979) developed the seminal NHPP-based SRM under the assumption that software bug-detection times are given by independent and identically distributed exponential random variables, and showed that it can be easily applied to assess the quantitative software reliability. Several authors also applied different software bug-detection time distributions such as the pareto distribution (Abdel-Ghaly *et al.* 1986), the lognormal and truncated normal distributions (Achcar and Niverthi 1998; Okamura *et al.* 2013), the log-logistic and truncated logistic distributions (Gokhale and Trivedi 1998; Ohba 1984), the Weibull distribution (Goel 1985; Ohishi *et al.* 2009), the gamma distributions (Yamada *et al.* 1983; Zhao and Xie 1996), the extreme-type distributions including the Gompertz distribution (Ohishi *et al.* 2009), among others.

In Software Reliability Assessment Tool on Spreadsheet (SRATS) (Okamura and Dohi 2013), eleven SRMs in the above references are implemented, where the maximum likelihood estimation is used to estimate the model parameters from the observed data. The model selection criterion such as Akaike information criterion (AIC) (Akaike 1973) is applied to determine an appropriate NHPP-based SRM with highest goodness-of-fit in the sense of information criterion. However, the best fit SRM based on the information criterion is not always the best prediction model for the future (unknown) data. Since the quantitative software reliability assessment is almost equivalent to prediction of the cumulative number of software bugs in future, it is pointed out that the model selection with the observed data may not work for the reliability assessment.

The present article concerns the dependency relationship between software bug prediction and information criteria, and investigates the applicability of software reliability models selected by the

 DOI: 10.1201/9781003491309-32

information criteria through empirical experiments with software bug count data. In addition to the well-known AIC (Akaike 1973), we use three well-known information criteria; Bayesian information criterion by Schwarz (1978), corrected AIC (AIC_C) by Sugiura (1978) and Hurvich and Tsai (1989), consistent AIC (CAIC) by Bozdogan (1987), to determine the best fit NHPP-based SRM, and apply them to determine the prediction model. Our experiments with actual software development project data show that the model selection works in the late system testing phase, but does not give good predictive performances in both the early and middle testing phases.

2 SOFTWARE RELIABILITY MODELS

In the system testing of a software product, let $N(t)$ be the cumulative number of software bugs detected by time $t\ (\geq 0)$, and assume that $N(t)$ follows a non-homogeneous Poisson process (NHPP) with the mean value function $\Lambda(t;\theta) = \int_0^t \lambda(x;\theta)dx$, where $\lambda(t;\theta)$ denotes the software intensity function and θ is the model parameter vector. The probability mass function of $N(t)$ is given by

$$\Pr\{N(t) = n\} = \frac{\{\Lambda(t;\theta)\}^n e^{-\Lambda(t;\theta)}}{n!}. \tag{1}$$

In Equation (1), $\Lambda(t;\theta) = aF(t;\alpha)$ and $\lambda(t;\theta) = af(t;\alpha)$ with $\theta \in (a,\alpha)$ are commonly assumed, where α is a non-negative real parameter vector, $a\ (> 0)$ is a real scalar parameter representing the total expected number of software bugs remaining in the software before the testing, $F(t;\alpha)$ and $f(t;\alpha) = dF(t;\alpha)/dt$ are the cumulative distribution function ($F(0;\alpha) = 0, F(\infty;\alpha) = 1$) and the probability density function of the time until each software bug is detected, respectively. Table 1 presents the representative eleven NHPP-based SRMs, which are implemented in SRATS (Okamura and Dohi 2013).

Table 1. Representative NHPP-based software reliability models.

Cumulative distribution function	$\Lambda(t;\theta),\ \theta \in (a,\alpha) = (a,b,c)$
Exp (Goel and Okumoto 1979)	$\Lambda(t;\theta) = a(1 - e^{-bt})$
Gamma (Yamada *et al.* 1983; Zhao and Xie 1996)	$\Lambda(t;\theta) = a\int_0^t \frac{c^b s^{b-1} e^{-cs}}{\Gamma(b)} ds$
Pareto (Abdel-Ghaly *et al.* 1986)	$\Lambda(t;\theta) = a\left(1 - \left(\frac{c}{t+c}\right)^b\right)$
TruncNormal (Okamura *et al.* 2013)	$\Lambda(t;\theta) = a\frac{F(t)-F(0)}{1-F(0)}$, $F(t;\alpha) = \frac{1}{\sqrt{2\pi}b}\int_t^{-\infty} e^{-\frac{(s-c)^2}{2b^2}} ds$
LogNormal (Achcar and Niverthi 1998; Okamura *et al.* 2013)	$\Lambda(t;\theta) = a\frac{1}{\sqrt{2\pi}b}\int_{\log(t)}^{-\infty} e^{-\frac{(s-c)^2}{2b^2}} ds$
TruncLogist (Ohba 1984)	$\Lambda(t;\theta) = a\frac{F(t;\alpha)-F(0;\alpha)}{1-F(0;\alpha)}$, $F(t;\alpha) = \frac{1}{1+e^{-\frac{t-c}{b}}}$
LogLogist (Gokhale and Trivedi 1998)	$\Lambda(t;\theta) = a\frac{1}{1+e^{-\frac{\log(t)-c}{b}}}$
TruncEVMax (Ohishi *et al.* 2009)	$\Lambda(t;\theta) = a\frac{F(t;\alpha)-F(0;\alpha)}{1-F(0;\alpha)}$, $F(t;\alpha) = e^{-e^{-\frac{t-c}{b}}}$
LogEVMax (Ohishi *et al.* 2009)	$\Lambda(t;\theta) = ae^{-e^{-\frac{\log(t)-c}{b}}}$
TruncEVMin (Ohishi *et al.* 2009)	$\Lambda(t;\theta) = a\frac{F(t;\alpha)-F(0;\alpha)}{1-F(0;\alpha)}$, $F(t;\alpha) = e^{-e^{-\frac{t-c}{b}}}$
LogEVMin (Goel 1985; Ohishi *et al.* 2009)	$\Lambda(t;\theta) = ae^{-e^{-\frac{\log(t)-c}{b}}}$

Once the mean value function is given, it is necessary to statistically estimate the unknown parameters $\theta \in (a, \alpha)$ from the software bug count data observed in the system testing. Suppose that n time-domain data, $\mathcal{D} = \{t_1, t_2, \ldots, t_n\}$, are observed, where t_i denotes the $i\,(= 1, 2, \ldots, n)$-th software bug-detection time. Then the log-likelihood function for the time-domain data is given by

$$\ln\mathcal{L}(\theta; \mathcal{D}) = \sum_{i=1}^{n} \ln\lambda(t_i; \theta) - \Lambda(t_n; \theta) \tag{2}$$

under the failure truncation at time t_n. The maximum likelihood estimation is reduced to the problem of finding the parameters θ (maximum likelihood estimate) satisfying

$$\widetilde{\theta} = \arg\max_{\theta} \ln\mathcal{L}(\theta; \mathcal{D}), \tag{3}$$

which maximizes the log-likelihood function in Equation (2). Finally, from Equation (1), quantitative software reliability for NHPP-based SRM is given by

$$R(u|t_n) = e^{-\{\Lambda(t_n+u;\widetilde{\theta}) - \Lambda(t_n;\widetilde{\theta})\}}, \tag{4}$$

where u is an arbitrary time-interval measured after the system testing is completed at time t_n.

3 INFORMATION CRITERIA

When the time-domain data $\mathcal{D}$ are given, we get the maximum likelihood estimate $\widetilde{\theta}$ and select an appropriate NHPP-based SRM for the future prediction, because the quantitative reliability $R(u|t_n)$ is a function of the future time u. In the frequentist approach based on Kullback-Leibler divergence, the model selection is made so as to minimize an approximate distance between the real model behind the underlying data and our model, say, SRM. Akaike (Akaike 1973) proposed AIC;

$$AIC = -2\ln\mathcal{L}(\widetilde{\theta}; \mathcal{D}) + 2\phi, \tag{5}$$

where $\ln\mathcal{L}(\widetilde{\theta}; \mathcal{D})$ represents the maximum log-likelihood, and ϕ is the model dimension denoting the number of arbitrary parameters in each NHPP-based SRM. A model with a smaller AIC is suitable better for the observed data.

The BIC (Schwarz 1978) is regarded as alternative information criterion for the AIC, where BIC is derived by maximizing the posterior probability instead of minimization of Kullback-Leibler divergence, and can be used for the model selection in the maximum likelihood estimation. Define

$$BIC = -2\ln\mathcal{L}(\widetilde{\theta}; \mathcal{D}) + \phi\ln(n), \tag{6}$$

where n is the number of data. The main feature of BIC is to adjust the penalty for model complexity on the sample size, and to enable more suitable model selection for large sample data. In fact, compared to AIC, it is known that BIC imposes a stricter penalty on complex models when the sample size is large, favoring the selection of simpler models.

Sugiura (1978) and, later, Hurvich and Tsai (1989) corrected a bias for the original AIC for regression and auto-regressive time series models, respectively, where the correction is of particular use when the sample size is small, or when the number of fitted parameters is a moderate to large fraction of the sample size. They defined $\mathrm{AIC_C}$ by

$$\mathrm{AIC_C} = AIC + \frac{2\phi^2 + 2\phi}{n - \phi - 1}, \tag{7}$$

where $n - \phi - 1$ 0 is required. In fact it is straightforward to see that AIC_C is reduced to AIC when $n \to \infty$.

Bozdogan (1987) also proposed CAIC as an extension of AIC, which makes AIC asymptotically consistent. Define

$$CAIC = -2\ln\mathcal{L}(\widetilde{\theta};\mathcal{D}) + \phi(\ln(n) + 1), \tag{8}$$

which tends to penalize over-parameterization more stringently, and is similar to BIC as well.

Based on the above four information criteria, we select the best goodness-of-fit NHPP-based SRM, and use them for the prediction. When the time-domain data $\mathcal{D}$ are given, we predict the mean value function $\Lambda(t_i;\widetilde{\theta})$ at time t_i $(i = n+1, n+2, \ldots, n+k)$ with the prediction length k $(= 1, 2, \ldots)$, in stead of the time sequence; $t_{n+1}, t_{n+2}, \ldots, t_{n+k}$, because the quantitative software reliability in Equation (4) is a function of the mean value function for an arbitrary time interval u. The predictive performance is measured by the prediction mean absolute error (PMAE):

$$PMAE = \frac{\sum_{i=n+1}^{n+k} |\Lambda(t_i;\widetilde{\theta}) - i|}{k}. \tag{9}$$

Since one cannot know the best NHPP-based SRM which minimizes the PMAE at the observation time t_n, as the best way, we select the best model with the minimum information criterion and apply it to the prediction.

Table 2. Software bug-count data sets.

Data Set	Total number of failures	Testing length (CPU time)	Source	Nature of system
DS1	54	108708	SYS2 (Musa 1979)	Real time CCS
DS2	38	67362	SYS3 (Musa 1979)	Real time CCS
DS3	136	88682	SYS1 (Musa 1979)	Real time CCS
DS4	53	52422	SYS4 (Musa 1979)	Real time CCS
DS5	73	5090	Project J5 (Lyu 1996)	Real time CCS
DS6	38	233700	S10 (Musa 1979)	Real time CCS
DS7	41	4312598	S27 (Musa 1979)	Military application
DS8	101	19572126	S17 (Musa 1979)	Real time CCS

CCS: Command and Control System

4 NUMERICAL EXPERIMENTS

4.1 *Experimental setup*

We investigate the predictive performances of the prediction models selected by four information criteria in Section 3. In Table refdataSetInfoTable we summarize eight bug-detection time data sets. In each data set, we set 20%, 50%, and 80% observation point of the whole data, estimate the model parameters and select the best fit NHPP-based SRM from eleven candidate models in Table 1. Next, we predict the cumulative number of software bugs detected during the remaining 80%, 50%, and 20% periods with the NHPP-based SRMs under four information criteria; AIC, AIC_C, BIC, and CAIC. Finally, the predictive performances are evaluated with the PMAE.

4.2 *Predictive performances*

In Tables 3, 4 and 5, we present the results on model selection, where each numerical value denotes the corresponding information criterion for each parametric model and the red colored value indicates the best model for each criterion. Table 6 presents the prediction results based on PMAE, where Best PMAE denotes the minimum PMAE value and its associated baseline model in Table 1. These models provide the minimum PMAE ex post facto, but cannot be known at each observation point. In the early testing phase with 20% data, it is seen that the best goodness-of-fit models with four information criteria were almost exponential distribution model except in three cases (Best AIC with DS3 and DS5, Best AIC_C with DS6) from Table 3, and that the model selection under four information criteria did not work comparing with the post hoc level on PMAE.

In the middle testing phase with 50% data in Table 4, we found that the results on model selection with four information criteria are almost same except in only one case (Best CAIC with DS7). It should be noted that we can select the model with best PMAE in advance based on the information criteria in some cases. This tendency became remarkable in the late testing phase (see Table 6), where four cases with DS1, DS2, DS4 and DS6 gave the same PMAE values as the Best PMAE results. Hence, as the system testing goes on and an enough number of software bug data are observed, the model selection could provide the best NHPP-based SRMs. Especially, we observed that both of AIC and AIC_C gave the Best PMAE results in DS1.

Table 3. Goodness-of-fit performances in early software testing phases (20%).

	DS1				DS2			
Model Name	AIC	AIC_C	BIC	CAIC	AIC	AIC_C	BIC	CAIC
Exp	154.98	156.48	155.78	157.78	105.54	107.94	105.70	107.70
Gamma	156.98	160.41	158.17	161.17	107.43	113.43	107.67	110.67
Pareto	156.99	160.42	158.18	161.18	106.89	112.89	107.12	110.12
TruncNormal	156.99	160.42	158.19	161.19	107.68	113.68	107.91	110.91
LogNormal	156.81	160.24	158.01	161.01	106.50	112.50	106.74	109.74
TruncLogist	156.99	160.42	158.18	161.18	107.58	113.58	107.81	110.81
LogLogist	156.98	160.41	158.18	161.18	107.12	113.12	107.36	110.36
TruncEVMax	156.99	160.42	158.18	161.18	107.57	113.57	107.81	110.81
LogEVMax	156.76	160.19	157.96	160.96	105.94	111.94	106.18	109.18
TruncEVMin	156.99	160.42	158.18	161.18	107.70	113.70	107.94	110.94
LogEVMin	156.98	160.41	158.17	161.17	107.35	113.35	107.59	110.59

	DS3				DS4			
Model Name	AIC	AIC_C	BIC	CAIC	AIC	AIC_C	BIC	CAIC
Exp	313.86	314.36	316.46	318.46	138.18	139.68	138.97	140.97
Gamma	313.78	314.82	317.66	320.66	138.84	142.27	140.03	143.03
Pareto	314.93	315.98	318.82	321.82	140.10	143.53	141.29	144.29
TruncNormal	316.10	317.14	319.99	322.99	140.22	143.65	141.41	144.41
LogNormal	314.11	315.15	318.00	321.00	139.32	142.75	140.52	143.52
TruncLogist	315.94	316.99	319.83	322.83	140.20	143.63	141.40	144.40
LogLogist	313.78	314.83	317.67	320.67	138.86	142.29	140.05	143.05
TruncEVMax	315.93	316.98	319.82	322.82	140.20	143.62	141.39	144.39
LogEVMax	314.60	315.64	318.49	321.49	139.69	143.12	140.88	143.88
TruncEVMin	316.20	317.25	320.09	323.09	140.23	143.66	141.42	144.42
LogEVMin	313.78	314.82	317.66	320.66	138.85	142.28	140.05	143.05

(continued)

Table 3. Continued

	DS5				DS6			
Model Name	AIC	AIC_C	BIC	CAIC	AIC	AIC_C	BIC	CAIC
Exp	133.73	134.73	135.14	137.14	148.95	151.35	149.11	151.11
Gamma	135.67	137.85	137.79	140.79	146.60	152.60	146.84	149.84
Pareto	135.73	137.91	137.86	140.86	151.22	157.22	151.46	154.46
TruncNormal	134.08	136.26	136.20	139.20	147.44	153.44	147.68	150.68
LogNormal	136.61	138.79	138.73	141.73	146.37	152.37	146.60	149.60
TruncLogist	133.32	135.50	135.44	138.44	147.80	153.80	148.04	151.04
LogLogist	136.01	138.19	138.13	141.13	146.77	152.77	147.01	150.01
TruncEVMax	134.18	136.36	136.30	139.30	146.71	152.71	146.95	149.95
LogEVMax	136.73	138.92	138.86	141.86	146.13	152.13	146.36	149.36
TruncEVMin	134.68	136.86	136.80	139.80	148.13	154.13	148.37	151.37
LogEVMin	135.47	137.66	137.60	140.60	147.02	153.02	147.26	150.26
	DS7				DS8			
Model Name	AIC	AIC_C	BIC	CAIC	AIC	AIC_C	BIC	CAIC
Exp	187.62	190.02	187.78	189.78	440.63	441.33	442.62	444.62
Gamma	189.62	195.62	189.86	192.86	441.70	443.20	444.69	447.69
Pareto	189.65	195.65	189.89	192.89	442.71	444.21	445.69	448.69
TruncNormal	189.55	195.55	189.79	192.79	441.35	442.85	444.34	447.34
LogNormal	189.60	195.60	189.84	192.84	443.84	445.34	446.83	449.83
TruncLogist	189.54	195.54	189.77	192.77	441.17	442.67	444.16	447.16
LogLogist	189.62	195.62	189.86	192.86	441.75	443.25	444.73	447.73
TruncEVMax	189.56	195.56	189.80	192.80	441.47	442.97	444.46	447.46
LogEVMax	189.59	195.59	189.82	192.82	445.32	446.82	448.30	451.30
TruncEVMin	189.53	195.53	189.77	192.77	441.15	442.65	444.14	447.14
LogEVMin	189.61	195.61	189.85	192.85	441.72	443.22	444.71	447.71

Table 4. Goodness-of-fit performances in middle software testing phases (50%).

	DS1				DS2			
Model Name	AIC	AIC_C	BIC	CAIC	AIC	AIC_C	BIC	CAIC
Exp	403.37	403.87	405.96	407.96	256.08	256.83	257.97	259.97
Gamma	405.32	406.37	409.21	412.21	257.40	259.00	260.24	263.24
Pareto	405.26	406.31	409.15	412.15	257.62	259.22	260.45	263.45
TruncNormal	405.47	406.51	409.36	412.36	258.20	259.80	261.04	264.04
LogNormal	404.96	406.01	408.85	411.85	257.07	258.67	259.90	262.90
TruncLogist	405.40	406.45	409.29	412.29	258.13	259.73	260.97	263.97
LogLogist	405.27	406.32	409.16	412.16	257.37	258.97	260.20	263.20
TruncEVMax	405.41	406.45	409.30	412.30	258.12	259.72	260.95	263.95
LogEVMax	404.87	405.92	408.76	411.76	256.98	258.58	259.81	262.81
TruncEVMin	405.48	406.52	409.37	412.37	258.20	259.80	261.04	264.04
LogEVMin	405.31	406.35	409.20	412.20	257.38	258.98	260.21	263.21

(continued)

Table 4. Continued

Model Name	DS3 AIC	DS3 AIC_C	DS3 BIC	DS3 CAIC	DS4 AIC	DS4 AIC_C	DS4 BIC	DS4 CAIC
Exp	866.50	866.68	870.94	872.94	334.77	335.29	337.29	339.29
Gamma	862.01	862.38	868.67	871.67	334.89	335.98	338.66	341.66
Pareto	866.29	866.66	872.94	875.94	336.70	337.79	340.47	343.47
TruncNormal	868.99	869.36	875.64	878.64	336.85	337.94	340.62	343.62
LogNormal	862.86	863.23	869.52	872.52	336.17	337.27	339.95	342.95
TruncLogist	868.69	869.07	875.35	878.35	336.83	337.92	340.60	343.60
LogLogist	862.05	862.42	868.71	871.71	334.92	336.01	338.70	341.70
TruncEVMax	868.62	868.99	875.28	878.28	336.82	337.91	340.59	343.59
LogEVMax	864.09	864.46	870.75	873.75	337.26	338.35	341.03	344.03
TruncEVMin	868.95	869.32	875.60	878.60	336.89	337.98	340.66	343.66
LogEVMin	862.04	862.41	868.70	871.70	334.91	336.00	338.68	341.68

Model Name	DS5 AIC	DS5 AIC_C	DS5 BIC	DS5 CAIC	DS6 AIC	DS6 AIC_C	DS6 BIC	DS6 CAIC
Exp	363.84	364.20	367.01	369.01	347.11	347.86	349.00	351.00
Gamma	364.97	365.72	369.72	372.72	345.98	347.58	348.81	351.81
Pareto	364.96	365.71	369.71	372.71	349.28	350.88	352.12	355.12
TruncNormal	366.24	366.99	370.99	373.99	348.22	349.82	351.06	354.06
LogNormal	365.29	366.04	370.04	373.04	345.14	346.74	347.98	350.98
TruncLogist	365.89	366.64	370.64	373.64	348.26	349.86	351.10	354.10
LogLogist	364.85	365.60	369.60	372.60	346.01	347.61	348.84	351.84
TruncEVMax	365.98	366.73	370.73	373.73	347.08	348.68	349.91	352.91
LogEVMax	365.49	366.24	370.24	373.24	344.60	346.20	347.44	350.44
TruncEVMin	366.22	366.97	370.97	373.97	348.67	350.27	351.50	354.50
LogEVMin	364.92	365.67	369.67	372.67	346.64	348.24	349.47	352.47

Model Name	DS7 AIC	DS7 AIC_C	DS7 BIC	DS7 CAIC	DS8 AIC	DS8 AIC_C	DS8 BIC	DS8 CAIC
Exp	447.10	447.81	449.09	451.09	1092.90	1093.16	1096.73	1098.73
Gamma	447.30	448.80	450.28	453.28	1094.69	1095.21	1100.43	1103.43
Pareto	449.45	450.95	452.44	455.44	1095.02	1095.54	1100.75	1103.75
TruncNormal	445.48	446.98	448.47	451.47	1094.43	1094.95	1100.16	1103.16
LogNormal	448.43	449.93	451.42	454.42	1098.89	1099.41	1104.63	1107.63
TruncLogist	445.33	446.83	448.32	451.32	1094.42	1094.94	1100.16	1103.16
LogLogist	447.24	448.74	450.23	453.23	1094.73	1095.25	1100.46	1103.46
TruncEVMax	445.59	447.09	448.58	451.58	1094.44	1094.96	1100.18	1103.18
LogEVMax	449.27	450.77	452.26	455.26	1102.56	1103.08	1108.30	1111.30
TruncEVMin	445.32	446.82	448.30	451.30	1094.42	1094.94	1100.16	1103.16
LogEVMin	447.22	448.72	450.20	453.20	1094.68	1095.21	1100.42	1103.42

Table 5. Goodness-of-fit performances in late software testing phases (80%).

	DS1				DS2			
Model Name	AIC	AIC_C	BIC	CAIC	AIC	AIC_C	BIC	CAIC
Exp	693.17	693.47	696.69	698.69	445.80	446.24	448.60	450.60
Gamma	693.18	693.79	698.46	701.46	445.44	446.37	449.65	452.65
Pareto	692.17	692.79	697.46	700.46	444.25	445.17	448.45	451.45
TruncNormal	695.85	696.47	701.14	704.14	448.93	449.85	453.13	456.13
LogNormal	691.85	692.47	697.14	700.14	443.98	444.91	448.19	451.19
TruncLogist	695.32	695.94	700.61	703.61	447.86	448.78	452.06	455.06
LogLogist	692.58	693.20	697.87	700.87	444.36	445.28	448.56	451.56
TruncEVMax	695.29	695.90	700.57	703.57	447.90	448.82	452.10	455.10
LogEVMax	691.68	692.29	696.96	699.96	443.89	444.82	448.10	451.10
TruncEVMin	695.98	696.60	701.27	704.27	448.62	449.54	452.82	455.82
LogEVMin	692.91	693.53	698.20	701.20	444.98	445.90	449.18	452.18

	DS3				DS4			
Model Name	AIC	AIC_C	BIC	CAIC	AIC	AIC_C	BIC	CAIC
Exp	1503.79	1503.91	1509.18	1511.18	565.50	565.81	568.97	570.97
Gamma	1494.48	1494.71	1502.55	1505.55	566.00	566.64	571.22	574.22
Pareto	1497.24	1497.47	1505.32	1508.32	567.15	567.78	572.36	575.36
TruncNormal	1507.89	1508.12	1515.97	1518.97	567.75	568.38	572.96	575.96
LogNormal	1495.22	1495.45	1503.30	1506.30	566.80	567.43	572.01	575.01
TruncLogist	1506.18	1506.41	1514.26	1517.26	567.59	568.22	572.81	575.81
LogLogist	1494.17	1494.40	1502.25	1505.25	566.13	566.76	571.34	574.34
TruncEVMax	1506.10	1506.33	1514.17	1517.17	567.57	568.20	572.78	575.78
LogEVMax	1495.77	1496.00	1503.84	1506.84	567.54	568.17	572.75	575.75
TruncEVMin	1507.68	1507.90	1515.75	1518.75	567.77	568.40	572.98	575.98
LogEVMin	1494.26	1494.49	1502.33	1505.33	566.09	566.72	571.30	574.30

	DS5				DS6			
Model Name	AIC	AIC_C	BIC	CAIC	AIC	AIC_C	BIC	CAIC
Exp	578.66	578.88	582.78	584.78	557.16	557.61	559.96	561.96
Gamma	577.54	577.98	583.72	586.72	555.27	556.19	559.47	562.47
Pareto	580.65	581.09	586.83	589.83	559.19	560.11	563.39	566.39
TruncNormal	580.76	581.20	586.94	589.94	557.96	558.89	562.17	565.17
LogNormal	579.69	580.13	585.87	588.87	554.06	554.98	558.26	561.26
TruncLogist	580.72	581.16	586.90	589.90	557.72	558.65	561.93	564.93
LogLogist	577.61	578.05	583.79	586.79	555.09	556.02	559.30	562.30
TruncEVMax	580.72	581.16	586.90	589.90	556.92	557.85	561.13	564.13
LogEVMax	581.43	581.87	587.61	590.61	553.47	554.39	557.68	560.68
TruncEVMin	580.76	581.20	586.94	589.94	558.40	559.32	562.61	565.61
LogEVMin	577.57	578.01	583.75	586.75	556.00	556.92	560.20	563.20

	DS7				DS8			
Model Name	AIC	AIC_C	BIC	CAIC	AIC	AIC_C	BIC	CAIC
Exp	795.91	796.31	798.90	800.90	1930.13	1930.29	1934.92	1936.92
Gamma	797.83	798.66	802.32	805.32	1927.55	1927.87	1934.74	1937.74
Pareto	796.37	797.19	800.86	803.86	1920.15	1920.46	1927.33	1930.33
TruncNormal	798.59	799.42	803.08	806.08	1940.08	1940.39	1947.26	1950.26
LogNormal	795.76	796.58	800.25	803.25	1935.30	1935.62	1942.49	1945.49

(*continued*)

Table 5. Continued

Model Name	DS7 AIC	DS7 AIC_C	DS7 BIC	DS7 CAIC	DS8 AIC	DS8 AIC_C	DS8 BIC	DS8 CAIC
TruncLogist	798.03	798.86	802.52	805.52	1932.40	1932.71	1939.58	1942.58
LogLogist	796.06	796.89	800.55	803.55	1920.26	1920.57	1927.44	1930.44
TruncEVMax	797.96	798.79	802.45	805.45	1932.42	1932.74	1939.61	1942.61
LogEVMax	795.82	796.64	800.30	803.30	1949.86	1950.17	1957.04	1960.04
TruncEVMin	798.61	799.44	803.10	806.10	1936.70	1937.01	1943.88	1946.88
LogEVMin	797.63	798.46	802.12	805.12	1925.23	1925.54	1932.41	1935.41

Table 6. Predictive performances in different software testing phases.

Early Software Testing Phase (20%)

Data Set	Best PMAE	Best AIC	Best AIC_C	Best BIC	Best CAIC
DS1	4.31 (LogEVMax)	14.02 (Exp)	14.02 (Exp)	14.02 (Exp)	14.02 (Exp)
DS2	8.65 (Pareto)	13.08 (Exp)	13.08 (Exp)	13.08 (Exp)	13.08 (Exp)
DS3	3.62 (LogLogist)	7.66 (Gamma)	45.46 (Exp)	45.46 (Exp)	45.46 (Exp)
DS4	3.95 (Gamma)	12.89 (Exp)	12.89 (Exp)	12.89 (Exp)	12.89 (Exp)
DS5	8.85 (LogNormal)	29.24 (TruncLogist)	25.34 (Exp)	25.34 (Exp)	25.34 (Exp)
DS6	2.27 (Pareto)	8.51 (LogEVMax)	4.87 (Exp)	8.51 (LogEVMax)	8.51 (LogEVMax)
DS7	3.77 (LogNormal)	12.38 (Exp)	12.38 (Exp)	12.38 (Exp)	12.38 (Exp)
DS8	16.03 (LogEVMax)	108.35 (Exp)	108.35 (Exp)	108.35 (Exp)	108.35 (Exp)

Middle Software Testing Phase (50%)

Data Set	Best PMAE	Best AIC	Best AIC_C	Best BIC	Best CAIC
DS1	1.65 (LogNormal)	8.01 (Exp)	8.01 (Exp)	8.01 (Exp)	8.01 (Exp)
DS2	2.78 (Exp)	2.78 (Exp)	2.78 (Exp)	2.78 (Exp)	2.78 (Exp)
DS3	3.15 (Pareto)	18.00 (Gamma)	18.00 (Gamma)	18.00 (Gamma)	18.00 (Gamma)
DS4	3.93 (LogEVMax)	6.80 (Exp)	6.80 (Exp)	6.80 (Exp)	6.80 (Exp)
DS5	14.44 (LogEVMax)	17.14 (Exp)	17.14 (Exp)	17.14 (Exp)	17.14 (Exp)
DS6	3.01 (LogEVMax)	3.01 (LogEVMax)	3.01 (LogEVMax)	3.01 (LogEVMax)	3.01 (LogEVMax)
DS7	41.78 (Pareto)	131.85 (TruncEVMin)	131.85 (TruncEVMin)	131.85 (TruncEVMin)	52.23 (Exp)
DS8	24.65 (TruncEVMin)	207.70 (Exp)	207.70 (Exp)	207.70 (Exp)	207.70 (Exp)

(continued)

Table 6. Continued

Late Software Testing Phase (80%)					
Data Set	Best PMAE	Best AIC	Best AIC_C	Best BIC	Best CAIC
DS1	2.38 (LogEVMax)	2.38 (LogEVMax)	2.38 (LogEVMax)	4.66 (Exp)	4.66 (Exp)
DS2	0.61 (LogEVMax)	0.61 (LogEVMax)	0.61 (LogEVMax)	0.61 (LogEVMax)	3.75 (Exp)
DS3	2.75 (LogEVMax)	4.47 (LogLogist)	4.47 (LogLogist)	4.47 (LogLogist)	4.47 (LogLogist)
DS4	1.60 (Exp)	1.60 (Exp)	1.60 (Exp)	1.60 (Exp)	1.60 (Exp)
DS5	2.94 (LogEVMax)	4.53 (Gamma)	4.53 (Gamma)	5.65 (Exp)	5.65 (Exp)
DS6	1.15 (LogEVMax)	1.15 (LogEVMax)	1.15 (LogEVMax)	1.15 (LogEVMax)	1.15 (LogEVMax)
DS7	2.28 (LogEVMax)	2.63 (LogNormal)	3.31 (Exp)	3.31 (Exp)	3.31 (Exp)
DS8	6.05 (LogEVMax)	9.93 (Pareto)	9.93 (Pareto)	9.93 (Pareto)	9.93 (Pareto)

5 CONCLUDING REMARKS

In this article, we have investigated the dependency relationship between software bug prediction and four information criteria; AIC, AIC_C, BIC, and CAIC, and confirmed that the model selection did not work especially in the early testing phase, and its effect could be improved as the testing went through. Among four information criteria, it has been shown that AIC tended to give different model selection results from the others, but both of AIC and AIC_C provided the performances on PMAE six cases. In future, we will also compare the other information criteria which do not always depend on only the maximum likelihood estimation, and investigate the dependency relationship with much more data sets.

REFERENCES

Abdel-Ghaly A. A., Chan P. Y. and Littlewood B. (1986), Evaluation of competing software reliability predictions, *IEEE Transactions on Software Engineering*, vol. SE-12, pp. 950–967.

Achcar J. A., Dey D. K., and Niverthi M. (1998), A Bayesian approach using nonhomogeneous Poisson processes for software reliability models, *Frontiers in Reliability*, A. P. Basu, S. K. Basu and S. Mukhopadyyay (eds.), pp. 1–18, World Scientific, Singapore.

Akaike H. (1973), Information theory and an extension of the maximum likelihood principle, *Proceedings of The 2nd International Symposium on Information Theory*, B. N. Petrov and F. Caski (eds.), Akadimiai Kiado, pp. 267–281, Budapest.

Bozdogan H. (1987), Model selection and Akaike's information criterion (AIC): The general theory and its analytical extensions, *Psychometrika*, vol. 52, pp. 345–370.

Goel A. L. (1985), Software reliability models: Assumptions, limitations and applicability, *IEEE Transactions on Software Engineering*, vol. SE-11, pp. 1411–1423, 1985.

Goel A. L., and Okumoto K. (1979), Time-dependent error-detection rate model for software reliability and other performance measures, *IEEE Transactions on Reliability*, vol. R-28, pp. 206–211.

Gokhale S. S., and Trivedi K. S. (1998), Log-logistic software reliability growth model, *Proceedings of The 3rd IEEE International High-Assurance Systems Engineering Symposium (HASE-1998)*, pp. 34–41, IEEE CPS.

Hurvich C. M., and Tsai C. L. (1989), Regression and time series model selection in small samples, *Biometrika*, vol. 76, pp. 297–307.

Lyu M. R. (ed.) (1996), *Handbook of Software Reliability Engineering*, McGraw-Hill, New York.

Musa J. D. (1979), *Software Reliability Data*, Technical Report in Rome Air Development Center, New Jersey.

Ohba M. (1984), Inflection S-shaped software reliability growth model, *Stochastic Models in Reliability Theory*, S. Osaki and Y. Hatoyama (eds.), pp. 144–162, Springer, Berlin/Heidelberg.

Ohishi K., Okamura H., and Dohi T. (2009), Gompertz software reliability model: Estimation algorithm and empirical validation, *Journal of Systems and Software*, vol. 82, pp. 535–543.

Okamura H., and Dohi T. (2013), SRATS: Software reliability assessment tool on spreadsheet, *Proceedings of The 2013 IEEE 24th International Symposium on Software Reliability Engineering (ISSRE 2013)*, pp. 100–107, IEEE CPS.

Okamura H., Dohi T., and Osaki S. (2013), Software reliability growth models with normal failure time distributions, *Reliability Engineering and System Safety*, vol. 116, pp. 135–141.

Schwarz G. (1978), Estimating the dimension of a model, *Annals of Statistics*, vol. 6, pp. 461–464.

Sugiura N. (1978), Further analysis of the data by Akaike's information criterion and the finite corrections, *Communications in Statistics*, vol. 7, pp. 13–26.

Yamada S., Ohba M., and Osaki S. (1983), S-shaped reliability growth modeling for software error detection, *IEEE Transactions on Reliability*, vol. R-32, pp. 475–484.

Zhao M., and Xie M. (1996), On maximum likelihood estimation for a general non-homogeneous Poisson process, *Scandinavian Journal of Statistics*, vol. 23, pp. 597–607.

Recent Advances in Reliability and Maintenance Modeling – Okamura, Inoue & Xiao (Eds)

Dynamic maintenance optimization of multi-indenture systems

Qian Qian Zhao & Zhenggeng Ye
School of Management Engineering, Zhengzhou University, Henan, China

Il Han Chung
Department of Integrated Safety Engineering, Ulsan College, Ulsan, South Korea

Won Young Yun
Department of Industrial Engineering, Pusan National University, Busan, South Korea

ABSTRACT: The paper deals with a dynamic maintenance optimization problem for multi-indenture systems with a modular structure. Components age over time and fail randomly, a discrete Weibull distribution is used to model the failure time with the aging process. The system's state is defined by a set of binary variables corresponding to the states of individual components, and failures of redundant components can be identified through inspections. Two maintenance strategies for maintenance objects are considered: component-level maintenance and module-level maintenance, different maintenance objects can adopt different maintenance actions accordingly. Since the state transitions and effective ages of all components in the system are independent of their history, the stochastic decision process of the maintenance strategy for minimizing the expected average cost is formulated as a finite discrete Markov decision process with a discrete state space. A reinforcement learning (RL) algorithm is developed to determine the optimal maintenance objects and corresponding maintenance actions based on the system state set over a finite time horizon. Numerical examples illustrate the applicability and effectiveness of the RL algorithm.

Keywords: multi-indenture systems, redundant structure, RL

1 INTRODUCTION

The maintenance policies are commonly designed and performed according to the specific operating environment and system structure of the target systems. The effectiveness of the maintenance strategy has a decisive impact on the performance of the system during its whole life cycle. Hence, maintaining performance requirements during the system operation and reducing economic losses caused by failures through effective maintenance activities are crucial tasks in reliability engineering. Especially for multi-component complex structure systems, it is necessary to fully consider the system structure, real system state, and maintainability of the system in the operation stage to dynamically determine the appropriate maintenance strategy. Generally, replacing units located at higher levels in the system structure could reduce maintenance time, but costly. Like the line-replaceable unit (LRU), modular units for aircraft, High-speed trains, or ships are designed for rapid replacement at the operation site [1,2]. The maintenance of the lowest level components mainly adopts repair methods because of their low cost, simplicity of maintenance, and spare parts acquisition. Depending on the units' type and their state, various maintenance actions, such as do nothing, minimal repair, imperfect repair, and replacement, could be considered over the system's lifetime. Most of the literature on the maintenance optimization problem of multi-component systems focuses on improving performance by establishing different maintenance policies, and developing customized algorithms

DOI: 10.1201/9781003491309-33

based on the policies, such as the exact optimal solution of dynamic programming [3] and the global optimal solution of heuristic algorithms [4]. These maintenance models are 'static' operation policies, where maintenance decisions (actions) are pre-determined and fixed based on an analysis of historical system failure patterns. The RL method trains the data through the interaction of the agent with the system environment and can handle a large number of state-action spaces. This makes the RL method more flexible and more widely applicable. Hu *et al.* [6] proposed a Q-learning algorithm for aircraft long-term maintenance decision optimization, which successfully learned maintenance strategies under different maintenance scenarios through the information of aircraft mission profiles. Compared with traditional methods, model-free algorithms in RL do not require the knowledge of the environment, the optimal strategy can be "learned" by iterating the experience trajectory path and its corresponding set of rewards and states. It also has significant computational efficiency advantages for large state-action spaces [5]. To handle systems with large state-action spaces, Yang *et al.* [7] addressed the condition-based maintenance problem for redundant systems with arbitrary structures through deep Q-learning. Huang *et al.* [8] proposed the deep Q-network to optimize the maintenance of a multi-component system with six machines and five buffers. Peng and Feng [9] modeled the condition-based maintenance problem as a discrete-time continuous state MDP, which is solved by reinforcement learning. Zhang and Si [10] presented the double deep Q network (DDQN) to optimize the maintenance of a multi-component system where both economic and stochastic dependencies between components were considered. Andriotis *et al.* [11] proposed a deep centralized multi-agent actor-critic algorithm. This study aims to address an optimization problem of dynamically determining the maintenance objects (components or modules) and corresponding maintenance actions for a multi-indenture system with a redundancy structure in a finite time horizon. The rest of the paper is organized as follows. Section 2 discusses the dynamic maintenance policy for multi-indenture systems and the optimization model and RL algorithm are presented. In Section 3, numerical examples are presented to illustrate the optimal maintenance models. Conclusions and future work are given in Section 4.

2 PROBLEM STATEMENT

Consider a multi-indenture system with a series-parallel modular structure composed of n statistically independent components. The units between the uppermost system and the lowermost component are referred to as modules. Note that the lowest component here refers to the lowest component that can be analyzed, rather than the physically lowest component in the actual structure. And all components have only two states: functioning and failed. The state of each module is affected by the state of its child units (which could be either modules or components) and the reliability structure (series or parallel) between them. Refer to Yun and Kim [12], system, modules, and components are collectively referred to as units, given a unit X, in the same failure path, the units located at an upper level are referred to as parent units, and the units located at a lower level are referred as offspring units. Units that share the same parent and are at the same level as unit X are referred to as sibling units. Figure 1 gives an example of the reliability structure of a three-indenture system consisting of five components, wherein the system consists of modules M_1 M_2 and M_3. M_3 as the parent

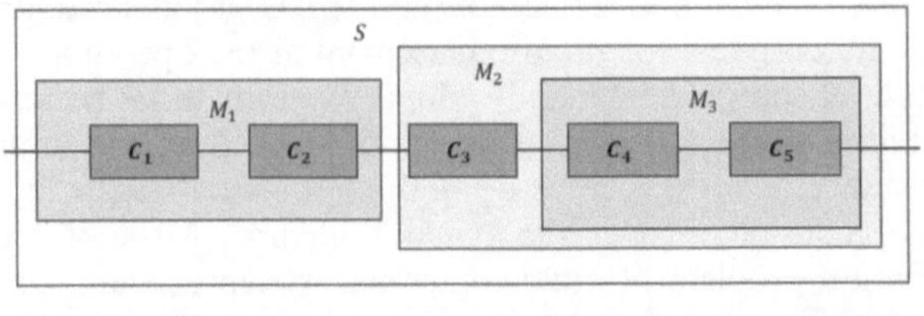

Figure 1. Example of a reliability structure of a three-indenture system with five components.

node, is composed of C_4 and C_5 in series, and as the offspring node connected with C_3 to form the module M_2. Module M_2 and module M_1 in series form system S.

The system's ages over time and experiences random failure. For the failure maintenance of modular systems, such as power systems, the system functions could be restored by replacing the components that directly cause the system failure or the higher-level modules along the failure path. These modules or components are defined as maintenance objects in this study. Two types of maintenance objects are considered: component-level maintenance and module-level maintenance. Component-level maintenance means that only components can be candidates for maintenance objects, while module-level maintenance allows units at any level (modules or components) for candidates for maintenance objects. Both preventive maintenance and corrective maintenance are considered for the aged system to maintain the required system functionality. Higher-level maintenance can delay system aging more effectively than lower-level maintenance, but it is relatively more expensive. The corresponding maintenance actions prioritize on-site replacements that are simple to perform and have a perfect effect, such as the maintenance of engine units. For the lowest level component, maintenance actions such as imperfect maintenance could be considered (for example, the belt can be adjusted for slight wear and replaced if necessary). In this study, the imperfect maintenance model proposed by Kijima *et al.* [13] is used to describe the effects of maintenance based on an age reduction effect. Consequently, several optional maintenance actions can be selected for each failed or aged component or module, including "doing nothing, imperfect repair, and preventive replacement and corrective replacement". Additionally, since the failure of the redundant units does not lead to immediate system failure, it is necessary to find the hidden failures of units through inspections. For multi-component systems that deteriorate over time, finding an appropriate maintenance strategy is a complex optimization problem. This study focuses on the multi-component system where each component ages over time and can be maintained individually or together. For progressively deteriorating systems, dynamic maintenance actions based on the degree of deterioration could provide a more cost-effective solution. We aim to determine the maintenance objects and their corresponding maintenance actions according to the actual degree of deterioration in the system at the inspection time points.

3 MAINTENANCE MODEL AND REINFORCEMENT LEARNING

The following assumptions are used in this study:

1) The time to failure of each component follows a discrete Weibull distribution.
2) Only one unit in a failure path can serve as the maintenance object, and the system level is not designated as a maintenance object.
3) The maintenance time is negligible.
4) Maintenance resources are always available.
5) The state of the system is known at any inspection time.

3.1 *Maintenance model of the multi-indenture system*

For n components in the system, the state of the ith component is defined by a binary variable $s_i(t), i \in \{1, 2, 3, \ldots, n\}$, where

$$s_i(t) = \begin{cases} 0 & \text{if component } i \text{ is functioning at time } t. \\ 1 & \text{if component } i \text{ has failed at time } t. \end{cases}$$

The state of the system at time t, $S_{sys}(t)$, can be represented by a set states of individual components. The total number of states is 2^n, the state vector of the system

$S_{sys}(t) = [s_1(t), s_2(t), \ldots, \ldots, s_n(t)]$, and $S_{sys}(t) \in \{[0,0,\ldots,0],[0,0,\ldots,1],\ldots,[1,1,\ldots,1]\}$. And then, the state of the system can finally be described by a binary function $\emptyset(S_{sys}(t)) = \emptyset(s_1(t), s_2(t), \ldots, \ldots, s_n(t))$, where

$$\emptyset(S_{sys}(t)) = \begin{cases} 0 & \text{if system is functioning at time } t. \\ 1 & \text{if system is in a failed state at time } t. \end{cases}$$

As mentioned in Section 2, we consider two types of maintenance policies: component-level maintenance and module-level maintenance, where only replacement is considered as a repair action of the module. The decision of maintenance actions is influenced not only by the maintenance object but also by the actual state of the system. The details are as follows:

(a) *For the maintenance of components:*

$$\begin{cases} \text{action set}\{\text{do nothing}(a_0), \text{preventive replacement}(a_1)\} & \text{if the system is working.} \\ \text{action set}\{\text{corrective replacement}(a_2), \text{imperfect repair}(a_3)\} & \text{if the system fails.} \end{cases}$$

(b) *For the maintenance of modules:*

$$\begin{cases} \text{action set}\{\text{do nothing}(a_0), \text{preventive replacement}(a_1)\} & \text{if the system is working.} \\ \text{action set}\{\text{corrective replacement}(a_2)\} & \text{if the system fails.} \end{cases}$$

Since the number of units involved in the failure path caused by each component varies, the available range of maintenance objects that can be selected also differs. For instance, in Figure 1, component C_5 has available maintenance objects M_2, M_3, and C_3, whereas component C_1 has available maintenance objects M_1 and C_1. The action set for each component consists of two parts of a sequential decision process: maintenance objects and maintenance actions for the selected maintenance objects. Moreover, there is a dependency between the "maintenance object" of components. For example, if module M_3 is selected as the maintenance object for preventive replacement in the maintenance decision for component C_4, then a maintenance decision is no longer needed for sibling component C_5. Therefore, we propose the definition of the action sets for component i (i = 1,2, … , n) and the algorithm decision process, as detailed below:

Step 1. Check whether the maintenance policy is component-level maintenance or module-level maintenance. if it is the component-level maintenance, proceed to Step 2. Otherwise, proceed to Step 3.

Step 2. Only components located at the lowest level can be selected as maintenance objects.
- 2.1 Identify the component i as the maintenance object.
- 2.2 Set the maintenance action $a_i(t)$ for component i at time t, $a_i(t) \in \{a_0, a_1, a_2, a_3\}$.

Step 3. All units can be selected as the maintenance objects. Identify the units involved in the failure path of component i and set these units as the available maintenance object sets.
- 3.1 Select one of the units as the maintenance object for failure path of component i,
- 3.2 Check whether the unit is a component or a module. If the unit is a component, proceed to Step 3.2.1. Otherwise, proceed to Step 3.2.2.
  - 3.2.1 Set the maintenance action $a_i(t)$ for component i at time t, $a_i(t) \in \{a_0, a_1, a_2, a_3\}$
  - 3.2.2 Set the maintenance action $a_i(t)$ for failure path of component i at time t, $a_i(t) \in \{a_0, a_1, a_2\}$

The action sets of the system at time t are defined as follows

$$A_{sys}(t) = [a_1(t), a_2(t), \ldots, \ldots, a_n(t)]$$

We consider a multi-component system in which each component ages over time within the system and fails independently and randomly. Assume that the time to failure of component i ($i = 1, \ldots, n$) follows a discrete Weibull distribution with parameters q_i and β_i. Let $k_i'(t)$ denote the effective age of component i after maintenance by a selected maintenance action. Since the Kijima type II model is adopted to model the effective age of imperfect repair, the $k_i'(t) + 1$ is expressed as $b \cdot k_i(t)$, where b denotes the age reduction factor, which indicates the maintenance effect of the component after selecting imperfect repair. Moreover, the value $b, 0 \leq b \leq 1$, and when $b = 0$ is equivalent to maintenance action of replacement, and $b = 1$ indicates no change in age, which is equivalent to maintenance action of "do nothing". Therefore, the failure rate function of the component is shown as follows:

$$r_i(t) = 1 - q_i^{k_i'(t)^{\beta_i} - \left(k_i'(t)-1\right)^{\beta_i}}, \quad \left(k_i'(t) = 1, 2, 3, \ldots\right)$$

If the "do nothing" action is selected at any time $t + 1$ for component i, it may remain in the same state with time t or transition to any of the next stages with a probability. The agent (decision maker) decides to take an action from the maintenance action set in each state, meaning that the state transition depends only on the current state and the selected action. Let $p(s_i(t+1)|s_i(t), a_i(t))$ denote the state transition probability of component i at time t, defined as follows.

$$p_i(s_i(t+1)|s_i(t), a_i(t)) = \begin{cases} r_i(t), & s_i(t) = 0, s_i(t+1) = 1 \\ 1 - r_i(t), & s_i(t) = 0, s_i(t+1) = 0 \end{cases}$$

Assuming that the components are independently identically distributed and fail randomly, the transition probability of the system state set from state set $[s_1(t), s_2(t), \ldots, s_n(t)]$ to state set $[s_1(t+1), s_2(t+1), \ldots, s_n(t+1)]$ is defined as follows.

$$p\left(S_{sys}(t+1)|, S_{sys}(t), A_{sys}(t)\right) = \prod_{i=1}^{n} p_i(s_i(t+1)|s_i(t), a_i(t))$$

3.2 *Cost analysis*

This study considers each time step as an inspection time, primarily used to identify the state of the redundant unit by time unit. The combination of all component states at time t represents the current state of the system, and maintenance actions are determined based on the current state information of the system. Therefore, we first consider the variable cost of maintenance actions corresponding to the selected maintenance objects along each failure path at each time t. For different maintenance policies, the maintenance action set can differ, leading to varying considerations to varying considerations for the associated maintenance costs. For the component level maintenance, when $s_i(t) = 0$, the action set includes the maintenance actions "Do nothing and Preventive maintenance (replacement)". Conversely, if $s_i(t) = 1$, the action set for component includes "Corrective replacement and Imperfect maintenance". For each component i, let c_{IR}^i denote the cost of imperfect repair, c_{PR}^i denote the cost of preventive replacement, and c_{CR}^i denote the cost of corrective replacement. Then, the maintenance cost $c_m^i(t)$ for component i at time t is given by

$$c_m^i(t) = \begin{cases} m_0(t) \cdot 0 + m_0(t) \cdot c_{PR}^i, & s_i(t) = 0 \\ m_1(t) \cdot c_{CR}^i + m_1(t) \cdot c_{IR}^i, & s_i(t) = 1 \end{cases}$$

where $m_0(t)$ is a binary variable used in the selection of maintenance actions for the component in an operating state at time t, $m_0(t) = 0$ indicates the action{'do nothing', 0}; $m_1(t)$ is a binary variable used in the selection of maintenance actions for the component fails, $m_0(t) = 0$ indicates the action{'corrective replacement', c^i_{CR}}.

As mentioned in the previous section, in the module-level maintenance policy, all units in the same failure path can be considered as alternative maintenance objects. Since the number of failure paths in the hierarchy equals the number of components at the lowest level, the decision to maintain objects needs to go through all components. Therefore, in this policy, the notation i refers to the failure path of component i, and the maintenance object and maintenance action for each failure path i are determined by the decision algorithm proposed in Section 3.1. Considering that the decision of the maintenance actions at each inspection time point would ultimately affect the state of the system and eventually lead to system failure, we introduce the expected penalty cost of the system failure into the reward function [13] to train of system failure into the reward function, which is the expected probability that the system fails at time $t+1$ given the state and action. For the costs spent by maintenance, we consider fixed administrative cost of maintenance of the system c^F_m, and total various maintenance cost $c^i_m(t)$ related to the maintenance action taken by component/ failure path i at time t. The expected total cost of the system at time t is estimated by:

$$E[c(t)] = p\big(\emptyset\big(S_{sys}(t+1)\big) = 1 | S_{sys}(t), A_{sys}(t)\big) \cdot c_f + \delta(t)c^F_m + \sum\nolimits_{i=1}^{n} c^i_m(t)$$

where $\delta(t)$ is a binary variable, indicating that a maintenance action is taken at time t. $p\big(\emptyset\big(S_{sys}(t+1)\big) = 1|, S_{sys}(t), A_{sys}(t)\big)$denote the probability of the system failure with the condition of the system state $S_{sys}(t)$, and the maintenance action $A_{sys}(t)$ decided.

The criterion of minimizing the average long-run cost rate is considered to determine the optimal maintenance decision at each inspection [10]. Thus, the expected total cost over a finite time T is given by:

$$E[C(T)] = \frac{\sum\limits_{t=1}^{T} E[c(t)]}{T}$$

3.3 *RL for dynamic maintenance decisions*

In this study, we propose a dynamic maintenance strategy for a multi-component system using a model-free RL approach. The goal of RL within a MDP framework is to maximize the expected cumulative reward. At any inspection time point, a trained agent can select the optimal maintenance action from the action set based on the current age and state of all components, thereby minimizing expected total costs. Then, we defined the negative value of the expected average maintenance cost as the reward function. The detailed steps of the RL algorithm are as follows:

Step 1: Set the number of iteration steps, $eps = 0$ and the inspection times $t = 0$. Initialize the learning rate α, exploration factor ε_0, finite time T, decaying factor φ and value of the action value function $Q\big(S_{sys}(t), A_{sys}(t)\big)$.

Step 2: Initialize the states and ages of each unit be 0. Calculate the exploration probability ε by using $\varepsilon = \varepsilon_0/\varphi^{eps}$.

Step 3: If $eps \leq \max(eps)$, input the initial state of the system. The explore action is selected randomly by ε, and the best action is selected by $(1-\varepsilon)$.

Step 4: Check whether the system life cycle is complete (the current time is T)

If $t < T$, the agent takes the maintenance action $A_{sys}(t)$ and calculate the corresponding immediate value of the $E[c(t)]$. Update the state of the system and component at time $t+1$.

And update the current value function $Q(S_{sys}(t), A_{sys}(t))$ following the rules:

$$UpdateQ(S_{sys}(t), A_{sys}(t))$$
$$\leftarrow (1-\alpha) \times Q(S_{sys}(t), A_{sys}(t)) + \alpha \times (R(S_{sys}(t), A_{sys}(t)) + \gamma \times \max Q(S_{sys}(t+1), \quad A_{sys}(t+1)))$$

Otherwise:

$$Update\ Q(S_{sys}(t), A_{sys}(t)) \leftarrow Q(S_{sys}(t), A_{sys}(t)) + R(S_{sys}(t), A_{sys}(t))$$

Step 5: Set $t = t + 1$, and calculate cumulative awards $E[c(t)]$. Repeat Step 4–5 until the $t \geq T$. Then, calculate expected average cost $E[C(T)]$
Step 6: Set $eps = eps + 1$, and repeat Steps 2–5 until the termination conditions Max(eps) is satisfied.

The parameter α refers to the learning rate, which is used to update the Q-value $Q(S_{sys}(t), A_{sys}(t))$. The exploration probability ε gradually decays following an exponential decaying trend during the learning process.

4 NUMERICAL EXPERIMENTS

The numerical examples are presented to demonstrate the effectiveness of the proposed method. We consider a two-indenture system with three components, M_1 is a parallel structure consisting of C_1 and C_2, as shown in Table 1, which gives the system structure and the input data of failure and maintenance of each unit. Also take into account a fixed inspection interval τ with a time unit (e.g., month, year,...). The goal is to find the best maintenance actions for a time horizon of 30-time units. For the RL, the parameter values of α and γ are set to 0.1 and 0.9, respectively. The exploration factor ε_0 and the decaying factor φ are empirically set to 1.0 and 1.0005, respectively. Failure cost (c_f) and the administrative cost of maintenance for the system are 10 and 15, respectively. The age reduction factor b is set to 0.5. To obtain steady-state solutions, we set the number of iterations to 5,0000 for each run. The RL algorithm is accomplished in Pycharm 2023 on a run on a PC with 11th Gen Intel(R) Core (TM) i5-1135G7 CPU @ 2.42 GHz and 16GB RAM.

Table 1. Input information of the two-indenture system.

		Failure rate parameters		Maintenance cost	
Parent unit	Unit	Shape parameters	scale parameters	Replacement cost	Repair cost
–	S	–	–	–	–
S	M_1	–	–	9	–
	C_3	0.9972	2.00	3	1.5
M_1	C_1	0.9968	2.89	8	4
	C_2	0.9968	2.89	8	4

To demonstrate the convergence of the proposed RL approach, we give the global learning curves for 30,000 epochs, and the expected average cost for the last 300 iterations. As shown in Figure 2, both strategies reach a steady state at around 10,000 epochs. Due to the randomness of failure occurrence and the selection of maintenance objects and their actions during the training process, strategies exhibit relatively large volatility even after convergence. At the last three hundred iterations, the mean and standard deviation of the cost rates of component-level maintenance are 15.95 and 0.92, respectively, while the mean and standard deviation for the module-level maintenance are 13.95 and 0.97, respectively. This indicates that the module-level maintenance strategy achieves a lower expected average cost, making it superior to the component-level maintenance strategy.

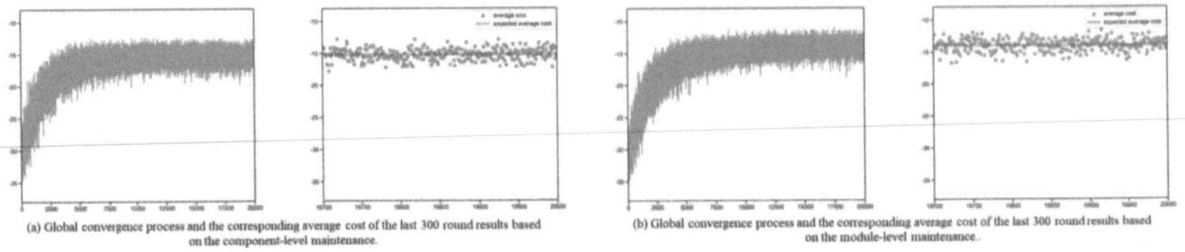

Figure 2. Learning curve under component-level and module-level maintenance.

Next, we discuss the effect of different maintenance objects on the optimal solutions with two maintenance strategies. Figures 3(a) and (b) give the optimal maintenance objects and the corresponding maintenance actions under different system states for component-level maintenance and module-level maintenance policies, respectively. The results for component level maintenance in Figure 3(a) show that to reduce the fixed administrative maintenance cost of the system, corrective replacements are performed on redundant components C_1 or C_2 in the early stages of system operation. Additionally, when the first random failure of component C_1 occurred, preventive replacement was opportunistically carried out on component C_3 to reduce the system failures from the failure of the component. For system failures caused by C_3 at inspection time $t = 20$ and $t = 28$, the average cost is minimized by selecting imperfect repair with lower cost and a corrective replacement with perfect maintenance at $t = 28$. In module-level maintenance, it is clear from Figure 3(b) that the focus of maintenance has shifted from components C_1 and C_2 to the more advanced preventive maintenance of module M_1. This is because, rather than maintaining individual components separately, module-level maintenance is more suitable for components with the same degree of aging. At $t = 16$, although the component of C_1 is in a failure state, system is working and to reduce the times of maintenance, it is maintained together with aging C_2 by preventive replacing the M_1 at time $t = 17$.

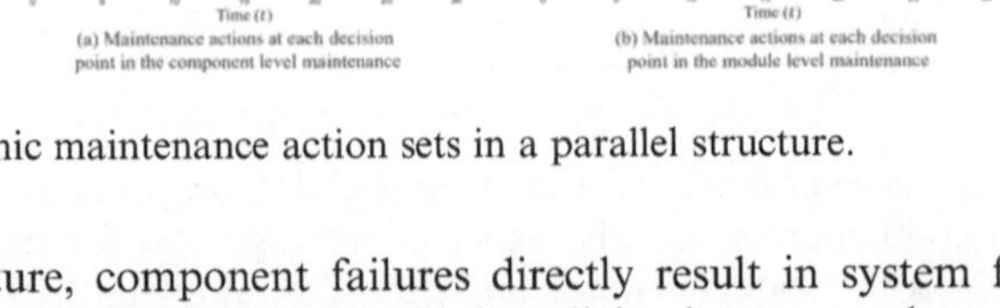

Figure 3. The dynamic maintenance action sets in a parallel structure.

In a series structure, component failures directly result in system failures. Hence, the average total cost of the system under both policies is greater than that of the parallel structure, 18.222 and 16.489 respectively. In component-level maintenance, as illustrated in Figure 4(a), to reduce the average cost, corrective replacement is not implemented, and we run the components to failures. Additionally, preventive replacement of other components is carried out simultaneously during the maintenance of failed components. The maintenance of modules at the upper level can improve the aging of the system to a great extent, thus reducing the number of system failures. Figure 4(b) shows that the maintenance of the failures of C_1 or C_2 is achieved by replacing module M_1. This is because, in this study, both the administrative cost of maintenance and the cost of failure loss are considered. To minimize the average cost, the agent always prefers module maintenance that reduces the number of component failures and the total number of maintenance actions required.

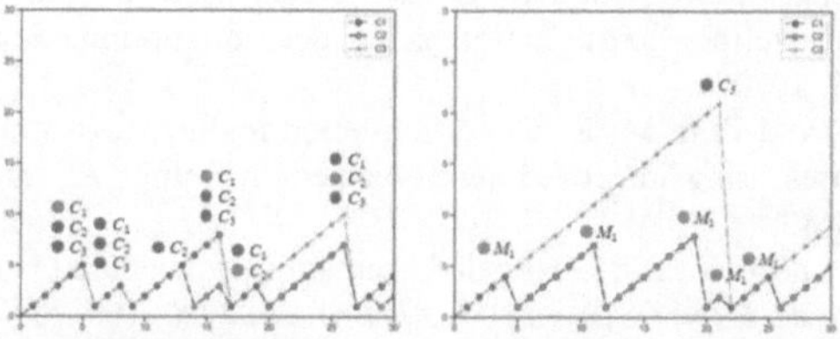

(a) Maintenance action decision of component level maintenance

(b) Maintenance action decision of module level maintenance

Figure 4. The optimal dynamic maintenance actions in a series structure.

5 CONCLUSIONS AND FUTURE WORK

In this paper, we considered the dynamic maintenance optimization problem of multi-component systems. All components deteriorate over time and their failure times are assumed to follow discrete Weibull distributions. The state of the system is a function (the system structure function) of states of components. Two different maintenance strategies are considered: component-level maintenance and modular-level maintenance to maintain the multi-component system correctively and preventively. The stochastic dynamic decision process is modeled as a finite discrete-time Markov decision process within a discrete state space. Through the RL algorithm, the dynamic maintenance objects of each specific aging state and the maintenance actions of the maintenance objects are obtained effectively. A numerical example with a two-indenture system demonstrates the feasibility and effectiveness of the proposed method. For further study, the problem of determining the optimal maintenance action for all components in a large discrete action space can be addressed using deep reinforcement learning methods. Furthermore, taking into account more appropriate variable failure penalty costs can help agents determine the optimal maintenance action more efficiently. Finally, the inspection can be considered as a maintenance action, and also as a decision variable for dynamic maintenance optimization.

ACKNOWLEDGMENT

This research was supported by the National Natural Science Foundation of China (Grant No. 72101240), the China Postdoctoral Science Foundation (Grant No. 2022M712860), the International Science and Technology Cooperation Project of Henan Province (Grant No. 242102521030)

REFERENCES

[1] Puig, J. P., and Basten, R. J. Defining line replaceable units. *European Journal of Operational Research*, 2015, 247(1): 310–320.

[2] Nguyen K. A., Do P, Grall A. Multi-level predictive maintenance for multi-component systems. *Reliability Engineering and System Safety*, 2015, 144: 83–94.

[3] Huang, Y., Yang, L., Tang, T., Gao, Z., Cao, F., and Li, K. Train speed profile optimization with on-board energy storage devices: A dynamic programming based approach. *Computers & Industrial Engineering*, 2018, 126, 149–164.

[4] Bülbül, P., Bayındır, Z. P., and Bakal, I. S. Exact and heuristic approaches for joint maintenance and spare parts planning. *Computers & Industrial Engineering*, 2019, Vol.129, 239–250.

[5] Yousefi, N., Tsianikas, S., and Coit, D. W. RL learning for dynamic condition-based maintenance of a system with individually repairable components. *Quality Engineering*, 2020, 32(3), 388–408.

[6] Hu, Y., Miao, X., Zhang, J., Liu, J and Pan, E. Reinforcement learning-driven maintenance strategy: A novel solution for long-term aircraft maintenance decision optimization. *Computers & Industrial Engineering*, 2021, 153, 107056.

[7] Yang, A., Qiu, Q. G., and Zhu, M. R. Condition-based maintenance strategy for redundant systems with arbitrary structures using improved reinforcement learning. *Reliability Engineering & System Safety*, 2022, 225, 108643.

[8] Huang, J., Chang, Q., and Arinez, J. Deep reinforcement learning based preventive maintenance policy for serial production lines. *Expert System with Application*, 2020, 160, 113701.

[9] Peng, S. L. Reinforcement learning with Gaussian processes for condition-based maintenance. *Computers & Industrial Engineering*, 2021, 158, 107321.

[10] Zhang, N., and Si, W. Deep reinforcement learning for condition-based maintenance planning of multi-component systems under dependent competing risks. *Reliability Engineering & System Safety*, 2020, 203, 107094

[11] Andriotis, C. P., and Papakonstantinou, K. G. Managing engineering systems with large state and action spaces through deep reinforcement learning. *Reliability Engineering & System Safety*, 2019, 191, 106483.

[12] Yun, W. Y., & Kim, J. W. Multi-level redundancy optimization in series systems. *Computers & Industrial Engineering*, 2004,46(2), 337–346.

[13] Kijima, M., Morimura, H., and Suzuki, Y. Periodical replacement problem without assuming minimal repair. *European Journal of Operational Research*, 1988, 37(2): 194–203.

[14] Zhang, P., Zhu, X. Y. and Xie, M. A model-based reinforcement learning approach for maintenance optimization of degrading systems in a large state space. *Computers and Industrial Engineering*, 2021, 161, 107622.

Recent Advances in Reliability and Maintenance Modeling – Okamura, Inoue & Xiao (Eds)

Analysis of fatigue behavior under multiaxial loading environment and calculation of low-cycle fatigue life in TC4 alloy

Junjie Zhu, Butong Li & Xufeng Zhao
College of Economics and Management, Nanjing University of Aeronautics and Astronautics, Nanjing, Jiangsu, PR China

ABSTRACT: Engineering components such as engine blades, turbofans, external parts, etc., are often subject-ed to complex loads in the serving environment. Fatigue failure of components under multiaxial loading will occur, causing a severe influence on operational safety. Centered on low-cycle fatigue under multiaxial loading conditions, we have developed a novel fatigue life prediction framework, which utilizes the physics-guided machine learning approach as a surrogate model for fatigue life prediction. We conducted preliminary experiments to obtain the material's mechanical properties and established reliable Finite Element Analysis (FEA) models based on these properties. Eventually, we can successfully predict low-cycle fatigue life in various loading environments.

Keywords: Fatigue life, finite element analysis, multiaxial loading, TC4 alloy

1 INTRODUCTION

Considerable fatigue damage occurs within the material under repeated cyclic loading; most will lead to multiaxial fatigue failures. Multiaxial fatigue failure is widespread in servicing engineering components such as engines, blades, pipes, and so forth under complex environments [1]. In the overall fatigue analysis of a component, three critical considerations have been identified as contributing to fatigue life: material properties, geometrical characteristics, and loading environment. The geometrical complexity of engineering components is such that, even if they are subjected to only a single fatigue load, the local stress-strain distribution of the structure is practically a multiaxial stress state [2]. Multiaxial fatigue life assessment under variable amplitude loads becomes problematic owing to the sophisticated environments in which the components are exposed. In order to study the nonlinear relationships in fatigue problems, researchers dedicated to finding out the trend of fatigue life with respect to multidimensional fatigue parameters by fitting equations [3]. In cases of multiaxial fatigue where only plain loading (e.g., proportional or constant amplitude cyclic loading) is considered, the yield criterion can be used as an alternative to the equivalent stress [13], namely equivalent stress-strain criteria. Multiple equivalence parameters for the out-phase and non-proportional cases were later defined based on the maximum prismatic hull [4] and minimum Frobenius' norm [5]. However, for more heterogeneous load cases, this approach gives an unconservative lifetime expectation. Subsequently, researchers focused on fatigue damage estimation methods based on cyclic plastic strain energy, which proved to be well-fitted to the multiaxial fatigue problem in low cycle fatigue (LCF) and high cycle fatigue (HCF) scenarios [6]. In the late fifties, Findley found that stress along the shear direction would significantly increase the rate of fatigue crack propagation, and accordingly, the critical plane method was proposed. The critical plane method is for assessing multiaxial fatigue on the basis of damage parameters calculated from stress and/or strain components, and it has been the prevalent method because of its effectiveness on multiaxial fatigue problems [7]. Further fatigue cracks in various cycle fatigue regimes can also be determined by the critical plane method [8].

DOI: 10.1201/9781003491309-34

The occurrence of fatigue and cracks is directly attributable to the stress-strain state of the material. The distribution of fatigue-related parameters remains ambiguous, with only built-up physical criteria. Finite element analysis (FEA) is a widely used method based on physical laws in fatigue analysis. The FEA method is capable of simulating fatigue conditions under compiled loading spectra, such as stress-strain distributions under a full multiaxial non-proportional loading [9]. The fatigue states on the surface can also be well characterized with regard to roughness and geometrical properties in the FEA of in-service components [10]. Apart from mechanical analysis, numerous studies have been focused on crack extension. The extended finite element method (XFEA) based on fracture mechanics can reveal the tendency of crack propagation in metallic parts [11], while insight into polysilicon fatigue under multiaxial loading gives the possibility to explore the texture anisotropy with the use of Poisson-Voronoi element [12]. Moreover, the stress-strain information output from FEA enables straightforward life assessment using fatigue analysis software such as Fe-Safe [13]. A few studies have been conducted combining Fe-Safe/Ansys nCode with the criterion described above to achieve reliable predictions of material fatigue life under cyclic multiaxial loading [14].

The topic of multiaxial fatigue life has been discussed in a number of literatures using statistical methods from the perspective of probabilistic fatigue life, fatigue strength. This paper analyzed fatigue behavior under multiaxial loading environment and calculated the low-cycle fatigue life. The primary works are as follows:

(1) The framework integrates physical laws, finite element analysis, which provides a novel method for greater efficiency and accuracy in low-cycle fatigue life calculation under multiaxial loading.
(2) The fatigue behaviors of alloy samples under various multiaxial loading paths are investigated. A highly reliable finite element model and preliminary database based on practical physical information and experiments have been established.

2 METHODOLOGY AND THEORETICAL FOUNDATION

2.1 *Preliminary tests*

Experiment-based pre-data preparation and law-of-physics-based data acquisition established a plausible fatigue reliability assessment. The multiaxial fatigue behavior of alloys is complicated and hard to analyze. Researchers have developed a large number of physically based numerical models and have attempted to find fatigue damage parameters that are consistent with the multiaxial fatigue failure mechanism and that are physically defined. In this work, tensile/torsional tests have been performed under a multitude of conditions, and the experimental data will be used to fit a multiaxial fatigue numerical model. In addition, experiments will be conducted to verify the performance and reliability of the fatigue life FEA model by correlating the calculated life with experimental life.

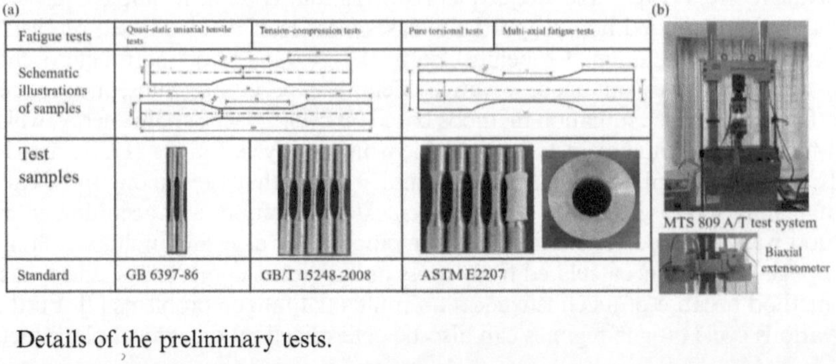

Figure 1. Details of the preliminary tests.

In order to investigate the fatigue mechanical properties of the materials, we will conduct quasi-static uniaxial tensile tests, tension-compression fatigue tests, and pure torsional fatigue tests. In this work, samples of three different geometries were fabricated from titanium alloy rods; sketches and details of the tests are shown in Figure 1a. The samples were subjected to the standard heat treatment, insulated at 800 degrees for 1.5 hours, and thereafter cooled in air. Fatigue tests were carried out on an MTS 809A/T testing system, as shown in Figure 1b. The amplitudes of axial and shear strains were measured using an axial-torsion extensometer. The fatigue lives, and stress-strain relation of alloys under cyclic loading were thereafter correlated by the Manson-Coffin equations:

$$\varepsilon_a = \varepsilon_{ea} + \varepsilon_{pa} = \frac{\sigma_f'}{E}(2N_f)^b + \varepsilon_f'(2N_f)^c \tag{1}$$

$$\gamma_a = \gamma_{ea} + \gamma_{pa} = \frac{\tau_f'}{G}(2N_f)^{t_0} + \gamma_f'(2N_f)^{c_0} \tag{2}$$

and Ramberg-Osgood equations:

$$\varepsilon_a = \frac{\Delta\varepsilon}{2} = \frac{\Delta\varepsilon_e}{2} + \frac{\Delta\varepsilon_p}{2} = \frac{\Delta\sigma}{2E} + \left(\frac{\Delta\sigma}{2K'}\right)^{1/n'} = \frac{\sigma_a}{E} + \left(\frac{\sigma_a}{K'}\right)^{1/n'} \tag{3}$$

$$\gamma_a = \frac{\Delta\gamma}{2} = \frac{\Delta\gamma_e}{2} + \frac{\Delta\gamma_p}{2} = \frac{\Delta\tau}{2G} + \left(\frac{\Delta\tau}{2K_0'}\right)^{1/n_0} = \frac{\tau_a}{G} + \left(\frac{\tau_a}{K_0'}\right)^{1/n_0} \tag{4}$$

with the acquired experimental data. Therefore, the fundamental mechanical properties of experimental alloy will be discussed, as elaborated in Section 3. Noteworthy, the Manson-Coffin equations were based on the von Mises equivalent strain parameter, and the strain magnitude on the critical plane will be introduced in the following section.

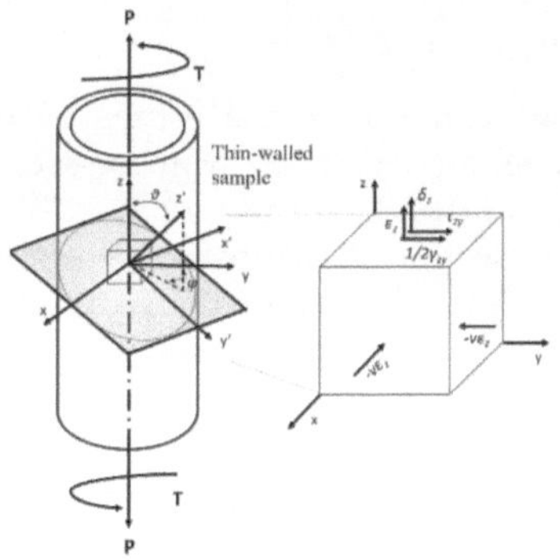

Figure 2. Physical states of a thin-walled tabular sample.

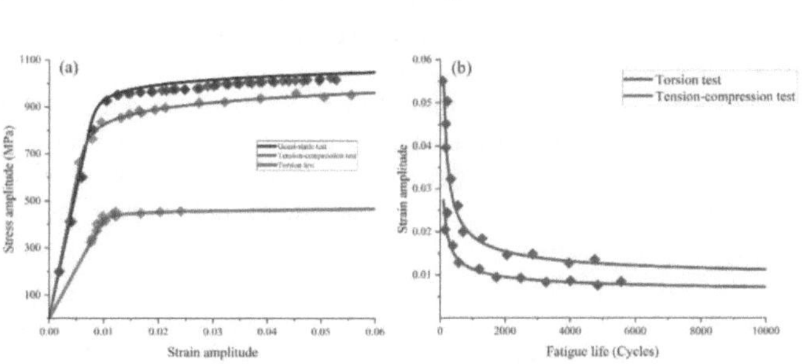

Figure 3. Curve fitting of the preliminary tests: (a) the Ramberg-Osgood equations ; (b) the Manson-Coffin equations.

2.2 *Critical plane model*

The critical plane is defined as the prevalent occurrence site of fatigue failure in metallic materials. For low-ratio (the ratio of wall and radius is less than 1:10) thin-walled specimens, the determination of the critical plane is critical for the simplification of fatigue [15], as shown in Figure 2. There have been studies on multiaxial fatigue life prediction models that are based on the critical plane theory [16]. Smith, Watson, and Topper investigated the fatigue of materials with tensile cracking under multiaxial loading and proposed SWT parameters [16]. Brown and Miller first proposed the model considering damage parameters related to maximum shear amplitude and normal strain on the maximum shear plane [17]:

$$\frac{1}{2}\gamma_{\max} = f(\varepsilon_n, N_f) \tag{5}$$

Kandil, Brown, and Miller (KBM) later introduced a simpler form based on the model, as follows [18]:

$$\begin{aligned}\frac{\Delta\gamma_{\max}}{2} + s_f\Delta\varepsilon_n = {} & [1+\nu_e+s(1-\nu_e)]\frac{\sigma_f'}{E}(2N_f)^b \\ & +[1+\nu_p+s_f(1-\nu_p)]\varepsilon_f'(2N_f)^c\end{aligned} \tag{6}$$

where s_f is the correction for fatigue conversion from uniaxial tension to torsional fatigue when the maximum shear strain amplitude is used as the fatigue damage parameter.

3 FATIGUE ANALYSIS OF THIN-WALLED SAMPLES UNDER MULTIAXIAL VARIABLE AMPLITUDE LOADING

3.1 *Fundamental fatigue properties of TC4 alloy*

Based on the preliminaries, we can obtain the estimated mechanical properties of the material by fitting the experimental results to Manson-Coffin equations and Ramberg-Osgood equations. Figure 3 illustrates the stress-strain profile and strain-life profile of the preliminary tests conducted in Section 2. Moreover, the relationship between Manson-Coffin equations and Ramberg-Osgood equations has a promising correlation with the test results. The fatigue properties of the material necessary for the simulations were eventually derived.

3.2 *Fatigue tests under multiaxial loading conditions*

In this section, we conducted multiaxial fatigue experiments on thin-walled samples of TC4 alloy. The experimental conditions are identical to the data accumulation procedure in Section 2 and the samples for multiaxial fatigue experiments are adopted in the same design scheme as the torsion fatigue tests shown in Figure 1. The tests were carried out at room temperature on a MTS809 A/T testing machine with displacement control in the axial direction and angle control in the torsional direction. Loading frequency is 0.1–1.0 Hz. The loading conditions are the primary considerations in this work. Therefore, we have developed nine loading paths in ascending order of simplicity to complexity in this section, as shown in Figure 4. The continuous loading will be terminated when cracks appear in the specimen, and the extensometer will be removed. In the case of the thin-walled samples in this study, it was assumed to be at the failure criterion when the cracks reached 10mm. The fatigue life of tests is summarized in Table 1. The results will be utilized to validate the results of the FEA simulation and subsequent predictions.

3.3 *Finite element simulation*

In view of the fact that the data-driven approach requires the support of a considerable amount of data, we introduced FEA for the analysis of multiaxial fatigue behavior. A 3-dimensional finite element model of the thin-walled TC4 sample was established by commercial software ABAQUS 2023 to simulate the process of fatigue failure. The main body of the finite element model is set as an incompatible mode eight-node brick element (C3D8I). Compared with the C3D8 element, the C3D8I element has an additional degree of freedom and is insensitive to shear locking behavior and bending behavior of the simulated structures. The rigid body binding is introduced at both ends of the sample to simulate the restraining effect of the fixture on the sample in practical tests. The end of the sample is constrained by two types of Displacement/Rotation boundary conditions (BC), and the degree of freedom of the node on the reference point (RP) is concentrated. The degrees of freedom of the RP are limited except for the Z axis and rotation around the Z axis ($U1 = U2 = 0$, $UR1 = UR2 = 0$). The employed meshing scheme of the model is demonstrated as follows: the mesh of the center section is more defined, and the parts of the ends that are not involved in fatigue behavior are rougher. The cyclic fatigue behavior of alloy material exhibits complicated Bauschinger effect and hardening. Under the consideration that it is difficult for bilinear or multilinear models to undertake the task of assessing the cyclic behavior of alloys, the stabilized combined harden model with Ramberg-Osgood plasticity correction was used in the simulation. The parameters of the employed model are derived from Section 3.1 and shown in Table 2. The loading spectrum will be defined in the Amplitudes and implemented into the numerical simulation. Eventually, the ABAQUS/Standard solver will be employed for calculations. It is noted that in the simulations, results of the single complete cycle will be calculated instead of full-time fatigue cycles (from the first cycle to the last cycle). The stress-strain results will be imported into fatigue post-processor Simulia Fe-Safe.

The fatigue life of the TC4 thin-walled samples is estimated using the commercial software Fe-Safe in this section. The necessary fatigue properties of the target material (TC4) in the Fe-Safe analysis are derived from Section 3.1 and also demonstrated in Table 2. After the setting up of the target material, the stress-strain history obtained from the simulation of ABAQUS is transferred as an ODB file and imported to Fe-Safe environment to evaluate the fatigue behavior. Surface roughness is not considered in this study and is defined as a default value $Kt = 1$. The selection of fatigue algorithms has a critical impact on the reliability of fatigue behavior assessment. Fe-Safe provides two fatigue evaluation methods, stress-controlled and strain-controlled, for loadings under different cyclic scenarios.

Table 1. The test results under multiaxial loading paths for titanium alloy TC4.

ε_a (%)	γ_a (%)	Path A	Path B	Path C	Path D	Path E	Path F	Path G	Path H	Path I
1	1	52723	32434	9352	8151	6877	8933	3597	10814	2439
		13614	2022	7097	2711	6287	9870	6501	1730	6485
		5808	8424	6670	10387	1039	11132	6683	1707	2291
		5260	1789	3349	1673	6958	11264	5140	7537	6927
		2851	2883	1800	12945	6749	2010	1319	4559	7876
		4909	3664	1737	22289	4354	8217	17587	5041	6655
		1377	4178	18030	3467	1299	1272	2213	2827	6747
		1007	950	1038	1056	10593	21434	6667	38089	6235
Average life excluding extremum		5637	3827	6447	8912	4448	7294	6574	3900	6163
Average life in logarithmic form		3.751	3.582	3.809	3.949	3.648	3.862	3.817	3.591	3.789

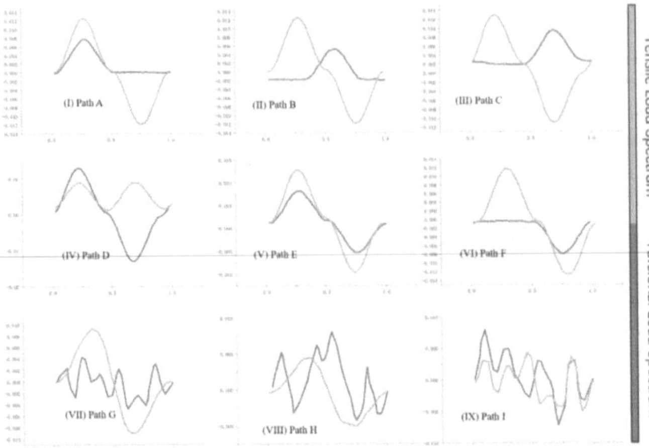

Figure 4. Multiaxial loading spectrums.

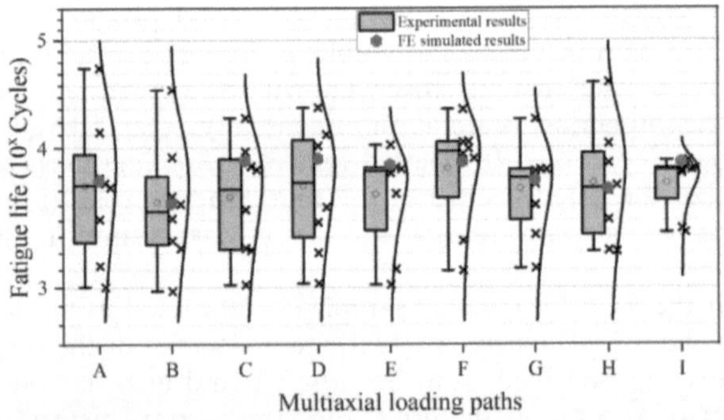

Figure 5. Comparisons between measured fatigue life and FE calculated fatigue life of TC4 samples under multiaxial loading paths.

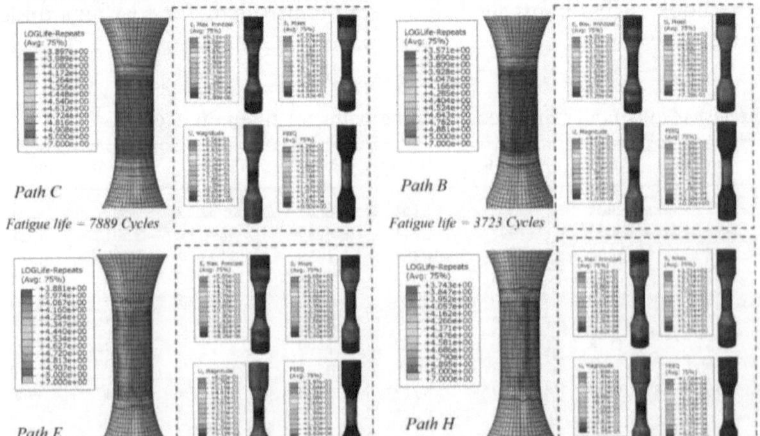

Figure 6. The simulation results and physical states of TC4 samples under multiaxial loadings.

Table 2. Mechanical properties for titanium alloy TC4 and other investigated materials.

Material	Mechanical and fatigue properties				
TC4 titanium alloy	E (Gpa)	σ_y (Mpa)	K'	n'	σ_f' (Mpa)
	108.6	936.1	1128	0.0535	1148
	G (Gpa)	τ_f' (Mpa)	γ_f'	v_e	ε_f'
	42.1	720	2.17	0.25	0.612
	b	c	b_0	c_0	
	−0.053	−0.652	−0.056	−0.74	

The results derived from the practical tests and corresponding FEA simulation are shown in Figure 5 for comparison. The X-shaped points in the figure are experimental data, and the blue points are calculated values by FEA simulation. There are fair agreements between the FE-calculated results and measured test results. The value discrepancies may stem from practical sources of uncertainties like reasonable errors in experiments. Under multiple experiments with the same loading path, the experimental values showed a normal distribution, which is shown in the form of a curve on the right side of the box plot. Moreover, it can be noted that the fatigue life is highly correlated with loading conditions. TC4 thin-walled structures have lower fatigue life under constant amplitude loading compared to non-proportional loading. This is particularly prominent under the simultaneous control of tensile and shear-type loads, attributed to material slippage with crack derivation. The distribution of strain and stress is further shown in the subgraph of Figure 6. The simulation results indicated that the stress and strain distributions on the specimens were identical to the fatigue life distributions, suggesting their contribution to the fatigue life variations. It can be assumed the proposed FEA model is proven to simulate multiaxial fatigue behavior, and it can be utilized as a reliable training data source for the data-driven surrogate model.

4 CONCLUSION

Fatigue life prediction under complex multiaxial loading is a critical issue in practical engineering problems. In this work, a reliable FEA model of fatigue behavior is proposed with a research perspective of fatigue life under different multiaxial loading conditions. Focusing on the TC4 alloy as the case study, we undertook the following works:

With the preliminary tests, we obtained the mechanical properties of the TC4 alloy well using the fitting of empirical formulas and developed a well-confident finite element model according to the derived information. The fatigue life prediction model combining ABAQUS\CAE and Fe-Safe proved to be a favorable fit for practical experimental results. In addition, the physical state under multiaxial loading conditions was analyzed to investigate the relationship between loading paths and fatigue life.

ACKNOWLEDGEMENT

This work was supported by the Aeronautical Science Foundation of China under Grant 2022Z057052002, the Key Laboratory of Intelligent Decision and Digital Operations, Ministry of Industrial and Information Technology under Grant NJ202307.

REFERENCES

[1] John E., Boxall J., Collins R., Bowman E. and Susmel L. (2024). Multiaxial fatigue of water pipe grey cast iron. *Int J Fatigue*;178:108002.

[2] Zhang D., Lv M. and Wen Z. (2023). A unified creep life prediction method and fracture mechanism of high-temperature alloys under multiaxial stress. *Eng Fail Anal*;149:107262.

[3] Santecchia E., Hamouda A. M. S., Musharavati F., Zalnezhad E., Cabibbo M., El Mehtedi M., et al. (2016). A Review on Fatigue Life Prediction Methods for Metals. *Advances in Materials Science and Engineering*;2016:9573524.

[4] Gonçalves C. A., Araújo J. A. and Mamiya E. N. (2005). Multiaxial fatigue: a stress based criterion for hard metals. *Int J Fatigue*;27:177–87.

[5] Zouain N., Mamiya E. N. and Comes F. (2006). Using enclosing ellipsoids in multiaxial fatigue strength criteria. *European Journal of Mechanics – A/Solids*;25:51–71.

[6] Itoh T., Nakamura H., Takanashi M. and Wu M. (2017). Multiaxial low cycle fatigue life of Ti-6Al-4V under non-proportional loading with mean strain. *Theoretical and Applied Fracture Mechanics*;90:165–73.

[7] Poczklán L., Polák J. and Kruml T. (2023). Comparison of critical plane models based on multiaxial low-cycle fatigue tests of 316L steel. *Int J Fatigue*;171:107569.

[8] Vantadori S., Carpinteri A., Ronchei C., Scorza D., Zanichelli A., Okamoto Y., et al. (2020). Lifetime estimation for 316 stainless steel specimens by using a critical plane approach. *Procedia Structural Integrity*;26:106–12.

[9] Duran E. T. (2022). Finite element based Multi-Axial low cycle fatigue analyses of Spot-Welded components and correlation with tests. *Eng Fail Anal*;132:105899.

[10] Jinlong W., Yuxin M., Wenjie P., Yongjie B. and Zeyu S. (2023). Evaluation of the effect of surface roughness parameters on fatigue of TC17 titanium alloy impeller using machine learning algorithm and finite element analysis. *Eng Fail Anal*;153:107586.

[11] Rashnooie R., Zeinoddini M., Ahmadpour F., Beheshti Aval S. B. and Chen T. (2023). A coupled XFEM fatigue modelling of crack growth, delamination and bridging in FRP strengthened metallic plates. *Eng Fract Mech*;279:109017.

[12] Xu R. and Komvopoulos K. (2023). Fatigue crack growth in polysilicon microstructures subjected to multi-axial loading: A Poisson–Voronoi-based finite element analysis. *Int J Fatigue* ;177:107898.

[13] Saadouki B., Sapanathan T., Pelca P. H., Elghorba M. and Rachik M. (2018). Fatigue damage in fieldshapers used during electromagnetic forming and welding processes at high frequency impulse current. *Int J Fatigue*;109:93–102.

[14] Ramakokovhu U., Desai D., Snedden G. and Jamiru T. (2021). Significance of residual stresses in fatigue life prediction of micro gas turbine blades. *Eng Fail Anal*;120:105092.

[15] Seyda J., Pejkowski Ł. and Skibicki D. (2020). The shear stress determination in tubular specimens under torsion in the elastic–plastic strain range from the perspective of fatigue analysis. *Materials*;13.

[16] Smith K. N. (1970). A Stress-Strain Function for the Fatigue of Metals.

[17] Brown M. W. and Miller K. J. (1973). A theory for fatigue failure under multiaxial stress-strain conditions. *Proceedings of the Institution of Mechanical Engineers*;187:745–55.

[18] Kandil F. M., Brown M. W. and Miller K. J. (1982). Biaxial Low-Cycle Fatigue Failure of 316 Stainless Steel at Elevated Temperatures.

Author index